T0182161

Birkhäuser

Advances in Mathematical Fluid Mechanics

Lecture Notes in Mathematical Fluid Mechanics

Lecture Notes in Mathematical Fluid Mechanics as a subseries of 'Advances in Mathematical Fluid Mechanics' is a forum for the publication of high quality monothematic work as well lectures on a new field or presentations of a new angle on the mathematical theory of fluid mechanics, with special regards to the Navier-Stokes equations and other significant viscous and inviscid fluid models.

In particular, mathematical aspects of computational methods and of applications to science and engineering are welcome as an important part of the theory as well as works in related areas of mathematics that have a direct bearing on fluid mechanics.

More information about this series at http://www.springer.com/series/15480

Ulrich Wilbrandt

Stokes–Darcy Equations

Analytic and Numerical Analysis

Ulrich Wilbrandt
Weierstrass Institute
for Applied Analysis and Stochastics
Berlin, Germany

ISSN 2297-0320 ISSN 2297-0339 (electronic)
Advances in Mathematical Fluid Mechanics
ISSN 2510-1374 ISSN 2510-1382 (electronic)
Lecture Notes in Mathematical Fluid Mechanics
ISBN 978-3-030-02903-6 ISBN 978-3-030-02904-3 (eBook)
https://doi.org/10.1007/978-3-030-02904-3

Library of Congress Control Number: 2018965415

Mathematics Subject Classification (2010): 46E35, 65J10, 65N12, 76D07, 76M10, 76S05

This book is published under the imprint Birkhäuser, www.birkhauser-science.com by the registered company Springer Nature Switzerland AG
The registered company address is: Gewerbestrasse 11, 6330 Cham, Switzerland

Acknowledgements

This research would not have been possible without the support of many people. First and foremost, I would like to express my great appreciation to my advisor Prof. Dr. Volker John for his constructive suggestions and useful critiques. His patience and enthusiasm always serve me as an invaluable guide.

Additionally, I would like to thank all the colleagues at the Weierstrass Institute for Applied Analysis and Stochastics, especially the members of the research group "Numerical Mathematics and Scientific Computing," which I am happy to be a part of. In particular I wish to acknowledge the help provided by Swetlana Giere and Alfonso Caiazzo, who shared an office with me, as well as Naveed Ahmed, Felix Anker, Clemens Bartsch, Laura Blank, Jürgen Fuhrmann, and Timo Streckenbach.

I furthermore wish to express my gratitude to my beloved families and friends, for their understanding and endless love, through the duration of my studies. I am particularly grateful for the assistance given by my parents, grandparents, and especially by my wife and my daughter who made this research possible.

This work originates from my dissertation at the "Freie Universität Berlin" which has been handed in 2018. I sincerely thank the reviewers Prof. Dr. Volker John and Prof. Dr. Paul Deuring for their time, as well as very helpful comments and corrections.

Contents

Chapter 1
Introduction

1.1 Motivation

Flows in domains which are partly occupied by a porous medium are of great interest and importance, noticeable examples include groundwater—surface water flow, as well as air and oil filters, blood filtration through vessel walls, and fuel cells. Simulations are often characterized by complex geometries and require the solution of large systems of equations. For this reason, efficient solvers are needed to tackle these types of problems in real-world scenarios. Since the domain of interest is composed of two parts, one allowing a free flow, and one being the porous matrix, two different models are used in the respective subdomains, namely Stokes and Darcy equations together with suitable coupling conditions on the common interface. The individual models are well known and tailored software is available to solve them. The coupled Stokes–Darcy model is somewhat different and it therefore is advisable to find solution strategies which use solutions to the individual models, rather than the coupled one. Inevitably, such strategies are iterative. In this monograph several such approaches are analyzed and their efficiency, especially with respect to the number of iterations, is shown theoretically as well as numerically.

It turns out that the straightforward definition of such an iterative algorithm already works very well, however, only for values of viscosity and hydraulic conductivity which are physically unrealistic. For realistic values, it fails. Alternative schemes have been developed but suffer from drawbacks as well. Therefore, it is an open problem to find algorithms which are efficient for a wide range of values of viscosity and hydraulic conductivity.

Several authors have studied the Stokes–Darcy coupled system and introduced algorithms which try to solve this problem. The following, while not a complete list, are important works on this subject: [DQ09, DQV07, CGHW11, JM00, Saf71, Ang11, CGHW10, CGH$^+$10, GOS11a, LSY02, RY05].

U. Wilbrandt, *Stokes–Darcy Equations*, Advances in Mathematical Fluid Mechanics,
https://doi.org/10.1007/978-3-030-02904-3_1

The analysis of the schemes introduced in the above mentioned literature considers an L^2 space on the interface. In order to develop a new algorithm however, the spaces and operators which contribute to the coupling have to be defined and studied in greater detail. It is essential to understand the notion of traces together with their kernels, image spaces, and existence of (right) inverses. This theory is particularly involved on domains which are not simple, e.g., with non-smooth interfaces. Therefore, a major part of this monograph is devoted to fully develop this theory with many proofs explicitly given.

1.2 Main Contributions

A new kind of Robin–Robin formulation of the coupled Stokes–Darcy problem is analyzed theoretically and numerically. In particular, it is clarified what the smoothness of the interface and the boundary conditions adjacent to it mean for the choice of appropriate spaces in the two subdomains as well as on the interface itself. These include subspaces of $H^{1/2}$ which are related to the so-called Lions–Magenes spaces and are introduced as the image spaces of suitable trace operators. Since the standard literature concerning the Stokes–Darcy coupling pays little attention to these theoretical details, in this monograph most statements can be found with complete, rigorous proofs.

The uniqueness and existence of solutions of the coupled Stokes–Darcy problem in the usual Neumann–Neumann formulation is given and shown to be equivalent to the newly introduced Robin–Robin formulation. Iterative schemes based on the former are analyzed with special focus on the effect of small viscosity and hydraulic conductivity. These include block-wise Gauss–Seidel, fixed point, as well as Steklov–Poincaré iterations. In this context, a modified Robin–Robin iteration from the literature is introduced, analyzed, and compared with iterations based on the new Robin–Robin formulation. Specifically, in the discrete setting, a closely related method called D-RR is proposed. All these formulations introduce Robin parameters which have to be chosen appropriately, such that the resulting algorithm is efficient. Even for the algorithms already proposed in the literature, different advice is given. The numerical studies in Chap. 8 show that a good choice is possible at least for the D-RR approach, which is therefore superior to previously existing approaches in the case of small viscosity and hydraulic conductivity. Thus, this monograph proposes the first subdomain iteration for the coupled Stokes–Darcy problem which is applicable in the practically relevant case.

The numerical studies support the given analysis for each of the introduced algorithms, using two standard examples from the literature. Finally, an example which is closer to geoscientific applications is considered to study the behavior of the new Robin–Robin iterations with respect to small viscosity and hydraulic conductivity, as well as the Robin parameters, which are introduced with this approach. This last example is furthermore used to get an impression of the effects of different Robin parameters in terms of the coupling conditions on the interface.

1.3 Outline

In order to define the involved operators in the Stokes–Darcy coupling, it is fundamental to exactly characterize traces in the context of Sobolev spaces. Starting with a chapter on notation and preliminary results, Chap. 2, Sobolev spaces and some of their important properties are studied in Chap. 3. The main result in this chapter is Theorem 3.4.5 showing that functions, which are smooth up to the boundary of a Lipschitz domain, are dense in certain Sobolev spaces. This is used in Chap. 4 to study trace operators on various domains, including their kernels and image spaces. The somewhat difficult Lions–Magenes spaces are introduced and characterized together with a careful analysis of surjectivity as well as existence of linear and continuous right inverses of traces. As a first application of these results, the two subproblems, Stokes and Darcy, are studied individually in Chap. 5. This includes existence and uniqueness theorems, as well as the definition of operators which are similar in nature to the ones needed in Chap. 6; where the coupling of Stokes and Darcy equations is studied in detail. Different formulations are derived and shown to be equivalent. The last, rather theoretical part, is Chap. 7 which proposes and analyzes several algorithms suitable to solve the coupled Stokes–Darcy problem. Finally, in Chap. 8 numerical examples are introduced and the efficiency of the studied algorithms is tested.

A rough overview of the dependencies among the chapters is given in Fig. 1.1. Readers who are well experienced with the theory of Stokes and Laplace problems, might directly start reading in Chap. 6. All chapters except the last also provide

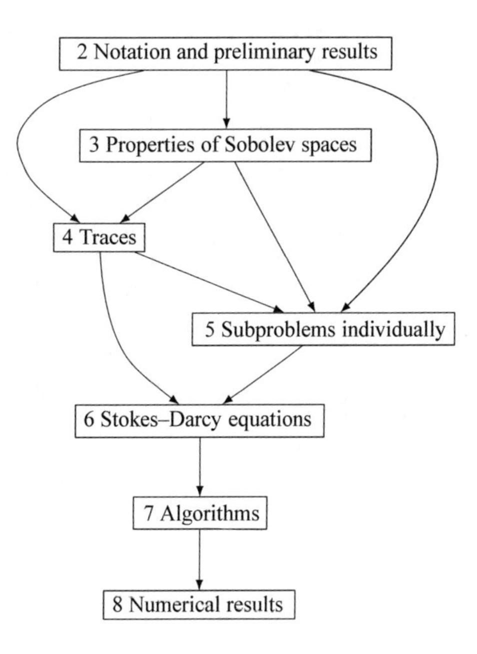

Fig. 1.1 A graph showing the dependencies among all chapters in this monograph. Note that implicit dependencies are not always shown

a graph to help understand the dependencies among the involved definitions, theorems, algorithms and so on. Finally, a list of notations used throughout the monograph follows.

1.4 List of Notations

Symbol	Meaning	Chapter/section
$a(\cdot,\cdot)$	Bilinear form used in the weak form of the Laplace and Stokes equation and in the formulation of the abstract saddle point problem	5.1.1, 5.2.1, 5.3.1
$a_{\mathrm{f}}(\cdot,\cdot)$	Stokes bilinear form used in the weak form of the Stokes–Darcy coupled problem	6.2
$a_{\mathrm{p}}(\cdot,\cdot)$	Darcy bilinear form used in the weak form of the Stokes–Darcy coupled problem	6.2
$a_{\mathrm{f}}^{\mathrm{R}}(\cdot,\cdot)$	Stokes bilinear form used in the Robin–Robin weak form of the Stokes–Darcy coupled problem	6.6.1
$a_{\mathrm{p}}^{\mathrm{R}}(\cdot,\cdot)$	Darcy bilinear form used in the Robin–Robin weak form of the Stokes–Darcy coupled problem	6.6.1
$\mathcal{A}$	Bilinear form used in the proof of existence and uniqueness of the Stokes–Darcy coupled system	6.3
α	Coercivity/positivity constant, the respective operator/bilinear form is in the index, for example α_a, also used as a parameter without index in the Beavers–Joseph–Saffman condition (6.3c)	5.1.2, 5.3.2, 6.2
$B(x_0,\alpha)$	Open ball around a point $x_0 \in X$ with radius α in a Banach space X, i.e., the set $B(x_0,\alpha) = \{x \in X \mid \|x - x_0\|_X < \alpha\}$	2.1, 3.1, 3.4.2, 4.3.1
$\mathcal{B}$	Bilinear form used in the proof of existence and uniqueness of the Stokes–Darcy coupled system	6.3
$b(\cdot,\cdot)$	Bilinear form used in the Stokes weak formulation and in the abstract saddle point theory	5.3.1, 5.2.1
$b_{\mathrm{f}}(\cdot,\cdot)$	Stokes bilinear form coupling the Stokes pressure and velocity space in the Stokes–Darcy coupling	6.2, 6.6.1
c	A continuity constant, typically with the operator or bilinear form as an index	
C	Generic constant, appears in many proofs	
C_1, C_2	Matrices coupling Stokes and Darcy subproblems	6.7, 7
$C^\infty(\Omega)$	Set of infinitely differentiable functions on the set Ω	2.2, 2.4, 3, 4
$\mathcal{D}(\Omega)$	Space of test functions on Ω, i.e., smooth functions with compact support in Ω, here Ω may be $\mathbb{R}^d$	2.4
D	Deformation tensor of a vector field, symmetric part of its gradient	3.5.1, 5.3, 6.1

(continued)

Symbol	Meaning	Chapter/section
D	Darcy matrix	6.7, 7
E	Extension operator, various occurrences, often with an index	3.3, 4, 7
E_f, E_p	Extension operators mapping variables on the interface Γ_I to the Stokes/Darcy subdomain	6.7, 7
$\mathcal{F}$	Linear form used in the proof of existence and uniqueness of the Stokes–Darcy coupled system	6.3
F	A Lipschitz transformation, also used as right-hand side in Theorem 2.1.13	3.2
f	Index indicating a quantity, set, or operator with respect to the Stokes (free flow) subdomain in the coupled Stokes–Darcy setting	6
$\mathcal{G}$	Linear form used in the proof of existence and uniqueness of the Stokes–Darcy coupled system	6.3
Γ_I	The interface separating the Stokes and Darcy subdomains, Ω_f and Ω_p	6.1
Γ	A part of the boundary $\partial\Omega$, often with an index $_\mathrm{N}$, $_\mathrm{D}$, or $_\mathrm{R}$ denoting Neumann, Dirichlet, or Robin boundary	6.1
γ	Parameter for Robin boundary conditions, in general a L^∞-function	5.1.1, 5.3.1
γ_f	Robin parameter for the Stokes subproblem	6.6
γ_p	Robin parameter for the Darcy subproblem	6.6
$H^k(\Omega)$	Sobolev space $W^{k,2}(\Omega)$, this is a Hilbert space	2.4
$H^s(\Omega)$	Sobolev space $W^{s,2}(\Omega)$, this is a Hilbert space	2.4.1
$\mathbf{H}^k(\Omega)$	Vector valued Sobolev space with each component in $H^k(\Omega)$, this is a Hilbert space	2.4
$\mathbf{H}^1_\Gamma(\Omega)$	Subspace of $H^1(\Omega)$ with vanishing trace on $\Gamma \subset \partial\Omega$, also denoted $W^{1,2}_\Gamma(\Omega)$	4.3
$H_\mathrm{f,N}$, $H_\mathrm{p,N}$	Operators taking Neumann data on the interface returning the Dirichlet data (trace) of the Stokes/Darcy solution, respectively; $H_\mathrm{f,N} = T_\mathrm{f} \circ K_\mathrm{f,N}$, $H_\mathrm{p,N} = T_\mathrm{p} \circ K_\mathrm{p,N}$	6.4.3
$H_\mathrm{f,D}$, $H_\mathrm{p,D}$	Operators taking Dirichlet data on the interface Γ_I returning the Neumann data of the Stokes/Darcy solution, respectively; $H_\mathrm{f,D} = T_\mathrm{f}^\mathrm{N} \circ K_\mathrm{f,D}$, $H_\mathrm{p,D} = T_\mathrm{p}^\mathrm{N} \circ K_\mathrm{p,D}$	6.4.3
$H_\mathrm{f}^{\gamma_\mathrm{f},\gamma_\mathrm{p}}$, $H_\mathrm{p}^{\gamma_\mathrm{p},\gamma_\mathrm{f}}$	Operators taking Robin data on the interface returning other Robin data of the Stokes/Darcy solution, respectively; $H_\mathrm{f}^{\gamma_\mathrm{f},\gamma_\mathrm{p}} = T^\mathrm{R}_{\mathrm{f},\gamma_\mathrm{p}} \circ K_\mathrm{f}^{\gamma_\mathrm{f}}$, $H_\mathrm{p}^{\gamma_\mathrm{p},\gamma_\mathrm{f}} = T^\mathrm{R}_{\mathrm{p},\gamma_\mathrm{f}} \circ K_\mathrm{p}^{\gamma_\mathrm{p}}$	6.6.5
Im	Image/range of an operator	2.1, 5.3.2, 6.4.2, 6.6.4
ker	Kernel of an operator, i.e., the preimage of {0} under that operator	2.1, 4.3, 5.3.2

(continued)

Symbol	Meaning	Chapter/section
K	Hydraulic conductivity tensor, assumed to be scalar in Algorithms 7 and 8	5.1, 6.1
K_{D}	Solution operator taking Dirichlet data on a part of the boundary, returning a Stokes/Darcy solution, respectively	5.1.3, 5.3.3
K_{R}	Solution operator taking Robin data on a part of the boundary, returning a Stokes/Darcy solution, respectively	5.1.3, 5.3.3
$K_{\mathrm{f,N}}$, $K_{\mathrm{p,N}}$	Solution operators taking Neumann data on the interface Γ_{I} and returning a Stokes/Darcy solution, respectively	6.4.2
$K_{\mathrm{f,D}}$, $K_{\mathrm{p,D}}$	Solution operators taking Dirichlet data on the interface Γ_{I} and returning a Stokes/Darcy solution, respectively	6.4.2
$K_{\mathrm{f}}^{\gamma_{\mathrm{f}}}$, $K_{\mathrm{p}}^{\gamma_{\mathrm{p}}}$	Solution operators taking Robin data on the interface Γ_{I} and returning a Stokes/Darcy solution, respectively	6.6.4
$L^p(\Omega)$	Lebesgue space, $1 \le p \le \infty$	2.3
$L_0^p(\Omega)$	Lebesgue space with zero mean value, $1 \le p \le \infty$	2.3
$L_{\mathrm{loc}}^1(\Omega)$	Lebesgue space of locally integrable functions	2.3
$L_{\mathrm{f,N}}$, $L_{\mathrm{p,N}}$	Linear operators taking Neumann data on the interface returning the Dirichlet data (trace) of the Stokes/Darcy solution, respectively	6.5
$L_{\mathrm{f,D}}$, $L_{\mathrm{p,D}}$	Linear operators taking Dirichlet data on the interface returning the Neumann data of the Stokes/Darcy solution, respectively	6.5
Λ_{f}	Trace space on the interface Γ_{I}, $\Lambda_{\mathrm{f}} = T_{\mathrm{f}}(\boldsymbol{V}_{\mathrm{f}}, Q_{\mathrm{f}})$	6.4.1
Λ_{p}	Trace space on the interface Γ_{I}, $\Lambda_{\mathrm{p}} = T_{\mathrm{p}}(Q_{\mathrm{p}})$	6.4.1
$\widetilde{\Lambda}_{\mathrm{f}}$	Trace space on the interface Γ_{I}, $\widetilde{\Lambda}_{\mathrm{f}} = T_{\mathrm{f}}\big(\widetilde{\boldsymbol{V}}_{\mathrm{f}} \times L^2(\Omega_{\mathrm{f}})\big)$	6.4.1
$\widetilde{\Lambda}_{\mathrm{p}}$	Trace space on the interface Γ_{I}, $\widetilde{\Lambda}_{\mathrm{p}} = T_{\mathrm{p}}\big(\widetilde{Q}_{\mathrm{p}}\big)$	6.4.1
ℓ	Right-hand side in a weak formulation, sometimes with an index	5.1.1, 5.3.1, 6.2
$\boldsymbol{n}$	The normal vector pointing out of Ω on $\partial\Omega$ and out of the Stokes subdomain on the interface Γ_{I}	6.1
ν	(Kinematic) viscosity	5.3, 6.1
Ω	Open and bounded subset of $\mathbb{R}^d$ with Lipschitz continuous boundary $\partial\Omega$	2.2, 5, 6.1
Ω_{f}	The Stokes subdomain	6.1
Ω_{p}	The Darcy subdomain	6.1
p_{f}	Stokes pressure solution component in the coupled Stokes–Darcy setting	6
p	Index indicating a quantity, set, or operator with respect to the Darcy (porous media) subdomain in the coupled Stokes–Darcy setting	6

(continued)

Symbol	Meaning	Chapter/section
φ_{p}	Darcy pressure (hydraulic head) solution component in the coupled Stokes–Darcy setting	6
Q_{f}	Stokes pressure test space in the coupled Stokes–Darcy setting, $Q_{\mathrm{f}} = L^2(\Omega_{\mathrm{f}})$	6.2
Q_{p}	Darcy pressure test space in the coupled Stokes–Darcy setting, $Q_{\mathrm{p}} = H^1_{\Gamma_{\mathrm{p,D}}}(\Omega_{\mathrm{p}})$	6.2
$\widetilde{Q}_{\mathrm{p}}$	Darcy pressure solution space in the coupled Stokes–Darcy setting, this is an affine linear space	6.4.1
R_{f}, R_{p}	Restriction operators which map the solutions of the Stokes/Darcy subproblems to the interface Γ_{I}	6.7, 7
$\mathbb{R}^d_+$	Half space	4.2
supp	The support of a function, i.e., the closure of all points which are not mapped to zero under that function	2.4, 3.1
S	Stokes matrix, saddle point structure	6.7, 7
T	Cauchy stress tensor	5.3, 6.1
T	Trace operator, often with indices	4, 6.1
T_{f}	Normal trace operator onto the interface Γ_{I} in the Stokes–Darcy coupling, Definition 6.2.3	6.2
T_{p}	Trace operator onto the interface Γ_{I} in the Stokes–Darcy coupling, Definition 6.2.3	6.2
$T_{\mathrm{f}}^{\mathrm{N}}$, $T_{\mathrm{p}}^{\mathrm{N}}$	Neumann data of a Stokes/Darcy solution on the interface Γ_{I}	6.4.2
T_{Γ}^{R}	Robin data of a solution on $\Gamma \subset \partial\Omega$	5.1.3, 5.3.3
$T_{\mathrm{f},\gamma}^{\mathrm{R}}$, $T_{\mathrm{p},\gamma}^{\mathrm{R}}$	Robin data of a Stokes/Darcy solution on Γ_{I}	6.6.4
$\boldsymbol{\tau}_i$	Tangential along the interface Γ_{I}	6.1
$\boldsymbol{u}_{\mathrm{f}}$	Stokes velocity solution component in the coupled Stokes–Darcy setting	6
$\boldsymbol{V}_{\mathrm{f}}$	Stokes velocity test space in the coupled Stokes–Darcy setting, $\boldsymbol{V}_{\mathrm{f}} = \mathbf{H}^1_{\Gamma_{\mathrm{f,D}}}(\Omega_{\mathrm{f}})$	6.2
$\widetilde{\boldsymbol{V}}_{\mathrm{f}}$	Stokes velocity solution space in the coupled Stokes–Darcy setting, this is an affine linear space	6.4.1
$W^{k,p}(\Omega)$	Sobolev space, $k \in \mathbb{N} \cup \{0\}$, $1 \leq p \leq \infty$	2.4
$W^{s,p}(\Omega)$	Sobolev space, $0 < s < 1$, $1 \leq p \leq \infty$	2.4.1
$W^{1,p}_{\Gamma}(\Omega)$	Sobolev space with vanishing trace on $\Gamma \subset \partial\Omega$, $1 \leq p \leq \infty$	4.3
$W^{s,p}_{00}$	Lions–Magenes space, $0 < s < 1$, $1 \leq p \leq \infty$	4.3.1
$(\cdot,\cdot)_X$	Inner product in the Hilbert space X	2.1
$\langle\cdot,\cdot\rangle_{X^*\times X}$	Dual product for a Banach space X and its dual X^*	2.1
$\Vert\cdot\Vert_X$	Norm in a Banach space X	2.1
$\partial\Omega$	The boundary of the domain Ω	2.2
∇	Nabla operator, first (weak) derivative	2.2, 2.4

Chapter 2
Notation and Preliminary Results

In this chapter most of the notation used in this monograph is introduced; in particular, Lipschitz domains on which the so-called Lebesgue and Sobolev spaces are defined, together with a few basic inequalities. Furthermore, the important theorem of Lax–Milgram is shown. To begin with, some definitions and results from functional analysis are stated.

All of the covered topics in this section can be found in many textbooks, including [AF03, Eva10, Gri85, Mac09, QV99, Tar07, Tri92].

2.1 Preliminaries from Functional Analysis

Let X be a Hilbert space with inner product $(\cdot, \cdot)_X : X \times X \to \mathbb{R}$ and corresponding norm $\|x\|_X := \sqrt{(x, x)_X}$. The inner product is linear in each component and a continuous map, which follows from the well known Cauchy–Schwarz inequality:

$$\text{for all } x, y \in X \text{ it is} \qquad (x, y)_X \leq \|x\|_X \|y\|_X. \tag{2.1}$$

Let Y be another Hilbert space with inner product $(\cdot, \cdot)_Y : Y \times Y \to \mathbb{R}$ and let the operator $T : X \to Y$ be linear and continuous. Note that such a linear operator $T : X \to Y$ is continuous if and only if there exists a constant $c_T \geq 0$ such that for all $x \in X$ it is $\|Tx\|_Y \leq c_T \|x\|_X$.

In the study of partial differential equations one often considers so-called bilinear forms on Hilbert spaces. These are maps $a : X \times Y \to \mathbb{R}$ which are bilinear, i.e., linear in each component. Similarly to linear operators, continuity of a is equivalent to the existence of a constant $c_a \geq 0$ such that for all elements $x \in X$, $y \in Y$ it is $|a(x, y)| \leq c_a \|x\|_X \|y\|_Y$. Furthermore, if the spaces X and Y coincide the bilinear

U. Wilbrandt, *Stokes–Darcy Equations*, Advances in Mathematical Fluid Mechanics,
https://doi.org/10.1007/978-3-030-02904-3_2

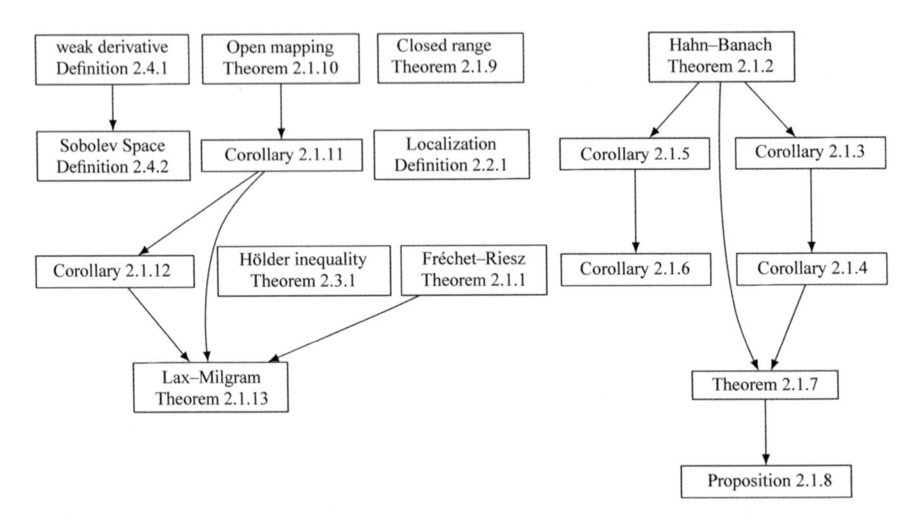

Fig. 2.1 A graph showing the dependencies among all theorems, definitions, corollaries, as well as propositions in Chap. 2. Note that implicit dependencies are not always shown

form a is called coercive[1] if there exists a constant $\alpha_a > 0$ such that for all $x \in X$ it is $a(x, x) \geq \alpha_a \|x\|_X^2$. The inner product is an example for such a coercive and continuous bilinear form.

Furthermore, in Hilbert spaces the parallelogram identity holds: For all $x, y \in X$ it is

$$\|x + y\|_X^2 + \|x - y\|_X^2 = 2\Big(\|x\|_X^2 + \|y\|_X^2\Big). \tag{2.2}$$

If on the other hand in a Banach space X with norm $\|\cdot\|_X$ the parallelogram law holds, then its norm is derived from an inner product, i.e., $\|x\|_X = \sqrt{(x, x)_X}$ for all $x \in X$ with

$$(x, y)_X := \frac{1}{4}\Big(\|x + y\|_X^2 - \|x - y\|_X^2\Big). \tag{2.3}$$

The set X^* of all linear and continuous functions mapping from X to $\mathbb{R}$ is called the dual of X. The action of an element $x^* \in X^*$ on an element $x \in X$ is written as $x^*(x) =: (x^*, x)_{X^* \times X}$ and often called the duality pairing between x^* and x. A norm on X^* is defined as $\|x^*\|_{X^*} = \sup_{\|x\|_X \leq 1} |x^*(x)|$, with which X^* is a Banach space. For sets $M \subset X$ and $N \subset X^*$, define the orthogonal complement $M^\perp \subset X$

[1]Sometimes also called elliptic; if the space is not clear it is often explicitly mentioned as X-coercivity or X-ellipticity.

and the annihilators $M^\circ \subset X^*$ and $N_\circ \subset X$ by

$$M^\perp := \{x \in X \mid \text{for all } x_m \in M\colon (x, x_m)_X = 0\},$$

$$M^\circ := \left\{x^* \in X^* \;\middle|\; \text{for all } x_m \in M\colon \left(x^*, x_m\right)_{X^* \times X} = 0\right\},$$

$$N_\circ := \left\{x \in X \;\middle|\; \text{for all } x_n^* \in N\colon \left(x_n^*, x\right)_{X^* \times X} = 0\right\}.$$

Note that $M^\perp$, M°, and $N_\circ$ are closed subspaces of X and X^*, respectively. If M is viewed as a subset of $X^{**} \supset X$ it is $M^\circ = M_\circ$. If additionally M is a subspace of X, define the quotient space X/M with the norm

$$\|\cdot\|_{X/M} = \inf_{x_m \in M} \|\cdot + x_m\|_X.$$

In the case of a single element set $N = \{x^*\}$, the annihilator $N_\circ$ is the kernel of x^*.

The definition of the dual X^* is used for Banach spaces X; however, Hilbert spaces admit an especially simple representation:

Theorem 2.1.1 (Fréchet–Riesz) *Let X be a Hilbert space. Then the map $Z : X \to X^*$, $x \mapsto (\cdot, x)_X$ is linear, continuous, bijective, and isometric.*

Proof By definition of an inner product Z is linear and the Cauchy–Schwarz inequality (2.1) implies $\|Zx\|_{X^*} \leq \|x\|_X$, i.e., continuity. For $x \in X$ $(x \neq 0)$ it is

$$\|Zx\|_{X^*} = \sup_{\|y\|_X \leq 1} \left|(Zx, y)_{X^* \times X}\right| = \sup_{\|y\|_X \leq 1} \left|(x, y)_X\right| \geq \left(x, \frac{x}{\|x\|_X}\right)_X = \|x\|_X$$

and hence Z is isometric and injective. Next, let $x^* \in X^* \setminus \{0\}$ be given. Denote the kernel of x^* by U which is a closed subspace of X. Let $u_\perp$ be a nonzero element of the orthogonal complement $U^\perp$ within X, hence $(x^*, u_\perp)_{X^* \times X} \neq 0$. Next, for any x in X define $x_U := x - \frac{(x^*, x)_{X^* \times X}}{(x^*, u_\perp)_{X^* \times X}} u_\perp \in U$. It holds

$$0 = (x_U, u_\perp)_X = (x, u_\perp)_X - \frac{(x^*, x)_{X^* \times X}}{(x^*, u_\perp)_{X^* \times X}} \|u_\perp\|_X^2$$

$$\implies \left(x^*, x\right)_{X^* \times X} = \left(x, \frac{(x^*, u_\perp)_{X^* \times X}}{\|u_\perp\|_X^2} u_\perp\right)_X.$$

Defining y to be the second argument in the last inner product in the previous equation shows $x^* = (\cdot, y)_X = Zy$, hence surjectivity of Z because x^* is arbitrary. □

Assigned to an operator $T : X \to Y$ is its dual[2] operator $T^* : Y^* \to X^*$ which is defined for all $y \in Y$ and all $x \in X$ as

$$\left(T^* y, x\right)_{X^* \times X} := (y, Tx)_{Y^* \times Y}.$$

In the case of Hilbert spaces the theorem of Fréchet–Riesz 2.1.1 admits the simplified definition of a Hilbert space dual operator $T' : Y \to X$ where in the above definition the duality pairings are replaced by inner products. The relation of the two types of dual operators in the Hilbert space setting is depicted in the following diagram:

$$\begin{array}{ccc} X & \xrightarrow{\;T\;} & Y \\ {\scriptstyle Z_X}\downarrow & & \downarrow{\scriptstyle Z_Y} \\ X^* & \xleftarrow[\;T^*\;]{} & Y^* \end{array} \qquad T' = Z_X^{-1} \circ T^* \circ Z_Y$$

The bidual T^{**} is then identified with T.

In the following, a few results are stated which are used to prove existence and uniqueness of solutions of saddle point problems in Sect. 5.2.

Theorem 2.1.2 (Hahn–Banach) *Let X be a normed linear space, $Y \subsetneq X$ a nontrivial subspace, and $y^* \in Y^*$ linear and continuous. Then there exists a norm-preserving extension $x^* \in X^*$, i.e., it holds $\|x^*\|_{X^*} = \|y^*\|_{Y^*}$ and*

$$x^*\big|_Y = y^*.$$

A proof of this fundamental theorem can be found for example in [Mac09, Chapter 3.1] where also the following conclusions can be found.

Corollary 2.1.3 *Let X be a nonempty, normed, linear space and $x \in X$. Then there exists a functional $x^* \in X^*$ with $\|x^*\|_{X^*} = 1$ and $(x^*, x)_{X^* \times X} = \|x\|_X$.*

Proof Define the subspace $M := \{\alpha x \mid \alpha \in \mathbb{R}\}$ of X and the functional x^* on M as $(x^*, \alpha x)_{M^* \times M} = \alpha \|x\|_X$. Then x^* has norm one and the desired value at x. Using the theorem of Hahn–Banach, Theorem 2.1.2, the functional x^* can be extended to all of X without changing its norm. □

[2] Also called adjoint.

Another consequence of the theorem of Hahn–Banach is the following corollary:

Corollary 2.1.4 *Let X be a normed linear space, Y a closed subspace, and $x_0 \in X \setminus Y$. Then there exists a separating functional $x^* \in X^*$ such that*

$$x^*\big|_Y = 0 \qquad \text{and} \qquad \left(x^*, x_0\right)_{X^* \times X} = 1.$$

Proof This proof is taken from the German book [Wer00, Korollar III.1.8]. Let $\omega : X \to X/Y =: Q$ be the quotient map. Then it is $\omega(x_0) \neq 0$ and $\omega(y) = 0$ for all $y \in Y$. According to Corollary 2.1.3 choose $\tilde{x}^* \in Q^*$ such that $\|\tilde{x}^*\|_{Q^*} = 1$ and $(\tilde{x}^*, \omega(x_0))_{Q^* \times Q} = \|\omega(x_0)\|_Q$. Finally define $x^* \in X^*$ for all $x \in X$ as

$$\left(x^*, x\right)_{X^* \times X} := \frac{1}{\|\omega(x_0)\|_Q} \left(\tilde{x}^*, \omega(x)\right)_{Q^* \times Q},$$

which has the desired properties. □

In case the subspace is not closed, there is a very similar result which is proved for example in [Şuh03], Corollary 6.8.4:

Corollary 2.1.5 *Let X be a normed linear space and Y a subspace of X. Further let $x_0 \in X$ be given such that* $\operatorname{dist}(x_0, Y) = \delta > 0$. *Then there exists a functional $x^* \in X^*$ such that $x^*\big|_Y = 0$, $(x^*, x_0)_{X^* \times X} = 1$, and $\|x^*\|_{X^*} = 1/\delta$.*

Proof Let Z be the subspace generated by the set $Y \cup \{x_0\}$, i.e., $z \in Z$ if and only if there are $y \in Y$ and $\lambda \in \mathbb{R}$ such that $z = y + \lambda x_0$. Then, on Z, define x^* to be

$$\left(x^*, y + \lambda x_0\right)_{Z^* \times Z} = \lambda$$

such that $x^*\big|_Y = 0$. For all $y \in Y$ and $\lambda \in \mathbb{R}$ it is $\|y + \lambda x_0\|_X = |\lambda| \left\| \frac{y}{\lambda} + x_0 \right\|_X \geq \delta|\lambda|$ and hence $(x^*, y + \lambda x_0)_{Z^* \times Z} \leq \|y + \lambda x_0\|_X / \delta$. This shows the bound $\|x^*\|_{Z^*} \leq 1/\delta$. On the other hand for all $y \in Y$ it is $1 = (x^*, x_0 - y)_{Z^* \times Z} \leq \|x^*\|_{Z^*} \|x_0 - y\|_X$ and this inequality therefore also holds for the infimum over Y, i.e., $\|x^*\|_{Z^*} \geq 1/\delta$. Together with the previous bound, the equality $\|x^*\|_{Z^*} = 1/\delta$ is shown. According to the theorem of Hahn–Banach, Theorem 2.1.2, there exists a norm-preserving extension of x^* which has the desired properties. □

The previous Corollary 2.1.5 admits a characterization of a dense subspace as follows:

Corollary 2.1.6 *Let X be a normed linear space and Y a subspace of X. If any functional $x^* \in X^*$ which vanishes on Y also vanishes on all of X, then Y is dense in X.*

Proof Suppose Y is not dense and let $x_0 \in X$ be given such that $\operatorname{dist}(x_0, Y) = \delta > 0$. According to Corollary 2.1.5 there exists a functional $x^* \in X^*$ which vanishes on Y and has a nonzero value at x_0. This contradicts the assumption that any such functional vanishes on all of X. □

In the following, a number of results are stated which are needed in later chapters.

Theorem 2.1.7 *Let X and Y be Banach spaces and $T : X \to Y$ be linear and continuous. Then it holds*

$$\left(\ker T^*\right)_\circ = \overline{\operatorname{Im} T}. \tag{2.4}$$

If X and Y are Hilbert spaces then additionally

$$(\ker T)^\circ = \overline{\operatorname{Im} T^*} \tag{2.5}$$

is true.

Proof Let $y = Tx \in \operatorname{Im} T$ be given. For any $y^* \in \ker T^*$ it holds

$$\left(y^*, y\right)_{Y^* \times Y} = \left(y^*, Tx\right)_{Y^* \times Y} = \left(T^* y^*, x\right)_{X^* \times X} = 0.$$

Therefore, $y \in (\ker T^*)_\circ$ and $\operatorname{Im} T \subset (\ker T^*)_\circ$. Since $(\ker T^*)_\circ$ is closed the inclusion $(\ker T^*)_\circ \supset \overline{\operatorname{Im} T}$ is verified. Given $y \notin \overline{\operatorname{Im} T}$ it must be shown that $y \notin (\ker T^*)_\circ$. Again since $\overline{\operatorname{Im} T}$ is a closed subspace, according to Corollary 2.1.4 of the Theorem of Hahn–Banach 2.1.2, there exists a $y^* \in Y^*$ such that $(y^*, \cdot)_{Y^* \times Y}$ is zero on $\overline{\operatorname{Im} T}$ and nonzero at y. The former reads $(y^*, Tx)_{Y^* \times Y} = 0$ for all $x \in X$, that means $y^* \in \ker T^*$. Due to $(y^*, y)_{Y^* \times Y} \neq 0$ it holds $y \notin (\ker T^*)_\circ$.

In case of a Hilbert space the bidual T^{**} can be identified with T and Eq. (2.4) with T^* instead of T yields (2.5). □

Proposition 2.1.8 *Let X and Y be Hilbert spaces and let $T : X \to Y$ be linear and continuous. Then it holds:*

a) Is T surjective then T^* is injective.
b) Is T^* surjective then T is injective.
c) Is T^* injective then $\operatorname{Im} T$ *is dense in Y.*
d) Is T injective then $\operatorname{Im} T^*$ *is dense in Y^*.*

Proof

a) Assume T is surjective and let $y^* \in \ker T^* \subset Y^*$ be given. Then it holds

$$\begin{aligned} 0 = T^* y^* &\iff 0 = \left(T^* y^*, x\right)_{X^* \times X} = \left(y^*, Tx\right)_{Y^* \times Y} \ \forall x \in X \\ &\overset{T \text{ surj.}}{\iff} 0 = \left(y^*, y\right)_{Y^* \times Y} \ \forall y \in Y. \end{aligned}$$

That means $y^* = 0$ which implies that the kernel of T^* is trivial, $\ker T^* = \{0\}$, i.e., T^* is injective.

b) Note that $T^{**} = T$ and apply a) to T^*.
c) Let T^* be injective, i.e., $\ker T^* = \{0\}$. Then it is $(\ker T^*)_\circ = Y$ and Eq. (2.4) completes the proof.
d) Note that $T^{**} = T$ and apply c) to T^*.

□

The following well known theorems can be found e.g. in [Wer00], Theorem IV.51.1, or in [Bre11], Theorem 2.19.

Theorem 2.1.9 (Closed Range Theorem) *Let X and Y be Banach spaces and let $T : X \to Y$ be linear and continuous. Then the following four statements are equivalent:*

(i) $\operatorname{Im} T$ *is closed,*
(ii) $\operatorname{Im} T = (\ker T^*)_\circ$,
(iii) $\operatorname{Im} T^*$ *is closed,*
(iv) $\operatorname{Im} T^* = (\ker T)^\circ$.

Theorem 2.1.10 (Open Mapping Theorem) *Let X and Y be Banach spaces and let $T : X \to Y$ be linear, continuous, and surjective. Then T is open (i.e., images of open sets are open under T).*

Proof See also in [DDE12], Theorem 1.12.

First note that for T to be open it suffices to show that[3]

$$\text{there exists a constant } \varepsilon > 0 \text{ such that } B(0, \varepsilon) \subset T(B(0, 1)). \tag{2.6}$$

In fact, let $O \subset X$ be an open set and $x \in O$, $y = Tx \in T(O)$. Furthermore, let r be small enough so that $B(x, r) \subset O$. Then it also is $T(B(x, r)) \subset T(O)$ and, using the linearity of T, $\frac{1}{r}T(B(0, 1)) + \{y\} \subset T(O)$. Now choose ε as in (2.6) and $\frac{1}{r}B(0, \varepsilon) = B(0, \varepsilon/r) \subset \frac{1}{r}T(B(0, 1))$. This means $B(y, \varepsilon/r) \subset T(O)$ is an open ball around y within the set $T(O)$. Hence, $T(O)$ is open and so is T.

Next, it is shown that

$$\text{there exists } \varepsilon > 0 \text{ such that } B(0, \varepsilon) \subset \overline{T(B(0, 1))}. \tag{2.7}$$

Writing X as the union of balls with increasing radius the surjectivity of T gives

$$Y = T(X) = \bigcup_{n \in \mathbb{N}} T(B(0, n)) = \bigcup_{n \in \mathbb{N}} \overline{T(B(0, n))}.$$

Because Y as a Banach space is also a Baire space, there exists $n_0 \in \mathbb{N}$ such that $\overline{T(B(0, n_0))}$ has nonempty interior. There therefore is a $y_0 \in \overline{T(B(0, n_0))}$ and an $\varepsilon_0 > 0$ such that $B(y_0, \varepsilon_0) \subset \overline{T(B(0, n_0))}$ and, equivalently, $B(y_0, \varepsilon_0/n_0) \subset \overline{T(B(0, 1))}$. Together with y_0 also $-y_0$ fulfills the latter inclusion because $T(B(0, 1))$ is symmetric and so is its closure. With this it is

$$B(0, \varepsilon_0/n_0) \subset \overline{T(B(0, 1))} + \{-y_0\} \subset \overline{T(B(0, 1))} + \overline{T(B(0, 1))} \subset \overline{T(B(0, 2))}.$$

Choosing $\varepsilon = \varepsilon_0/2n_0$ results in (2.7).

---

[3]Here and in what follows $B(x_0, \alpha) \subset X$ denotes the ball around x_0 with radius α, i.e., $B(x_0, \alpha) = \{x \in X \mid \|x - x_0\|_X < \alpha\}$. Furthermore, for sets $M, N \subset X$ and a real number a the product aM and the sum $M + N$ have to be understood element-wise: $aM = \{ax \mid x \in M\}$, $M + N = \{m + n \mid m \in M, n \in N\}$.

In the last part of the proof let ε be as in (2.7) and the goal is to show that with this ε it is $B(0,\varepsilon) \subset \overline{T(B(0,1))}$. Let $y \in B(0,\varepsilon)$ and $\|y\|_Y < \varepsilon_0 < \varepsilon$. Then define $\tilde{y} := \frac{\varepsilon}{\varepsilon_0} y \in B(0,\varepsilon)$. Since $\tilde{y} \in \overline{T(B(0,1))}$ and according to statement (2.7), there exists $y_0 = Tx_0 \in T(B(0,1))$ with $\|\tilde{y} - y_0\|_Y < \alpha\varepsilon$, where $0 < \alpha < 1$ is chosen to be smaller than $1 - \frac{\varepsilon_0}{\varepsilon}$, i.e., $\frac{\varepsilon_0}{\varepsilon}\frac{1}{1-\alpha} < 1$. Because $\frac{1}{\alpha}(\tilde{y} - y_0) \in B(0,\varepsilon)$, again due to (2.7), there exists $y_1 = Tx_1 \in T(B(1,0))$ with $\left\|\frac{\tilde{y}-y_0}{\alpha} - y_1\right\|_Y < \alpha\varepsilon$, or equivalently $\|\tilde{y} - (y_0 + \alpha y_1)\|_Y < \alpha^2\varepsilon$. This in turn means $\frac{1}{\alpha^2}(\tilde{y} - y_0 - \alpha y_1) \in B(0,\varepsilon)$ and inductively a sequence $y_i = Tx_i$ with $x_i \in B(0,1)$ can be constructed with

$$\left\|\tilde{y} - T\left(\sum_{i=0}^{n} \alpha^i x_i\right)\right\|_Y < \alpha^{n+1}\varepsilon.$$

The sequence $(\sum_{i=0}^{n} \alpha^i x_i)_{n\in\mathbb{N}}$ is a Cauchy sequence[4] and therefore converges to, say, $\tilde{x} = \sum_{i=0}^{\infty} \alpha^i x_i$. Due to the construction it is $T\tilde{x} = \tilde{y}$. Finally set $x := \frac{\varepsilon_0}{\varepsilon}\tilde{x}$ so that $Tx = y$ and

$$\|x\|_X = \frac{\varepsilon_0}{\varepsilon}\|\tilde{x}\|_X \le \frac{\varepsilon_0}{\varepsilon}\sum_{i=0}^{\infty} \alpha^i \|x_i\|_X \le \frac{\varepsilon_0}{\varepsilon}\sum_{i=0}^{\infty} \alpha^i = \frac{\varepsilon_0}{\varepsilon}\frac{1}{1-\alpha} < 1,$$

i.e., $x \in B(0,1)$ and $y \in T(B(0,1))$. Thus statement (2.6) is shown. □

Corollary 2.1.11 *Let X and Y be Banach spaces and let $T : X \to Y$ be linear, continuous, and bijective. Then the inverse operator T^{-1} is continuous.*

Proof First note that the inverse operator is in fact linear. The inverse T^{-1} is continuous if and only if preimages of open sets are open under T^{-1} (this is a topological definition of continuous functions), i.e., if for any arbitrary open set $O \subset X$ the preimage $\left(T^{-1}\right)^{-1}(O) \subset Y$ is open. According to the open mapping theorem, Theorem 2.1.10, the operator T is open. Therefore, $\left(T^{-1}\right)^{-1}(O) = T(O)$ is an open set, and hence T^{-1} continuous. □

Corollary 2.1.12 *Let X and Y be Banach spaces and let $T : X \to Y$ be linear, continuous, and injective. Then the image of T is closed if and only if T is bounded away from zero, i.e. there exists a constant $c > 0$ such that* $\inf_x \|Tx\|_Y/\|x\|_X \ge c$.[5]

Proof Assume $\operatorname{Im} T$ is closed. Then $\operatorname{Im} T$ is a Banach space and $T : X \to \operatorname{Im} T$ is bijective. According to Corollary 2.1.11 of the open mapping theorem the inverse T^{-1} is continuous. This implies $\left\|T^{-1}y\right\|_X \le \left\|T^{-1}\right\| \|y\|_Y$ for all $y \in \operatorname{Im} T$. Writing

[4] Note that $\left\|\sum_{i=n}^{m} \alpha^i x_i\right\| \le \sum_{i=n}^{m} \alpha^i = \frac{\alpha^{n+1}-\alpha^{m+1}}{1-\alpha} \xrightarrow{n,m\to\infty} 0$.

[5] In this work it is always assumed that infima or suprema are taken with respect to a set for which the following expression is defined. In this case, this means $x \in X$ and $x \neq 0$.

$y = Tx$, this reads $\|x\|_X \leq c\|Tx\|_Y$ for all $x \in X$, with $c = \left\|T^{-1}\right\| > 0$. This proves the boundedness of T away from zero.

Now assume there exists $c > 0$ such that $\|Tx\|_Y \geq c\|x\|_X$ for all $x \in X$. Let $y_n = Tx_n \in \operatorname{Im} T$ be a converging sequence in the image of T, say $y_n \to y \in Y$. Then the assumption implies $\|y_n - y_m\|_Y = \|T(x_n - x_m)\|_Y \geq c\|x_n - x_m\|_X$. Therefore, x_n is as well a Cauchy sequence in the complete space X. Denote its limit by x. Then continuity of T proves $y \in \operatorname{Im} T$: $Tx = T(\lim x_n) = \lim Tx_n = \lim y_n = y$. □

Finally, the important theorem of Lax and Milgram is stated. It is very helpful in the analysis of elliptic problems.

Theorem 2.1.13 (Lax–Milgram) *Let X be a Hilbert-space, $a(\cdot,\cdot) : X \times X \to \mathbb{R}$ a continuous and coercive bilinear form, and $F : X \to \mathbb{R}$ linear and continuous. Then there exists a unique $u \in X$ such that for all $v \in X$ it holds*

$$a(u, v) = F(v).$$

Proof For a fixed $w \in X$ the map $v \mapsto a(w, v)$ is an element of the dual X^*. According to the theorem of Fréchet–Riesz 2.1.1 there is a unique $Tw \in X$ such that $(Tw, v)_X = a(w, v)$ holds for all $v \in X$. The map $w \mapsto Tw$ is linear and continuous because a is in its first component. Furthermore, it is injective: $Tw = 0 \implies a(w, w) = 0 \implies w = 0$, because a is assumed to be coercive. Due to Corollary 2.1.12 its image is closed. If T was not surjective, let $y \in (\operatorname{Im} T)^{\perp}$, $y \neq 0$. Then for all $x \in X$ it is $0 = (Tx, y) = a(x, y)$ and for $x = y$ the coercivity of a implies the contradiction $y = 0$. Consequently, T is bijective as well and, according to Corollary 2.1.11, the inverse T^{-1} is continuous. Then define $u = T^{-1}(Z^{-1}F)$ with the operator Z from the theorem of Fréchet–Riesz 2.1.1. The continuity constant of T^{-1} is the coercivity constant α_a of a which gives a bound on u:

$$\|u\|_X \leq \frac{1}{\alpha_a}\|F\|_{X^*}.$$

Similarly it is $\|F\|_{X^*} \leq c_a\|u\|_X$, where c_a is the continuity constant of a (and T). □

2.2 Suitable Domains

In order to appropriately introduce Stokes and Darcy as well as other partial differential equations a solution space is needed. It turns out that special subspaces of the Lebesgue spaces, the so-called Sobolev spaces, together with the concept of weak derivatives are suitable. These spaces are defined on a domain $\Omega \subset \mathbb{R}^d$. In the entire monograph the set Ω is open, bounded, and connected. The space dimension

d is either 2 or 3, however many proofs in the following chapters hold for arbitrary d. The boundary of Ω is denoted by $\partial\Omega$ and assumed to be Lipschitz continuous. In fact, in most situations it is enough if a weaker condition is fulfilled, namely the cone condition. For simplicity here only Lipschitz domains are considered.

A domain Ω is said to be Lipschitz if, for each $\boldsymbol{x}_0 \in \partial\Omega$, the coordinate system can be rotated by a map $R : \mathbb{R}^d \to \mathbb{R}^d$ in such a way that within a neighborhood U of $\boldsymbol{x}_0$ the boundary is described by a Lipschitz continuous function $h : V \to \mathbb{R}$, $V \subset \mathbb{R}^{d-1}$. Elements in $\mathbb{R}^{d-1}$ are denoted with an apostrophe, e.g. $\boldsymbol{y}' \in V$, and elements in $\mathbb{R}^d$ as $\boldsymbol{y} = (\boldsymbol{y}', y_d)$. More specifically R shall map the graph of h onto the boundary of Ω within U, i.e., $R(V \times h(V)) = U \cap \partial\Omega$. For any point $\boldsymbol{x} \in U$ there exists a $\boldsymbol{y}' \in V$ and a $y_d \in \mathbb{R}$, $|y_d| < r$, such that

$$\boldsymbol{x} = R(\boldsymbol{y}', y_d + h(\boldsymbol{y}')).$$

In other words it is

$$U = R(W) \qquad \text{with} \qquad W = \left\{ \boldsymbol{y} = (\boldsymbol{y}', y_d + h(\boldsymbol{y}')) \in \mathbb{R}^d \;\middle|\; \boldsymbol{y}' \in V, |y_d| < r \right\}.$$

Here the boundary shall be characterized by $y_d = 0$, the interior $U \cap \Omega$ by $y_n > 0$ and the exterior $U \setminus \overline{\Omega}$ by $y_n < 0$. In Fig. 2.2 a typical such situation is sketched. Since the boundary is compact, finitely many such neighborhoods $\Omega_1, \ldots, \Omega_N$ cover a neighborhood $U_{\partial\Omega}$ of $\partial\Omega$ with a uniform parameter $r > 0$. Associated to each Ω_i is a rotation R_i so that $\Omega_i = R_i(W_i)$. In order to cover all of Ω, additionally

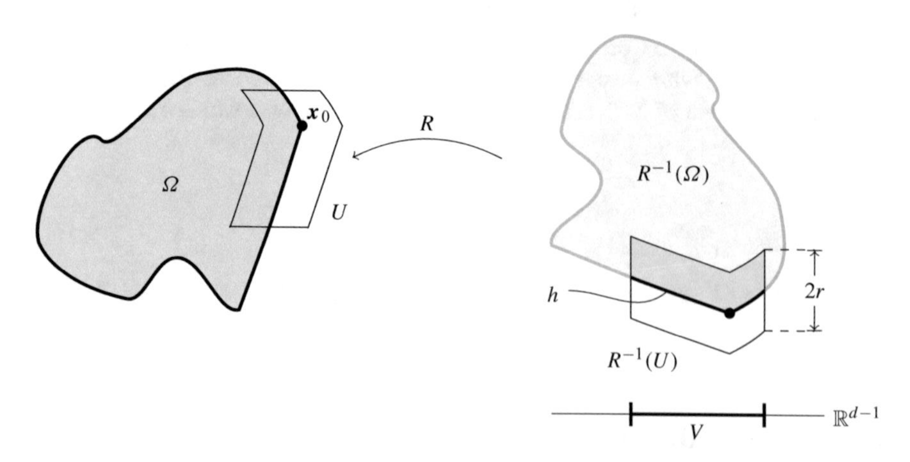

Fig. 2.2 Sketch of some relevant objects in a localization in two space dimensions. The domain Ω is rotated by R such that a local part of the boundary can be represented as a graph of a function h

define the open set $\Omega_0 := \{\boldsymbol{x} \in \Omega \mid \operatorname{dist}(\boldsymbol{x}, \partial\Omega) > r/2\}$. Then indeed it is

$$\Omega \subset \bigcup_{i=0}^{N} \Omega_i.$$

Adapted to the Ω_i, $i = 0, \ldots, N$, define a partition of unity $\phi_0, \ldots, \phi_N \in C^\infty(U_{\partial\Omega} \cup \Omega)$ with

$$\phi_i \in C_0^\infty(\Omega_i), \qquad \sum_{i=0}^{N} \phi_i(\boldsymbol{x}) = 1 \quad \text{for all } \boldsymbol{x} \in U_{\partial\Omega}.$$

The rotations R_i are usually not explicitly mentioned because they are linear and the notation becomes less verbose, in particular ∇R_i is constant, has determinant one, and preserves the norm in $\mathbb{R}^d$, i.e., $|\nabla R_i \boldsymbol{x}| = |\boldsymbol{x}|$. Hence, a change of variables using R_i does not alter the value of integrals.

Definition 2.2.1 (Localization) The triples (V_i, h_i, ϕ_i), $i = 0, \ldots, N$, constructed above are referred to as a (Lipschitz) localization of (the boundary of) Ω.

Such a localization is of importance to define Lebesgue spaces on the boundary. It is possible to allow only special sets for the V_i, for example the unit ball around zero in $\mathbb{R}^{d-1}$. This can facilitate the analysis but is not strictly necessary.

2.3 Lebesgue Spaces

The majority of the spaces appearing in this monograph are based on the Lebesgue spaces which in turn depend on the Lebesgue measure on $\mathbb{R}^d$. Introductions to measure theory and integration can be found in, e.g., [EG92, Fol99], and in many other text books. Important results include Lebesgue's dominated convergence theorem, the theorem of Fubini, and the fact that measurable functions can be approximated by continuous ones. The Lebesgue spaces are defined as

$$L^p(\Omega) := \big\{ f : \Omega \to \mathbb{R} \;\big|\; f \text{ is Lebesgue-measurable and } \|f\|_{L^p(\Omega)} < \infty \big\},$$

with $1 \le p \le \infty$ and the norm

$$\|f\|_{L^p(\Omega)} = \begin{cases} \left(\int_\Omega |f(\boldsymbol{x})|^p \,\mathrm{d}\boldsymbol{x}\right)^{1/p} & \text{if } p < \infty, \\ \operatorname{ess\,sup}_{\boldsymbol{x}\in\Omega} |f(\boldsymbol{x})| & \text{if } p = \infty. \end{cases}$$

A short notation for the integral is sometimes used if the dependence of the integrand on the variable is clear: $\int_\Omega f = \int_\Omega f(\boldsymbol{x})\, d\boldsymbol{x}$. Strictly speaking $L^p(\Omega)$ consists of equivalence classes of functions, rather than functions by themselves. However in this monograph this distinction is not made, that means any statement about a function in a Lebesgue space holds for all members of its equivalence class. Many text books introduce the Lebesgue spaces and show important properties which are not proved here, see, e.g., [Jos05, Wil13, AF03]. For example it can be shown that the Lebesgue spaces with the given norm are complete, i.e., they are Banach spaces. Furthermore, the continuous functions $\mathcal{C}(\Omega)$ are dense in L^p. Since the considered domain is bounded, the Lebesgue spaces are nested: For $p > q$ it is $L^p(\Omega) \subset L^q(\Omega)$, which follows from the following frequently used theorem.[6]

Theorem 2.3.1 (Hölder Inequality) *Let* $1 \leq p, q \leq \infty$ *such that* $\frac{1}{p} + \frac{1}{q} = 1$ *with the convention* $\frac{1}{\infty} = 0$. *Further let* $f \in L^p(\Omega)$ *and* $g \in L^q(\Omega)$ *be given. Then the product* fg *is in* $L^1(\Omega)$ *and it is*

$$\|fg\|_{L^1(\Omega)} \leq \|f\|_{L^p(\Omega)} \|g\|_{L^q(\Omega)}.$$

Proof In case p or q is ∞ the Hölder inequality follows directly

$$\|fg\|_{L^1(\Omega)} = \int_\Omega |f(\boldsymbol{x})g(\boldsymbol{x})|\, d\boldsymbol{x} \leq \|f\|_{L^\infty(\Omega)} \int_\Omega |g(\boldsymbol{x})|\, d\boldsymbol{x} = \|f\|_{L^\infty(\Omega)} \|g\|_{L^1(\Omega)}.$$

In case $p, q < \infty$ the proof is taken from [AF03, Theorem 2.4]. Because the exponential function is strictly convex (its second derivative is strictly positive), given $A, B \in \mathbb{R}$, it is

$$e^{A/p+B/q} \leq \frac{e^A}{p} + \frac{e^B}{q}.$$

For $a, b > 0$ set $A = \ln(a^p)$ and $B = \ln(b^q)$ yielding the so-called Young inequality

$$ab \leq \frac{a^p}{p} + \frac{b^q}{q}. \tag{2.8}$$

If f or g is zero the Hölder inequality is trivially true, otherwise it follows from the Young inequality setting $a = |f(\boldsymbol{x})|/\|f\|_{L^p(\Omega)}$ and $b = |g(\boldsymbol{x})|/\|g\|_{L^q(\Omega)}$ and integrating over the domain Ω. □

Of special interest is the space $L^2(\Omega)$, which is even a Hilbert space with inner product $(\cdot,\cdot)_0 : L^2(\Omega) \times L^2(\Omega) \to \mathbb{R}$, defined by

$$(f, g)_0 = \int_\Omega f(\boldsymbol{x})g(\boldsymbol{x})\, \mathrm{d}\boldsymbol{x} \qquad \text{for all } f, g \in L^2(\Omega).$$

[6] In the case $p = q = 2$ the Hölder inequality is the Cauchy–Schwarz inequality (2.1).

For clarity the norm is denoted by $\|\cdot\|_0 := \|\cdot\|_{L^2(\Omega)}$. If only a subset $M \subset \Omega$ is considered, denote by $(\cdot,\cdot)_{0,M}$ and $\|\cdot\|_{0,M}$ the corresponding inner product and norm, respectively. The same symbols are used in the case of vectors, matrices, and higher order tensors, i.e., $(\boldsymbol{f},\boldsymbol{g})_0$ is defined also for $\boldsymbol{f},\boldsymbol{g} : \Omega \to \mathbb{R}^d$ or $\boldsymbol{f},\boldsymbol{g} : \Omega \to \mathbb{R}^{d\times d}$ as the sum of the inner products over all components. The norm is then defined in the standard way: $\|\boldsymbol{f}\|_0 = \sqrt{(\boldsymbol{f},\boldsymbol{f})_0}$.

Functions in $L^p(\Omega)$ which have a vanishing mean value are often considered in the context of Stokes and Darcy equations:

$$L_0^p(\Omega) := \left\{ f \in L^p(\Omega) \,\middle|\, \int_\Omega f(\boldsymbol{x})\,\mathrm{d}\boldsymbol{x} = 0 \right\}. \tag{2.9}$$

A larger space which is used to define weak derivatives in the next section is

$$L^1_{\mathrm{loc}}(\Omega) = \left\{ f \in L^1(\Omega) \,\middle|\, f \in L^1(\Omega_0) \text{ for all } \Omega_0 \subset \overline{\Omega}_0 \subset \Omega \right\}$$

and it consists of locally integrable functions.

2.4 Sobolev Spaces

Sobolev spaces are subspaces of the Lebesgue spaces $L^p(\Omega)$ which consist of weakly differentiable functions up to a certain degree. A weak derivative is a generalization of the classical one. Its definition is partly based on the space of test functions

$$\mathcal{D}(\Omega) := \left\{ f \in C^\infty(\Omega) \mid \mathrm{supp(f)} \subset \Omega \text{ is compact} \right\},$$

where $\mathrm{supp}(f) := \overline{\{\boldsymbol{x} \mid f(\boldsymbol{x}) \neq 0\}}$ is the support of f. It can be shown that $\mathcal{D}(\Omega)$ is dense in $L^p(\Omega)$, $1 \leq p < \infty$, see e.g., [AF03, Corollary 2.30]. Further results regarding density of smooth functions are shown in Sect. 3.4. The following definition of weak derivatives uses a so-called multiindex $\alpha \in \mathbb{N}_0^d$, where $\mathbb{N}_0 = \mathbb{N} \cup \{0\}$. Its order is defined as $|\alpha| = \sum_{i=1}^d \alpha_i$, with the components α_i of α.

Definition 2.4.1 (Weak Derivative) Let $\alpha \in \mathbb{N}_0^d$ be a multiindex and $f \in L^1_{\mathrm{loc}}(\Omega)$. A function $g \in L^1_{\mathrm{loc}}(\Omega)$ is called the αth weak derivative of f if for all test functions $v \in \mathcal{D}(\Omega)$ it is[7]

$$\int_\Omega f(\boldsymbol{x}) D^\alpha v(\boldsymbol{x})\,\mathrm{d}\boldsymbol{x} = (-1)^{|\alpha|} \int_\Omega g(\boldsymbol{x}) v(\boldsymbol{x})\,\mathrm{d}\boldsymbol{x}.$$

[7] The derivative $D^\alpha v$ is defined as $D^\alpha v = \frac{\partial^{|\alpha|} v}{\partial^{\alpha_1} x_1 \ldots \partial^{\alpha_d} x_d}$.

It can be shown that such a weak derivative, if it exists, is unique. The usual rules such as the product and composition rule as well as the linearity of the differentiation remain valid in the weak sense. Furthermore, a (classically) differentiable function is weakly differentiable with the same derivative, which is hence denoted with the same symbol $g = D^\alpha f$, moreover ∇ is used for the weak gradient as well. The term

$$(-1)^{|\alpha|} \int_\Omega f(\boldsymbol{x}) D^\alpha v(\boldsymbol{x}) \,\mathrm{d}\boldsymbol{x}$$

is called distributional derivative of f even if f is not weakly differentiable. It has to be understood as a functional on $\mathcal{D}(\Omega)$.

Definition 2.4.2 (Sobolev Space) Let $1 \leq p \leq \infty$ and $k \in \mathbb{N}_0 := \mathbb{N} \cup \{0\}$ be given. Then

$$W^{k,p}(\Omega) := \left\{ f \in L^p(\Omega) \;\middle|\; \begin{array}{l} \text{for all multiindices } \alpha \in \mathbb{N}_0^d,\ |\alpha| \leq k, \text{ there exists} \\ D^\alpha f \text{ in the weak sense and } D^\alpha f \in L^p(\Omega) \end{array} \right\}$$

is called a Sobolev space. Its norm $\|\cdot\|_{W^{k,p}(\Omega)}$ is defined by

$$\|f\|_{W^{k,p}(\Omega)} = \left(\sum_{|\alpha| \leq k} \left\| D^\alpha f \right\|_{L^p(\Omega)}^p \right)^{\frac{1}{p}} \qquad \text{for all } f \in W^{k,p}(\Omega).$$

Furthermore, define the semi-norm

$$|f|_{W^{k,p}(\Omega)} = \left(\sum_{|\alpha| = k} \left\| D^\alpha f \right\|_{L^p(\Omega)}^p \right)^{\frac{1}{p}} \qquad \text{for all } f \in W^{k,p}(\Omega).$$

Remark 2.4.3 Sobolev spaces are introduced in different ways by many authors. Most notably is the definition as the completion of the set $\{f \in C^\infty(\Omega) \mid \|f\|_{W^{k,p}(\Omega)} < \infty\}$ with respect to the norm $\|\cdot\|_{W^{k,p}(\Omega)}$. These definitions coincide with the one given above, see Theorem 3.4.1 in Sect. 3.4.1. Many more related spaces with special properties are studied and used for particular problems. See for example [AF03] and [Gri85].

It can be shown that the Sobolev spaces are complete, i.e., Banach spaces, see for example [AF03, Theorem 3.3]. In case $p = 2$, $H^k(\Omega) := W^{k,2}(\Omega)$ is even a Hilbert space with inner product $(\cdot,\cdot)_k : H^k(\Omega) \times H^k(\Omega) \to \mathbb{R}$, defined by

$$(f, g)_k = \sum_{|\alpha| \leq k} \left(D^\alpha f, D^\alpha g \right)_0 \qquad \text{for all } f, g \in H^k(\Omega).$$

The norm is denoted by $\|\cdot\|_k = \|\cdot\|_{H^k(\Omega)}$, the semi-norm by $|\cdot|_k = |\cdot|_{H^k(\Omega)}$. As before the notation $\|\cdot\|_{k,M}$ and $(\cdot,\cdot)_{k,M}$ is used for the norm and inner product where all integrals are taken over $M \subset \Omega$ instead of Ω. Note that $W^{0,p}(\Omega) = L^p(\Omega)$ and $H^0(\Omega) = L^2(\Omega)$. The spaces consisting of vector valued functions are written in bold face, for example

$$\mathbf{H}^k(\Omega) := \left(H^k(\Omega)\right)^d.$$

2.4.1 *Fractional Order Sobolev Spaces*

So far only Sobolev spaces of nonnegative integer order are introduced, namely $W^{k,p}(\Omega)$ with $k \in \mathbb{N}_0$. It is however meaningful to define intermediate spaces $W^{s,p}(\Omega) \subset W^{0,p}(\Omega) = L^p(\Omega)$ with real $0 < s < 1$. Functions in $W^{s,p}(\Omega)$ also have a finite Sobolev–Slobodeckij semi-norm, which is defined in Eq. (2.10) below. They can be understood as a generalization of Hölder spaces, in the sense that $W^{s,\infty}(\Omega) = C^{0,s}(\Omega)$, [DNPV12], see also [Eva10] in the case of $s = 1$ and a C^1 boundary. Fractional order Sobolev spaces can as well be defined through interpolation methods, see [BS08] or [AF03].

The Sobolev–Slobodeckij semi-norm is given as

$$|v|_{W^{s,p}(\Omega)} = \left(\iint_{\Omega\times\Omega} \frac{|v(\boldsymbol{x}) - v(\boldsymbol{y})|^p}{|\boldsymbol{x} - \boldsymbol{y}|^{sp+d}} \,\mathrm{d}\boldsymbol{x}\,\mathrm{d}\boldsymbol{y}\right)^{1/p}. \tag{2.10}$$

It is also called Aronszajn or Gagliardo semi-norm and is introduced by these three authors. Then define

$$W^{s,p}(\Omega) := \left\{v \in L^p(\Omega) \;\middle|\; |v|_{W^{s,p}(\Omega)} < \infty\right\}$$

which is a Banach space with the norm

$$\|v\|_{W^{s,p}(\Omega)} = \left(\|v\|^p_{L^p(\Omega)} + |v|^p_{W^{s,p}(\Omega)}\right)^{1/p},$$

see [DDE12], Proposition 4.24. In case $p = 2$, $H^s(\Omega) = W^{s,2}(\Omega)$ is a Hilbert space with the inner product

$$(u, v)_{H^s(\Omega)} = (u, v)_{0,\Omega} + \iint_{\Omega\times\Omega} \frac{(u(\boldsymbol{x}) - u(\boldsymbol{y}))\,(v(\boldsymbol{x}) - v(\boldsymbol{y}))}{|\boldsymbol{x} - \boldsymbol{y}|^{2s+d}} \,\mathrm{d}\boldsymbol{x}\,\mathrm{d}\boldsymbol{y}.$$

Equivalently, on all of $\mathbb{R}^d$ the fractional order Sobolev space $H^s(\mathbb{R}^d)$ can be conveniently defined using the Fourier transform, see [DDE12], Section 4.2, and also [Tar07, DNPV12].

Remark 2.4.4 (Sobolev Spaces with Negative Index) So far in this monograph the space $W^{s,p}(\Omega)$ is defined for $s \in [0,1]$. An extension to arbitrary positive s is discussed in Remark 3.4.7. Additionally, it is possible to define the above space for negative s. In $\mathbb{R}^d$, these are the duals of corresponding spaces with positive index: for $s < 0$ it is $W^{s,p}(\mathbb{R}^d) := \left(W^{-s,p'}(\mathbb{R}^d)\right)^*$ with $1 = 1/p + 1/p'$. For a domain $\Omega \neq \mathbb{R}^d$, one instead uses a suitable subspace of $W^{-s,p'}(\Omega)$ in the definition of $W^{s,p}(\Omega)$. In this monograph Sobolev spaces with negative index are not studied any further, but more information on them can be found in the literature, see for example [AF03, Eva98] ($p = 2$), or [BC84].

2.5 Lebesgue Spaces on the Boundary

Consider a localization (V_i, h_i, ϕ_i), $i = 0, \ldots, N$, of Ω as in Definition 2.2.1. A function $u : \partial\Omega \to \mathbb{R}$ is called measurable on $\partial\Omega$ if the functions $u_i : V_i \subset \mathbb{R}^{d-1} \to \mathbb{R}$, $i > 0$, defined as

$$u_i(\boldsymbol{y}') = \phi_i\big(\boldsymbol{y}', h_i(\boldsymbol{y}')\big)\, u\big(\boldsymbol{y}', h_i(\boldsymbol{y}')\big)$$

are measurable. Furthermore, u is integrable if additionally the integrals

$$\int_{\partial\Omega} u_i(\boldsymbol{x})\,\mathrm{d}\sigma_{\boldsymbol{x}} = \int_{V_i} u_i(\boldsymbol{y}')\sqrt{1 + |Dh_i(\boldsymbol{y}')|^2}\,\mathrm{d}\boldsymbol{y}'$$

exist in the Lebesgue sense. As Lipschitz continuous functions, the mappings h_i are differentiable almost everywhere (Rademacher's theorem) and

$$\|\nabla h_i\|_{L^\infty(V_i)} \le |h_i|_{C^{0,1}(V_i)},$$

see Theorem 6 in Section 5.8.3 of [Eva98].[8] The term on the right-hand side is the semi-norm

$$|h_i|_{C^{0,1}(V_i)} = \sup_{\substack{\boldsymbol{x},\boldsymbol{y} \in V_i \\ \boldsymbol{x} \neq \boldsymbol{y}}} \frac{h_i(\boldsymbol{x}) - h_i(\boldsymbol{y})}{|\boldsymbol{x} - \boldsymbol{y}|}.$$

Therefore, the integration of u_i over $\partial\Omega$ is well defined and the integral of u on the boundary is then

$$\int_{\partial\Omega} u(\boldsymbol{x})\,\mathrm{d}\sigma_{\boldsymbol{x}} = \sum_{i=1}^{N} \int_{\partial\Omega} u_i(\boldsymbol{x})\,\mathrm{d}\sigma_{\boldsymbol{x}}.$$

[8] Some generalization on Rademacher's theorem can be found in [EG92].

Remark 2.5.1 On the boundary $\partial\Omega$ it is important to define the outer unit normal vector $\boldsymbol{n} : \partial\Omega \to S^d = \{\boldsymbol{x} \in \mathbb{R}^d \mid |\boldsymbol{x}| = 1\}$. Using the localization as above it is defined as

$$\boldsymbol{n}(\boldsymbol{y}) = \frac{1}{\sqrt{1 + |\nabla h_i(\boldsymbol{y}')|^2}} \begin{pmatrix} \nabla h_i(\boldsymbol{y}') \\ -1 \end{pmatrix},$$

where i is taken to be such that $\boldsymbol{y} = (\boldsymbol{y}', h_i(\boldsymbol{y}')) \in \Omega_i \cap \partial\Omega$.

Finally, the Lebesgue space on the boundary $\partial\Omega$ is defined similarly as before for Ω:

$$L^p(\partial\Omega) = \{f : \partial\Omega \to \mathbb{R} \mid f \text{ is integrable and } \|f\|_{L^p(\partial\Omega)} < \infty\},$$

with the norm

$$\|f\|^p_{L^p(\partial\Omega)} = \int_{\partial\Omega} |f(\boldsymbol{x})|^p \,\mathrm{d}\sigma_{\boldsymbol{x}}$$

and the usual convention in the case of $p = \infty$.

Remark 2.5.2 Since Lipschitz functions are almost everywhere differentiable, many classic results involving integrals of differentiable functions extend to Lipschitz functions. This includes a change of variables (substitution) under the integral as well as the fundamental theorem of calculus.

2.6 Sobolev Spaces on the Boundary

Having introduced Lebesgue spaces on the boundary, the Sobolev–Slobodeckij spaces on the boundary can be defined analogously to what is done for domains:

$$W^{s,p}(\partial\Omega) = \{v \in L^p(\partial\Omega) \mid |v|_{W^{s,p}(\partial\Omega)} < \infty\},$$

with the Sobolev–Slobodeckij semi-norm

$$|v|^p_{W^{s,p}(\partial\Omega)} = \iint_{\partial\Omega\times\partial\Omega} \frac{|v(\boldsymbol{x}) - v(\boldsymbol{y})|^p}{|\boldsymbol{x} - \boldsymbol{y}|^{sp+d-1}} \,\mathrm{d}\sigma_{\boldsymbol{x}} \,\mathrm{d}\sigma_{\boldsymbol{y}}.$$

Again, restrictions to parts of the boundary are possible.

Chapter 3
Properties of Sobolev Spaces

This chapter collects a number of important properties of Sobolev spaces. Almost every claim is provided together with a proof. The main result is the density of $\mathcal{D}(\overline{\Omega})$ in $W^{s,p}(\Omega)$ and proved in Theorem 3.4.5. The necessary tools to establish these proofs are introduced and intermediate results are presented in the following subsections. This entire chapter can be viewed as a preparation for subsequent ones on traces, Chap. 4, and to meaningfully define the weak forms of some partial differential equations in Chap. 5.

3.1 Mollifications

Essential parts in many proofs in this chapter are based on mollifications, which are convolutions with an appropriately scaled smooth function, which has compact support. This smooth function is called the standard mollifier and is introduced in the following definition. Afterwards, the mollification is defined and its main properties are shown, namely that it provides a technique to approximate integrable functions. This is introduced in many textbooks. e.g., [Eva98], Appendix C.4, and the second chapter in [AF03].

Definition 3.1.1 (Standard Mollifier) Define $\eta \in C^\infty(\mathbb{R}^d)$ as

$$\eta(\boldsymbol{x}) = \begin{cases} C\mathrm{e}^{1/(|\boldsymbol{x}|^2-1)} & \text{if } |\boldsymbol{x}| < 1 \\ 0 & \text{else,} \end{cases}$$

where C is chosen so that $\int_{\mathbb{R}^d} \eta(\boldsymbol{x})\,\mathrm{d}\boldsymbol{x} = 1$. Then the *standard mollifier*

$$\eta_\varepsilon(\boldsymbol{x}) := \varepsilon^{-d}\eta(\boldsymbol{x}/\varepsilon)$$

U. Wilbrandt, *Stokes–Darcy Equations*, Advances in Mathematical Fluid Mechanics,
https://doi.org/10.1007/978-3-030-02904-3_3

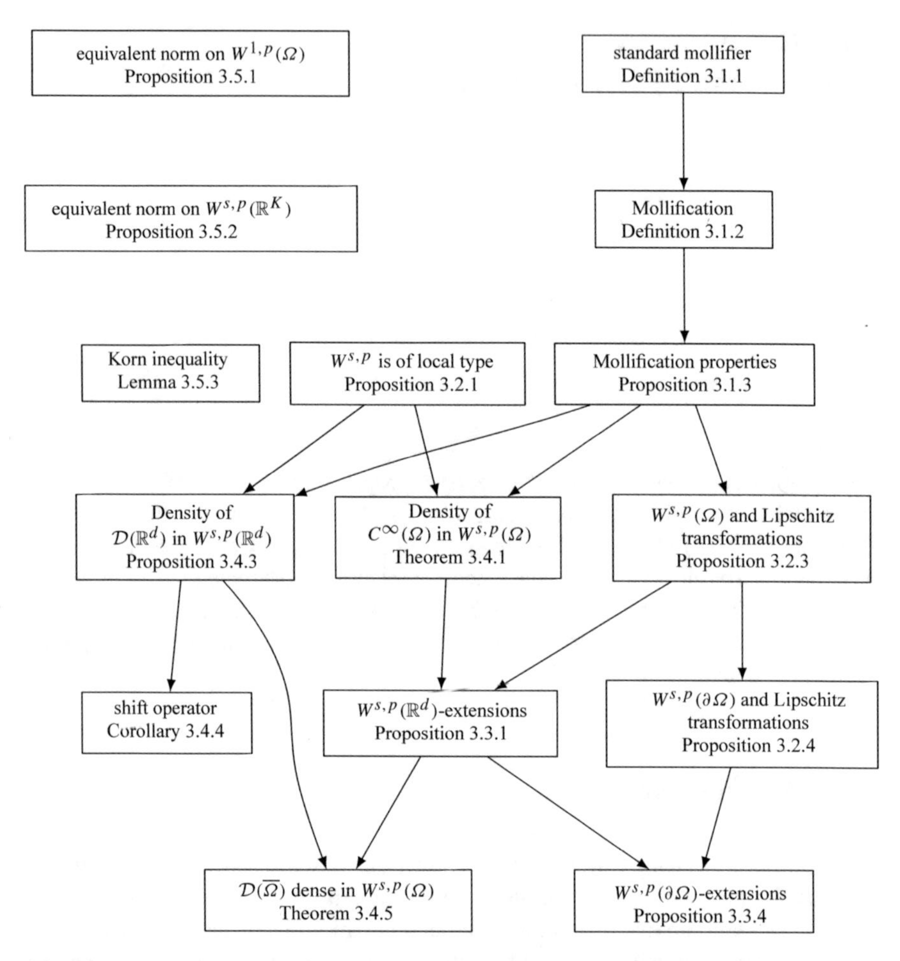

Fig. 3.1 A graph showing the dependencies among all theorems, definitions, corollaries, lemmas as well as propositions in Chap. 3. Note that implicit dependencies are not always shown

is also in $C^\infty(\mathbb{R}^d)$ and it holds

$$\int_{\mathbb{R}^d} \eta_\varepsilon(\boldsymbol{x})\,\mathrm{d}\boldsymbol{x} = 1, \qquad \operatorname{supp}(\eta_\varepsilon) = \overline{B(0,\varepsilon)} = \left\{\boldsymbol{x} \in \mathbb{R}^d \;\middle|\; |\boldsymbol{x}| \le \varepsilon\right\}.$$

Definition 3.1.2 (Mollification) For $\varepsilon > 0$ and a given $f \in L^1_{\mathrm{loc}}(\Omega)$ its *mollification* $f_\varepsilon := \eta_\varepsilon * f$, with the standard mollifier η_ε as in Definition 3.1.1, is defined on the slightly smaller set $\Omega_\varepsilon = \{\boldsymbol{x} \in \Omega \mid \operatorname{dist}(\boldsymbol{x}, \partial\Omega) > \varepsilon\}$.[1] That means

[1]In this definition Ω may be $\mathbb{R}^d$, then $\Omega_\varepsilon = \Omega$ for all ε.

for each $\boldsymbol{x} \in \Omega_\varepsilon$ it is

$$f_\varepsilon(\boldsymbol{x}) = \int_\Omega \eta_\varepsilon(\boldsymbol{x} - \boldsymbol{y}) f(\boldsymbol{y}) \,\mathrm{d}\boldsymbol{y} = \int_{B(0,\varepsilon)} \eta_\varepsilon(\boldsymbol{y}) f(\boldsymbol{x} - \boldsymbol{y}) \,\mathrm{d}\boldsymbol{y}.$$

The set Ω_ε is chosen such that f is only evaluated inside of Ω in the above integral. Extending f by zero outside of Ω results in a mollification on all of Ω or even $\mathbb{R}^d$. The same notation f is used for this extension if no confusion is possible.

The advantage of mollifications is that they are smooth approximations which is the statement of the following theorem, see [Eva98], Theorem 6 in Appendix C and Theorem 1 in Chapter 5.3 as well as [AF03], Theorem 2.29.

Proposition 3.1.3 (Mollification Properties) *Let $u \in L^1_{\mathrm{loc}}(\Omega)$ be given and u_ε its mollification as in Definition 3.1.2. Then it holds*

i) $u_\varepsilon \in \mathcal{C}^\infty(\Omega_\varepsilon)$ *and for* $\alpha \in \mathbb{N}^d$ *it is* $D^\alpha u_\varepsilon = D^\alpha(\eta_\varepsilon * u) = (D^\alpha \eta_\varepsilon) * u$,
ii) *if* $\operatorname{supp}(u) \subset \Omega$ *and* $0 < \varepsilon < \varrho := \operatorname{dist}(\operatorname{supp}(u), \partial\Omega)$, *then* $u_\varepsilon \in \mathcal{D}(\Omega)$,
iii) *if* $u \in L^p_{\mathrm{loc}}(\Omega)$, $p \geq 1$, *then* $u_\varepsilon \to u$ *in* $L^p_{\mathrm{loc}}(\Omega)$ *for* $\varepsilon \to 0$,
iv) *if* $u \in W^{k,p}_{\mathrm{loc}}(\Omega)$, $p \geq 1$, $k \in \mathbb{N}$, *then* $u_\varepsilon \to u$ *in* $W^{k,p}_{\mathrm{loc}}(\Omega)$ *for* $\varepsilon \to 0$,
v) *if* $u \in W^{s,p}_{\mathrm{loc}}(\Omega)$, $p \geq 1$, $s \in (0, 1)$, *then* $u_\varepsilon \to u$ *in* $W^{s,p}_{\mathrm{loc}}(\Omega)$ *for* $\varepsilon \to 0$.

Proof

i) Let $\boldsymbol{x} \in \Omega_\varepsilon$, $i = 1, \ldots, d$, and $h > 0$ small enough so that $\boldsymbol{x} + h\mathbf{e}_i \in \Omega_\varepsilon$. Then it is

$$\frac{u_\varepsilon(\boldsymbol{x} + h\mathbf{e}_i) - u_\varepsilon(\boldsymbol{x})}{h} = \frac{1}{\varepsilon^d} \int_\Omega \frac{1}{h}\left(\eta\left(\frac{\boldsymbol{x} + h\mathbf{e}_i - \boldsymbol{y}}{\varepsilon}\right) - \eta\left(\frac{\boldsymbol{x} - \boldsymbol{y}}{\varepsilon}\right)\right) u(\boldsymbol{y}) \,\mathrm{d}\boldsymbol{y}.$$

The support of the integrand above is a compact subset of Ω in which

$$\frac{1}{h}\left(\eta\left(\frac{\boldsymbol{x} + h\mathbf{e}_i - \boldsymbol{y}}{\varepsilon}\right) - \eta\left(\frac{\boldsymbol{x} - \boldsymbol{y}}{\varepsilon}\right)\right) \to \frac{1}{\varepsilon}\frac{\partial \eta}{\partial x_i}\left(\frac{\boldsymbol{x} - \boldsymbol{y}}{\varepsilon}\right) = \varepsilon^d \frac{\partial \eta_\varepsilon}{\partial x_i}(\boldsymbol{x} - \boldsymbol{y})$$

converges uniformly for $h \to 0$. According to the dominated convergence theorem the derivative in $\mathbf{e}_i$ direction of u_ε exists and equals

$$\frac{\partial u_\varepsilon}{\partial x_i}(\boldsymbol{x}) = \frac{\partial \eta_\varepsilon}{\partial x_i} * u(\boldsymbol{x}) = \int_\Omega \frac{\partial \eta_\varepsilon}{\partial x_i}(\boldsymbol{x} - \boldsymbol{y}) u(\boldsymbol{y}) \,\mathrm{d}\boldsymbol{y}.$$

Similarly, it can be shown that any derivative of u_ε exists and for all multiindices $\alpha \in \mathbb{N}_0^d$ it is $D^\alpha u_\varepsilon = D^\alpha(\eta_\varepsilon * u) = D^\alpha \eta_\varepsilon * u$. Therefore, $u_\varepsilon \in \mathcal{C}^\infty(\Omega_\varepsilon)$.

ii) According to *i)*, u_ε is an element of $\mathcal{D}(\Omega_\varepsilon)$. Due to the choice of ε, its extension by zero to all of Ω fulfills $\operatorname{supp}(u) \subset \operatorname{supp}(u_\varepsilon) \subset \overline{\Omega_{\varrho-\varepsilon}} \subset \Omega$ and hence is contained in $\mathcal{D}(\Omega)$.

iii) Let $U \subset \Omega$ be an open set whose closure is also contained in Ω, i.e., $\overline{U} \subset \Omega$, and choose $\varepsilon > 0$ such that $\varepsilon < \operatorname{dist}(U, \partial\Omega)$. Then, using Hölder's inequality, Theorem 2.3.1, it is

$$\begin{aligned} |u_\varepsilon(\boldsymbol{x})| &= \left| \int_{B(0,\varepsilon)} \eta_\varepsilon(\boldsymbol{y}) u(\boldsymbol{x} - \boldsymbol{y}) \, \mathrm{d}\boldsymbol{y} \right| \\ &\le \left(\int_{B(0,\varepsilon)} \eta_\varepsilon(\boldsymbol{y}) \, \mathrm{d}\boldsymbol{y} \right)^{1-1/p} \left(\int_{B(0,\varepsilon)} \eta_\varepsilon(\boldsymbol{y}) |u(\boldsymbol{x} - \boldsymbol{y})|^p \, \mathrm{d}\boldsymbol{y} \right)^{1/p} \end{aligned}$$

for any $\boldsymbol{x} \in U$. The first factor above is one and therefore the L^p-norm of u_ε can be bounded by

$$\begin{aligned} \|u_\varepsilon\|_{L^p(U)}^p = \int_U |u_\varepsilon(\boldsymbol{x})|^p \, \mathrm{d}\boldsymbol{x} &\le \int_U \int_{B(0,\varepsilon)} \eta_\varepsilon(\boldsymbol{y}) |u(\boldsymbol{x} - \boldsymbol{y})|^p \, \mathrm{d}\boldsymbol{y} \, \mathrm{d}\boldsymbol{x} \\ &= \int_{B(0,\varepsilon)} \eta_\varepsilon(\boldsymbol{y}) \int_U |u(\boldsymbol{x} - \boldsymbol{y})|^p \, \mathrm{d}\boldsymbol{x} \, \mathrm{d}\boldsymbol{y} \\ &\le \int_{B(0,\varepsilon)} \eta_\varepsilon(\boldsymbol{y}) \int_V |u(\boldsymbol{x})|^p \, \mathrm{d}\boldsymbol{x} \, \mathrm{d}\boldsymbol{y} \\ &= \|u\|_{L^p(V)}^p, \end{aligned}$$

where Fubini's theorem is applied and $V = \left\{ \boldsymbol{y} \in \mathbb{R}^d \;\middle|\; \operatorname{dist}(U, \boldsymbol{y}) < \varepsilon \right\} \subset \Omega$. Next, let $\delta > 0$ be given and choose a continuous $v \in C(V)$ such that

$$\|u - v\|_{L^p(V)} \le \delta/3$$

and hence $\|u_\varepsilon - v_\varepsilon\|_{L^p(U)} = \|(u - v)_\varepsilon\|_{L^p(U)} \le \|u - v\|_{L^p(V)} \le \delta/3$. Furthermore, for $\boldsymbol{x} \in U$ it is

$$|v_\varepsilon - v|(\boldsymbol{x}) = \int_{B(\boldsymbol{x},\varepsilon)} \eta_\varepsilon(\boldsymbol{x} - \boldsymbol{y}) |v(\boldsymbol{y}) - v(\boldsymbol{x})| \, \mathrm{d}\boldsymbol{y} \le \sup_{\boldsymbol{y} \in B(\boldsymbol{x},\varepsilon)} |v(\boldsymbol{y}) - v(\boldsymbol{x})|,$$

which tends to zero independently of $\boldsymbol{x}$ because v is uniformly continuous on $U \subsetneq V$. Therefore, upon possibly further reducing ε it is $\|v_\varepsilon - v\|_{L^p(V)} \le \delta/3$. Combining all these results yields

$$\|u_\varepsilon - u\|_{L^p(U)} \le \|u_\varepsilon - v_\varepsilon\|_{L^p(U)} + \|v_\varepsilon - v\|_{L^p(U)} + \|v - u\|_{L^p(U)} \le \delta.$$

iv) Let $u \in W^{k,p}_{\mathrm{loc}}$ and U an open set whose closure is contained in Ω. According to *iii)*, u_ε converges to u in $L^p(U)$. Additionally, the derivatives $D^\alpha u_\varepsilon$, $|\alpha| \le k$, belong to $L^p(U)$ and similarly to *i)* it is $D^\alpha u_\varepsilon = \eta_\varepsilon * D^\alpha u$. Using again *iii)* also $D^\alpha u_\varepsilon$ converges to $D^\alpha u$ in $L^p(U)$, which completes the proof.

v) This step is shown in a very similar manner as *iii)* but in $\mathbb{R}^{2d}$. As before, let U be an open set such that its closure is a subset of Ω and $V = \{\boldsymbol{y} \in \mathbb{R}^d \mid \operatorname{dist}(U, \boldsymbol{y}) < \varepsilon\} \subset \Omega$, i.e., $U \subset V$. Let $\delta > 0$ be given. For all $\boldsymbol{x}, \boldsymbol{y} \in V$ denote

$$f_u(\boldsymbol{x}, \boldsymbol{y}) = \frac{u(\boldsymbol{x}) - u(\boldsymbol{y})}{|\boldsymbol{x} - \boldsymbol{y}|^{s+d/p}},$$

so that $|f_u(\boldsymbol{x}, \boldsymbol{y})|^p$ is the integrand in the Sobolev–Slobodeckij semi-norm, see Eq. (2.10). Since $u \in W^{s,p}(V)$ it is $f_u \in (L^p(V))^2$. Now choose $g \in C(V^2)$ close to f_u such that $\|f_u - g\|_{(L^p(V))^2} \leq \delta/3$. Instead of a regular mollification in $\mathbb{R}^{2d}$, a modified one is used. For this purpose define $\widetilde{\eta} : \mathbb{R}^{2d} \to \mathbb{R}$ using the regular mollification η in $\mathbb{R}^d$:

$$\widetilde{\eta}(\boldsymbol{x}, \boldsymbol{y}) = \eta(\boldsymbol{x})\eta(\boldsymbol{y}).$$

The properties of η are inherited with slight modifications:

- $\operatorname{supp}(\widetilde{\eta}) = B(0, 1)^2 \subset B(0, \sqrt{2}) \subset \mathbb{R}^{2d}$ and
- $\int_{\mathbb{R}^{2d}} \widetilde{\eta}(\boldsymbol{x}, \boldsymbol{y}) \, \mathrm{d}\boldsymbol{x} \, \mathrm{d}\boldsymbol{y} = \left(\int_{B(0,1)} \eta(\boldsymbol{x}) \, \mathrm{d}\boldsymbol{x}\right)^2 = 1.$

The scaled version $\widetilde{\eta}_\varepsilon(\boldsymbol{x}, \boldsymbol{y}) = \varepsilon^{-2d}\widetilde{\eta}(\boldsymbol{x}/\varepsilon, \boldsymbol{y}/\varepsilon) = \eta_\varepsilon(\boldsymbol{x})\eta_\varepsilon(\boldsymbol{y})$ is then a mollifier in $\mathbb{R}^{2d}$ and therefore $g_\varepsilon := \widetilde{\eta}_\varepsilon * g$ an approximation to g. In particular, it is

$$\begin{aligned} (g_\varepsilon - g)(\boldsymbol{x}, \boldsymbol{y}) &= \iint_{B(0,\varepsilon)^2} \eta_\varepsilon(\boldsymbol{z})\eta_\varepsilon(\boldsymbol{w})(g(\boldsymbol{x} - \boldsymbol{z}, \boldsymbol{y} - \boldsymbol{w}) - g(\boldsymbol{x}, \boldsymbol{y})) \, \mathrm{d}\boldsymbol{z} \, \mathrm{d}\boldsymbol{w} \\ &\leq \sup_{(\boldsymbol{z},\boldsymbol{w}) \in B(0,\varepsilon)^2} |g(\boldsymbol{x} - \boldsymbol{z}, \boldsymbol{y} - \boldsymbol{w}) - g(\boldsymbol{x}, \boldsymbol{y})|, \end{aligned}$$

which tends to zero uniformly, i.e., independently of $(\boldsymbol{x}, \boldsymbol{y}) \in U^2$, because g is uniformly continuous on the closure of U^2 as it is continuous on the larger set V^2. This implies $\|g_\varepsilon - g\|_{(L^p(V))^2} \leq \delta/3$ for a small enough ε. Furthermore, f_{u_ε} and g_ε are close as well. First note that for $\boldsymbol{x}, \boldsymbol{y} \in U$ it is

$$\begin{aligned} &f_{u_\varepsilon}(\boldsymbol{x}, \boldsymbol{y}) - g_\varepsilon(\boldsymbol{x}, \boldsymbol{y}) \\ &\quad = \frac{(\eta_\varepsilon * u)(\boldsymbol{x}) - (\eta_\varepsilon * u)(\boldsymbol{y})}{|\boldsymbol{x} - \boldsymbol{y}|^{s+d/p}} - (\widetilde{\eta}_\varepsilon * g)(\boldsymbol{x}, \boldsymbol{y}) \\ &\quad = \int_{B(0,\varepsilon)} \eta_\varepsilon(\boldsymbol{z}) \frac{u(\boldsymbol{x} - \boldsymbol{z}) - u(\boldsymbol{y} - \boldsymbol{z})}{|\boldsymbol{x} - \boldsymbol{y}|^{s+d/p}} \, \mathrm{d}\boldsymbol{z} \end{aligned}$$

$$
\begin{aligned}
&- \iint_{B(0,\varepsilon)^2} \eta_\varepsilon(z)\eta_\varepsilon(\boldsymbol{w}) g(\boldsymbol{x}-z, \boldsymbol{y}-\boldsymbol{w})\, \mathrm{d}z\, \mathrm{d}\boldsymbol{w} \\
&= \iint_{B(0,\varepsilon)^2} \eta_\varepsilon(z)\eta_\varepsilon(\boldsymbol{w}) \left(\frac{u(\boldsymbol{x}-z) - u(\boldsymbol{y}-z)}{|\boldsymbol{x}-\boldsymbol{y}|^{s+d/p}} - g(\boldsymbol{x}-z, \boldsymbol{y}-\boldsymbol{w}) \right) \mathrm{d}z\, \mathrm{d}\boldsymbol{w} \\
&\leq \left(\iint_{B(0,\varepsilon)^2} \eta_\varepsilon(z)\eta_\varepsilon(\boldsymbol{w})\, \mathrm{d}z\, \mathrm{d}\boldsymbol{w} \right)^{1-1/p} \\
&\quad \left(\iint_{B(0,\varepsilon)^2} \eta_\varepsilon(z)\eta_\varepsilon(\boldsymbol{w}) \left| \frac{u(\boldsymbol{x}-z) - u(\boldsymbol{y}-z)}{|\boldsymbol{x}-\boldsymbol{y}|^{s+d/p}} - g(\boldsymbol{x}-z, \boldsymbol{y}-\boldsymbol{w}) \right|^p \mathrm{d}z\, \mathrm{d}\boldsymbol{w} \right)^{1/p}
\end{aligned}
$$

where Hölder's inequality, Theorem 2.3.1, is applied. The first factor above is one. Hence, the L^p-norm of $f_{u_\varepsilon} - g_\varepsilon$ is bounded by

$$
\begin{aligned}
&\left\| f_{u_\varepsilon} - g_\varepsilon \right\|^p_{(L^p(U))^2} \\
&\leq \iint_{U^2} \iint_{B(0,\varepsilon)^2} \eta_\varepsilon(z)\eta_\varepsilon(\boldsymbol{w}) \left| \frac{u(\boldsymbol{x}-z) - u(\boldsymbol{y}-z)}{|\boldsymbol{x}-\boldsymbol{y}|^{s+d/p}} - g(\boldsymbol{x}-z, \boldsymbol{y}-\boldsymbol{w}) \right|^p \mathrm{d}z\, \mathrm{d}\boldsymbol{w}\, \mathrm{d}\boldsymbol{x}\, \mathrm{d}\boldsymbol{y} \\
&\leq \iint_{V^2} \left| \frac{u(\boldsymbol{x}) - u(\boldsymbol{y})}{|\boldsymbol{x}-\boldsymbol{y}|^{s+d/p}} - g(\boldsymbol{x}, \boldsymbol{y}) \right|^p \mathrm{d}\boldsymbol{x}\, \mathrm{d}\boldsymbol{y} \\
&= \| f_u - g \|^p_{(L^p(V))^2} \leq \delta/3,
\end{aligned}
$$

where Fubini's theorem and the fact $U^2 \subset V^2$ are applied similarly to *iii)*. Finally, the results can be combined (using *ii)* for the L^p-part)

$$
\begin{aligned}
\|u_\varepsilon - u\|^p_{W^{s,p}(U)} &= \|u_\varepsilon - u\|^p_{L^p(U)} + \left\| f_{u_\varepsilon} - f_u \right\|^p_{(L^p(U))^2} \\
&\leq \delta + \left\| f_{u_\varepsilon} - g_\varepsilon \right\|^p_{(L^p(U))^2} + \|g_\varepsilon - g\|^p_{(L^p(U))^2} + \|g - f_u\|^p_{(L^p(U))^2} \\
&\leq 2\delta.
\end{aligned}
$$

□

Remark 3.1.4 The above properties remain valid in the case $\Omega = \mathbb{R}^d$.

3.2 Products with Smooth Functions and Lipschitz Transformations

The Hölder inequality, Theorem 2.3.1, shows that the product of two functions in suitable Lebesgue spaces is in L^1. In the following it is shown that the product of a function in a Sobolev space with a smooth one is in that same Sobolev space. Sometimes such spaces are called to be of local type, see [DDE12],

Proposition 4.26. A proof for unbounded domains can also be found in [DNPV12], Lemma 5.3.

Proposition 3.2.1 (**$W^{s,p}$ is of Local Type**) *Let $\Omega \subset \mathbb{R}^d$ be a bounded domain and $p > 1$, as well as $s \in [0,1]$. Then for any $u \in W^{s,p}(\Omega)$ and $\varphi \in \mathcal{D}(\Omega)$ their product φu is also in $W^{s,p}(\Omega)$ and it holds*

$$\|\varphi u\|_{W^{s,p}(\Omega)} \le c\|u\|_{W^{s,p}(\Omega)}$$

for a constant c which depends on $\|\varphi\|_\infty$ and, if $s > 0$, also on $\|\nabla\varphi\|_\infty$.

Proof In case $s = 0$, i.e., $u \in W^{0,p}(\Omega) = L^p(\Omega)$ the claim follows directly from the Hölder inequality, Theorem 2.3.1, because $\varphi^p \in L^\infty(\Omega)$ and $u^p \in L^1(\Omega)$:

$$\|\varphi u\|^p_{L^p(\Omega)} = \left\|\varphi^p u^p\right\|_{L^1(\Omega)} \le \left\|\varphi^p\right\|_{L^\infty(\Omega)}\left\|u^p\right\|_{L^1(\Omega)} = \left\|\varphi^p\right\|_{L^\infty(\Omega)}\|u\|^p_{L^p(\Omega)}.$$

For $s = 1$ the same reasoning shows that $\varphi\nabla u$ is in $(L^p(\Omega))^d$. Similarly it is $\nabla\varphi \in (\mathcal{D}(\Omega))^d$ and hence $\nabla\varphi u$ is also in $(L^p(\Omega))^d$. Therefore, $\nabla(\varphi u) = \nabla\varphi u + \varphi\nabla u$ is an element of $(L^p(\Omega))^d$, i.e., $\varphi u \in W^{1,p}(\Omega)$. In the case of intermediate $s \in (0,1)$, observe that the Sobolev–Slobodeckij semi-norm can be written in two terms using the inequality $(a+b)^p \le 2^{p-1}(a^p + b^p)$:

$$\begin{aligned}
|\varphi u|^p_{W^{s,p}(\Omega)} &= \iint_{\Omega^2} \frac{|(\varphi u)(\boldsymbol{x}) - (\varphi u)(\boldsymbol{y})|^p}{|\boldsymbol{x}-\boldsymbol{y}|^{sp+d}}\,\mathrm{d}\boldsymbol{x}\,\mathrm{d}\boldsymbol{y} \\
&\le 2^{p-1}\iint_{\Omega^2} \frac{|\varphi(\boldsymbol{x})(u(\boldsymbol{x}) - u(\boldsymbol{y}))|^p}{|\boldsymbol{x}-\boldsymbol{y}|^{sp+d}}\,\mathrm{d}\boldsymbol{x}\,\mathrm{d}\boldsymbol{y} \\
&\quad + 2^{p-1}\iint_{\Omega^2} \frac{|(\varphi(\boldsymbol{x}) - \varphi(\boldsymbol{y}))u(\boldsymbol{y})|^p}{|\boldsymbol{x}-\boldsymbol{y}|^{sp+d}}\,\mathrm{d}\boldsymbol{x}\,\mathrm{d}\boldsymbol{y}.
\end{aligned}$$

The first term can be bounded by the product of $\|\varphi\|^p_{L^\infty(\Omega)}$ and $|u|^p_{W^{s,p}(\Omega)}$. In order to bound the second term note that, due to the mean value theorem, it is $\varphi(\boldsymbol{x}) - \varphi(\boldsymbol{y}) \le \|\nabla\varphi\|_{L^\infty(\Omega)}|\boldsymbol{x}-\boldsymbol{y}|$ and therefore

$$\begin{aligned}
&\int_\Omega\int_\Omega \frac{|(\varphi(\boldsymbol{x}) - \varphi(\boldsymbol{y}))u(\boldsymbol{y})|^p}{|\boldsymbol{x}-\boldsymbol{y}|^{sp+d}}\,\mathrm{d}\boldsymbol{x}\,\mathrm{d}\boldsymbol{y} \\
&\qquad\le \|\nabla\varphi\|^p_{L^\infty(\Omega)}\int_\Omega |u(\boldsymbol{y})|^p \int_\Omega \frac{1}{|\boldsymbol{x}-\boldsymbol{y}|^{(s-1)p+d}}\,\mathrm{d}\boldsymbol{x}\,\mathrm{d}\boldsymbol{y}.
\end{aligned}$$

The inner integral is finite because $|x - y| \le \varrho := \operatorname{diam}(\Omega)$ and $(s-1)p + d < d$. In fact, using polar coordinates it can be bounded by $\omega_d\varrho^{p(1-s)}/(p(1-s))$, where ω_d is the measure of the unit ball's surface (i.e., the unit sphere). Concluding, $|\varphi u|_{W^{s,p}(\Omega)} \le c\|u\|_{W^{s,p}(\Omega)}$ where c depends on $\|\varphi\|^p_{L^\infty(\Omega)}$, $\|\nabla\varphi\|^p_{L^\infty(\Omega)}$, ϱ, s, p, and ω_d. □

Remark 3.2.2 In case a function $u \in W^{s,p}(\Gamma)$, $\Gamma \subset \partial\Omega$, $0 \le s < 1$, is multiplied with a smooth function ϕ, the same proof as in the previous Proposition 3.2.1 also shows that $u\phi \in W^{s,p}(\Gamma)$.

Many proofs involving Sobolev spaces on a bounded domain $\Omega \subset \mathbb{R}^d$ use a localization as introduced in Definition 2.2.1 to show the claim locally first. In a second step the local results are suitably combined to conclude the statement. The previous Proposition 3.2.1 shows that the products of the smooth functions ϕ_i, which formed the partition of unity, with a $W^{s,p}$-function u essentially keep the norm of the factor u. The transformations to the local setting from a global one and vice versa have to preserve the necessary norms, too. This is the core statement of the following two propositions:

Proposition 3.2.3 **($W^{s,p}(\Omega)$ and Lipschitz Transformations)** *Let Ω and Ω' be two Lipschitz domains. Furthermore, let $F : \Omega' \to \Omega$ be a bijective Lipschitz map whose inverse is also Lipschitz continuous. The composition $u \circ F : \Omega' \to \mathbb{R}$ of $u : \Omega \to \mathbb{R}$ inherits its properties from u as follows: If $u \in W^{s,p}(\Omega)$, $s \in [0, 1]$, then $u \circ F$ is in $W^{s,p}(\Omega')$ and there exists a constant c such that*

$$\|u \circ F\|_{W^{s,p}(\Omega')} \le c\|u\|_{W^{s,p}(\Omega)}.$$

Proof A similar proof for $s = 0$ and $s = 1$ can be found in [DDE12], Lemma 2.22 and 2.21.

The transformation formula (change of variables) yields

$$\|u \circ F\|^p_{L^p(\Omega')} = \int_{\Omega'} |u(F(\boldsymbol{y}))|^p \,\mathrm{d}\boldsymbol{y} = \int_{\Omega} |u(\boldsymbol{x})|^p \left|\det(\nabla F^{-1}(\boldsymbol{x}))\right| \mathrm{d}\boldsymbol{x}.$$

Since the determinant is a continuous operator, the above integral can be bounded by $c\|u\|^p_{L^p(\Omega)}$, where c depends on $\left\|\nabla F^{-1}\right\|_{L^\infty(\Omega)}$. In the case of $s = 1$ the same reasoning for the derivative shows the desired bound. However, it remains to show that $u \circ F$ is in $W^{1,p}(\Omega')$. Let $\phi \in \mathcal{D}(\Omega')$ be given and let $\varepsilon < \operatorname{dist}\bigl(\partial\Omega, F^{-1}(\operatorname{supp}(\phi))\bigr)$. The mollification u_ε of u is in $W^{1,p}_{\mathrm{loc}}(\Omega_\varepsilon)$, see Proposition 3.1.3. Since u_ε is smooth it is

$$-\int_{\Omega'} (u_\varepsilon \circ F)(\boldsymbol{x})\nabla\phi(\boldsymbol{x})\,\mathrm{d}\boldsymbol{x} = \int_{\Omega'} \nabla(u_\varepsilon \circ F)(\boldsymbol{x})\phi(\boldsymbol{x})\,\mathrm{d}\boldsymbol{x}$$
$$= \int_{\Omega'} \nabla u_\varepsilon(F(\boldsymbol{x})) \cdot \nabla F(\boldsymbol{x})\phi(\boldsymbol{x})\,\mathrm{d}\boldsymbol{x}.$$

Taking the limit $\varepsilon \to 0$ shows that $(\nabla u \circ F) \cdot \nabla F$ is the weak derivative of $u \circ F$.

Assume now $0 < s < 1$. Then again using a change of variables the Sobolev–Slobodeckij semi-norm can be bounded as well:

$$
\begin{aligned}
&|u \circ F|_{W^{s,p}(\Omega')} \\
&= \iint_{\Omega'^2} \frac{|u(F(y)) - u(F(y'))|^p}{|y - y'|^{sp+d}} \, dy \, dy' \\
&\le L_F^{sp+d} \iint_{\Omega'^2} \frac{|u(F(y)) - u(F(y'))|^p}{|F(y) - F(y')|^{sp+d}} \, dy \, dy' \\
&= L_F^{sp+d} \iint_{\Omega^2} \frac{|u(x) - u(x')|^p}{|x - x'|^{sp+d}} \left|\det(DF^{-1}(x))\right| \left|\det(DF^{-1}(x'))\right| \, dx \, dx' \\
&= c|u|_{W^{s,p}(\Omega)},
\end{aligned}
$$

where L_F is the Lipschitz constant of F and the constant c depends on L_F and on $\|\nabla F^{-1}\|_{L^\infty(\Omega)}$. □

The previous result, Proposition 3.2.3, extends to the case of a function u defined on the boundary of a domain Ω. Then it is $\|u \circ F\|_{W^{s,p}(\partial\Omega')} \le c\|u\|_{W^{s,p}(\partial\Omega)}$:

Proposition 3.2.4 ($W^{s,p}(\partial\Omega)$ and Lipschitz Transformations) *Let Ω, Ω', and F be as in Proposition 3.2.3 and $u \in W^{s,p}(\partial\Omega)$, $s \in [0, 1)$. Then the composition $u \circ F \in W^{s,p}(\partial\Omega')$ and there exists a constant c such that*

$$
\|u \circ F\|_{W^{s,p}(\partial\Omega')} \le c\|u\|_{W^{s,p}(\partial\Omega)}.
$$

Proof Consider a localization (V_i', h_i', ϕ_i') of Ω' as in Definition 2.2.1 and let $\{\Omega_i'\}_{i=0}^N$ be the corresponding covering of Ω'. Then the transformed sets $\Omega_i = F(\Omega_i')$ cover Ω and the transformed functions $\phi_i = \phi_i' \circ F^{-1}$ form a partition of unity adapted to Ω_i. Note however that the ϕ_i are no longer in $C_0^\infty(\Omega_i)$ but only Lipschitz continuous with compact support, $\phi_i \in C_0^{0,1}(\Omega_i)$, the following argument still holds though. Alternatively, they can be approximated uniformly with smooth functions.

Next, consider the L^p-norm of $u \circ F$ on $\partial\Omega'$:

$$
\begin{aligned}
\|u \circ F\|_{L^p(\partial\Omega')}^p &= \int_{\partial\Omega'} |u(F(x))|^p \, d\sigma_x = \sum_{i=1}^N \int_{\partial\Omega' \cap \Omega_i'} \phi_i'(x)|u(F(x))|^p \, d\sigma_x \\
&= \sum_{i=1}^N \int_{\partial\Omega' \cap \Omega_i'} \phi_i(F(x))|u(F(x))|^p \, d\sigma_x \\
&= \sum_{i=1}^N \int_{V_i'} \phi_i(F(x, h_i'(x))) \left|u(F(x, h_i'(x)))\right|^p \sqrt{1 + \left|\nabla h_i'(x)\right|^2} \, dx.
\end{aligned}
$$

For each $\boldsymbol{y} \in V_i \subset \mathbb{R}^{d-1}$ there exists an $\boldsymbol{x} \in V_i' \subset \mathbb{R}^{d-1}$ such that $F(\boldsymbol{x}, h_i'(\boldsymbol{x})) = (\boldsymbol{y}, h_i(\boldsymbol{y}))$ and it is unique because F is bijective. Therefore, the map $\kappa_i : V_i \to V_i'$, $\boldsymbol{y} \mapsto \boldsymbol{x}$ is well defined and bijective, even Lipschitz continuous:

$$\begin{aligned}
\left|\kappa_i(\boldsymbol{y}_1) - \kappa_i(\boldsymbol{y}_2)\right| &\le \left|(\kappa_i(\boldsymbol{y}_1), h_i'(\kappa_i(\boldsymbol{y}_1))) - (\kappa_i(\boldsymbol{y}_2), h_i'(\kappa_i(\boldsymbol{y}_2)))\right| \\
&= \left|F^{-1}(\boldsymbol{y}_1, h_i(\boldsymbol{y}_1)) - F^{-1}(\boldsymbol{y}_2, h_i(\boldsymbol{y}_2))\right| \\
&\le L_{F^{-1}}\left|(\boldsymbol{y}_1, h_i(\boldsymbol{y}_1)) - (\boldsymbol{y}_2, h_i(\boldsymbol{y}_2))\right| \\
&= L_{F^{-1}}\sqrt{\left|\boldsymbol{y}_1 - \boldsymbol{y}_2\right|^2 + \left|h_i(\boldsymbol{y}_1) - h_i(\boldsymbol{y}_2)\right|^2} \\
&\le L_{F^{-1}}\sqrt{1 + L_{h_i}^2}\,\left|\boldsymbol{y}_1 - \boldsymbol{y}_2\right|,
\end{aligned}$$

where L are the Lipschitz constants of the functions in the subscript, i.e., it is $L_{\kappa_i} \le L_{F^{-1}}\sqrt{1 + L_{h_i}^2}$. In particular the determinant of the derivative of κ_i is bounded by a constant C which only depends on the domain Ω and the transformation F^{-1}, $|\det(\nabla\kappa_i)| \le C$. In the integrals over $V_i' \subset \mathbb{R}^{d-1}$ above a change of variables $\boldsymbol{x} = \kappa_i(\boldsymbol{y})$ is done. For this purpose it is necessary to understand what happens to the term in the square root above. In fact, it is the norm of a suitable vector whose behavior under the desired change of variables is easier to study. Let $g_i : \mathbb{R}^d \to \mathbb{R}$ be such that $\partial\Omega \cap \Omega_i = g_i^{-1}(\{0\})$, i.e., $g_i(\boldsymbol{y}, \xi) = h_i(\boldsymbol{y}) - \xi$ and $g_i(\boldsymbol{y}, h_i(\boldsymbol{y})) = 0$ for all $\boldsymbol{y} \in V_i$. Similarly define g_i' and note that it is $g_i' = g_i \circ F$. With this notation it is

$$\begin{aligned}
\sqrt{1 + \left|\nabla h_i'(\boldsymbol{x})\right|^2} = \left|\nabla g_i'(\boldsymbol{x}, h_i'(\boldsymbol{x}))\right| &= \left|\nabla g_i(F(\boldsymbol{x}, h_i'(\boldsymbol{x})))\nabla F(\boldsymbol{x}, h_i'(\boldsymbol{x}))\right| \\
&\le \|\nabla F\|_{L^\infty(\Omega')}\left|\nabla g_i(F(\boldsymbol{y}, h_i'(\boldsymbol{y})))\right|.
\end{aligned}$$

Consequently, using the two identities $F\big(\kappa_i(\boldsymbol{y}), h_i'(\kappa_i(\boldsymbol{y}))\big) = (\boldsymbol{y}, h_i(\boldsymbol{y}))$ and $|\nabla g_i(\boldsymbol{y}, h_i(\boldsymbol{y}))| = \sqrt{1 + |\nabla h_i(\boldsymbol{y})|^2}$, a change of variables $\boldsymbol{x} = \kappa_i(\boldsymbol{y})$ in the L^p-norm of $u \circ F$ on $\partial\Omega'$ yields

$$\begin{aligned}
&\|u \circ F\|_{L^p(\partial\Omega')}^p \\
&\quad \le C\|\nabla F\|_{L^\infty(\Omega')}\sum_{i=1}^N \int_{V_i} \phi_i(\boldsymbol{y}, h_i(\boldsymbol{y}))|u(\boldsymbol{y}, h_i(\boldsymbol{y}))|^p\sqrt{1 + |\nabla h_i(\boldsymbol{y})|^2}\,\mathrm{d}\boldsymbol{y} \\
&\quad = C\|\nabla F\|_{L^\infty(\Omega')}\sum_{i=1}^N \int_{\partial\Omega\cap\Omega_i} \phi_i(\boldsymbol{y}, h_i(\boldsymbol{y}))|u(\boldsymbol{y}, h_i(\boldsymbol{y}))|^p\,\mathrm{d}\sigma_{\boldsymbol{y}} \\
&\quad = C\|\nabla F\|_{L^\infty(\Omega')}\|u\|_{L^p(\partial\Omega)}^p.
\end{aligned}$$

Next, consider the Sobolev-Slobodeckij semi-norm of u:

$$\iint_{\partial\Omega'^2} \frac{|u(F(\boldsymbol{x})) - u(F(\boldsymbol{y}))|^p}{|\boldsymbol{x}-\boldsymbol{y}|^{sp+d-1}} \,\mathrm{d}\sigma_{\boldsymbol{x}}\,\mathrm{d}\sigma_{\boldsymbol{y}}$$

$$= \sum_{i=1}^{N}\sum_{j=1}^{N} \iint_{V_i'^2} \phi_i(F(\boldsymbol{x}, h_i'(\boldsymbol{x})))\phi_j(F(\boldsymbol{y}, h_i'(\boldsymbol{y})))$$

$$\frac{\left|u(F(\boldsymbol{x}, h_i'(\boldsymbol{x}))) - u(F(\boldsymbol{y}, h_i'(\boldsymbol{y})))\right|^p}{\left|(\boldsymbol{x}, h_i'(\boldsymbol{x})) - (\boldsymbol{y}, h_i'(\boldsymbol{y}))\right|^{sp+d-1}}$$

$$\sqrt{1+\left|\nabla h_i'(\boldsymbol{x})\right|^2}\sqrt{1+\left|\nabla h_i'(\boldsymbol{y})\right|^2}\,\mathrm{d}\boldsymbol{x}\,\mathrm{d}\boldsymbol{y}.$$

Again, changing variables using κ_i and κ_j as before and using the Lipschitz continuity of the transformation F, i.e., $\left|(\boldsymbol{x}, h_i'(\boldsymbol{x})) - (\boldsymbol{y}, h_i'(\boldsymbol{y}))\right|^{-1} \leq L_F\left|F(\boldsymbol{x}, h_i'(\boldsymbol{x})) - F(\boldsymbol{y}, h_i'(\boldsymbol{y}))\right|^{-1}$, gives the desired result for the Sobolev-Slobodeckij semi-norm, $|u \circ F|_{W^{s,p}(\partial\Omega')} \leq c|u|_{W^{s,p}(\partial\Omega)}$. Summarizing, there is a constant c depending on the domain Ω and the transformation F such that $\|u \circ F\|_{W^{s,p}(\partial\Omega')} \leq c\|u\|_{W^{s,p}(\partial\Omega)}$. □

3.3 Extension from Ω to $\mathbb{R}^d$

The following proposition states that functions defined on a Lipschitz domain can be continuously extended to all of $\mathbb{R}^d$. The proof is taken from [DDE12], Proposition 4.43 and [DNPV12], Theorem 5.4, for $0 < s < 1$. The case $s = 1$ is shown in [DDE12], Proposition 2.70, as well as in [Eva98], Theorem 1 in Section 5.4.

Proposition 3.3.1 ($W^{s,p}(\mathbb{R}^d)$-Extensions) *Let $\Omega \subset \mathbb{R}^d$ be Lipschitz, $p > 1$ and $s \in [0, 1]$. Then there exists a linear and continuous extension operator $E : W^{s,p}(\Omega) \to W^{s,p}(\mathbb{R}^d)$, i.e., for all $u \in W^{s,p}(\Omega)$ and almost every $\boldsymbol{x} \in \Omega$ it is $Eu(\boldsymbol{x}) = u(\boldsymbol{x})$.*

Proof Note that for $s = 0$ a function $u \in W^{0,p}(\Omega) = L^p(\Omega)$ can be extended by zero outside of Ω. Then the norm is not changed, $\|u\|_{L^p(\mathbb{R}^d)} = \|u\|_{L^p(\Omega)}$.

Let (V_i, h_i, ϕ_i) be a localization of Ω, see Definition 2.2.1, and $u \in W^{s,p}(\Omega)$. Then the product $u\phi_i$ is in $W^{s,p}(\Omega)$ as well with

$$\|u\phi_i\|_{W^{s,p}(\Omega)} \leq c\|u\|_{W^{s,p}(\Omega)},$$

see Proposition 3.2.1. For any point $\boldsymbol{x} = (\boldsymbol{x}', x_d)$ of Ω_i outside of Ω, i.e., $x_d < h_i(\boldsymbol{x}')$, define its reflexion $P(\boldsymbol{x}) = (\boldsymbol{x}', 2h_i(\boldsymbol{x}') - x_d)$. Next, define the local extension $\widetilde{u\phi_i}$ of $u\phi_i$ as

$$\widetilde{u\phi_i}(\boldsymbol{x}, x_d) = \begin{cases} 0 & \text{if } (\boldsymbol{x}', x_d) \notin \Omega_i, \\ (u\phi_i)(\boldsymbol{x}', x_d) & \text{if } x_d \geq h_i(\boldsymbol{x}'), \\ (u\phi_i)(P(\boldsymbol{x}', x_d)) & \text{if } x_d < h_i(\boldsymbol{x}'). \end{cases}$$

Clearly, $\widetilde{u\phi_i}$ is in $L^p(\mathbb{R}^d)$ with $\|\widetilde{u\phi_i}\|_{L^p(\mathbb{R}^d)} \leq 2\|u\phi_i\|_{L^p(\Omega)}$. The cases $s = 1$ and $0 < s < 1$ are treated separately.

$0 < s < 1$: In order to show that this local extension is in fact in $W^{s,p}(\mathbb{R}^d)$, the double integral in the Sobolev–Slobodeckij semi-norm $|\widetilde{u\phi_i}|^p_{W^{s,p}(\mathbb{R}^d)}$ is split into four parts J_1, J_2, J_3, and J_4 which in turn are double integrals over the sets $\Omega \times \Omega$, $\Omega \times \Omega_c$, $\Omega_c \times \Omega$, and $\Omega_c \times \Omega_c$ with $\Omega_c = \mathbb{R}^d \setminus \Omega$. The first part J_1 is $|u\phi_i|_{W^{s,p}(\Omega)}$. The second and the third are equal due to the Theorem of Fubini and can be bounded as follows. For $\boldsymbol{x} \in \Omega$ and $\boldsymbol{y} \in \Omega_c$ (i.e., $x_d \geq h_i(\boldsymbol{x}')$ and $y_d \leq h_i(\boldsymbol{y}')$) it is[2]

$$\begin{aligned} x_d - 2h_i(\boldsymbol{y}') + y_d &\leq x_d - y_d \leq |\boldsymbol{x} - \boldsymbol{y}|, \\ x_d - 2h_i(\boldsymbol{y}') + y_d &\geq 2(h_i(\boldsymbol{x}') - h_i(\boldsymbol{y}')) - x_d + y_d \\ &\geq -2\|\nabla h_i\|_\infty |\boldsymbol{x}' - \boldsymbol{y}'| - |x_d - y_d| \\ &\geq -(1 + 2\|\nabla h_i\|_\infty)|\boldsymbol{x} - \boldsymbol{y}|. \end{aligned}$$

It therefore is $|\boldsymbol{x} - P(\boldsymbol{y})| = |\boldsymbol{x}' - \boldsymbol{y}'| + |x_d - 2h_i(\boldsymbol{y}') + y_d| \leq C|\boldsymbol{x} - \boldsymbol{y}|$ where C depends on $\|\nabla h_i\|_\infty$ which in turn can be bounded independently of i. Then with a different C it is

$$\begin{aligned} J_2 &= \int_\Omega \int_{\Omega_c} \frac{|\widetilde{u\phi_i}(\boldsymbol{x}) - \widetilde{u\phi_i}(\boldsymbol{y})|^p}{|\boldsymbol{x} - \boldsymbol{y}|^{sp+d}} \, \mathrm{d}\boldsymbol{x} \, \mathrm{d}\boldsymbol{y} \leq C \int_\Omega \int_{\Omega_c} \frac{|\widetilde{u\phi_i}(\boldsymbol{x}) - \widetilde{u\phi_i}(\boldsymbol{y})|^p}{|\boldsymbol{x} - P(\boldsymbol{y})|^{sp+d}} \, \mathrm{d}\boldsymbol{x} \, \mathrm{d}\boldsymbol{y} \\ &= C \int_\Omega \int_{\Omega_c} \frac{|(u\phi_i)(\boldsymbol{x}) - (u\phi_i)(P(\boldsymbol{y}))|^p}{|\boldsymbol{x} - P(\boldsymbol{y})|^{sp+d}} \, \mathrm{d}\boldsymbol{x} \, \mathrm{d}\boldsymbol{y} \\ &= C \iint_{\Omega^2} \frac{|(u\phi_i)(\boldsymbol{x}) - (u\phi_i)(\boldsymbol{y})|^p}{|\boldsymbol{x} - \boldsymbol{y}|^{sp+d}} \, \mathrm{d}\boldsymbol{x} \, \mathrm{d}\boldsymbol{y}, \end{aligned}$$

where in the last step a change of variables is done. Hence, $J_2 = J_3 \leq CJ_1$. In the last integral J_4 note that for $\boldsymbol{x} \in \Omega_c$ and $\boldsymbol{y} \in \Omega_c$ it is

$$\begin{aligned} |P(\boldsymbol{x}) - P(\boldsymbol{y})| &= |\boldsymbol{x}' - \boldsymbol{y}'| + |2h_i(\boldsymbol{x}') - x_d - 2h_i(\boldsymbol{y}') + y_d| \\ &\leq (1 + 2\|\nabla h_i\|_\infty)|\boldsymbol{x} - \boldsymbol{y}|. \end{aligned}$$

[2]For simplicity here the norm on $\mathbb{R}^d$ is the 1-norm, i.e., $|\boldsymbol{x}| = \sum |x_i|$.

Hence, the integral can be bounded, similarly as above, by CJ_1. Together it is shown that $\widetilde{u\phi_i} \in W^{s,p}(\mathbb{R}^d)$ and that $|\widetilde{u\phi_i}|_{W^{s,p}(\mathbb{R}^d)} \le 4C|u\phi_i|_{W^{s,p}(\Omega\cap\Omega_i)}$. Combining these results, it is

$$\|\widetilde{u\phi_i}\|^p_{W^{s,p}(\mathbb{R}^d)} \le 2\|u\phi_i\|^p_{L^p(\Omega)} + 4C|u\phi_i|^p_{W^{s,p}(\Omega\cap\Omega_i)}.$$

Finally, these local extensions are combined to a global one, namely $Eu = \sum_{i=0}^N \widetilde{u\phi_i}$, which is linear and continuous:

$$\|Eu\|_{W^{s,p}(\mathbb{R}^d)} \le \sum_{i=0}^N \|\widetilde{u\phi_i}\|_{W^{s,p}(\mathbb{R}^d)} \le C\|u\|_{W^{s,p}(\Omega)}$$

with a constant C which does not depend on u.

$s = 1$: First, a result very similar to the definition of weak derivatives is shown, see also Lemma 2.58 in [DDE12]. With $v \in W^{1,p}(\mathbb{R}^d_+)$, $\mathbb{R}^d_+$ is the half space $\{\boldsymbol{x} = (\boldsymbol{x}', x_d) \in \mathbb{R}^d \mid x_d > 0\}$, and $\varphi \in \mathcal{D}(\mathbb{R}^d)$[3] it is

$$\int_{\mathbb{R}^d_+} \frac{\partial v}{\partial x_j}(\boldsymbol{x})\varphi(\boldsymbol{x})\,\mathrm{d}\boldsymbol{x} = -\int_{\mathbb{R}^d_+} v(\boldsymbol{x})\frac{\partial \varphi}{\partial x_j}\,\mathrm{d}\boldsymbol{x} \tag{3.1}$$

for $1 \le j < d$ and this equation is also true for $j = d$ provided that $\varphi(\boldsymbol{x}', 0) = 0$ for all $\boldsymbol{x}' \in \mathbb{R}^{d-1}$. To see this, let $v_n \in C^\infty(\mathbb{R}^d_+) \cap W^{1,p}(\mathbb{R}^d_+)$ converge[4] to v in $W^{1,p}(\mathbb{R}^d_+)$. For the v_n the stated equation (3.1) is true (integration by parts for $i = d$) and taking the limit also shows it for v.

For convenience denote $v := u\phi_i$ and transform the space $\mathbb{R}^d$ such that the boundary of Ω is straightened out, i.e., define the bi-Lipschitz continuous transformation $F : \mathbb{R}^d \to \mathbb{R}^d$ to be $F(\boldsymbol{x}', x_d) = (\boldsymbol{x}', x_d + h_i(\boldsymbol{x}'))$. Then, according to Proposition 3.2.3, the function $w := v \circ F$ is in $W^{1,p}(\mathbb{R}^d_+)$. Assume that $\widehat{w} := \widetilde{v} \circ F$ is a $W^{1,p}$-extension of w with $\|\widehat{w}\|_{W^{1,p}(\mathbb{R}^d)} \le 2\|w\|_{W^{1,p}(\mathbb{R}^d_+)}$, then Proposition 3.2.3 states that $\widetilde{v} \in W^{1,p}(\mathbb{R}^d)$ and

$$\|\widetilde{v}\|_{W^{1,p}(\mathbb{R}^d)} \le \tilde{c}\|\widehat{w}\|_{W^{1,p}(\mathbb{R}^d)} \le 2\tilde{c}\|w\|_{W^{1,p}(\mathbb{R}^d_+)} \le c\|v\|_{W^{1,p}(\Omega_i\cap\Omega)},$$

where c is a constant depending on $\|h_i\|_\infty$. It remains to show that $\widehat{w}$ is a $W^{1,p}$ extension of w. Note that it is

$$\widehat{w}(\boldsymbol{y}', y_d) = \begin{cases} w(\boldsymbol{y}', y_d) & \text{for } y_d \ge 0 \\ w(\boldsymbol{y}', -y_d) & \text{for } y_d < 0. \end{cases}$$

[3]In contrast to the definition of the weak derivative, the test function φ is not in $\mathcal{D}(\mathbb{R}^d_+)$ but in $\mathcal{D}(\mathbb{R}^d)$.

[4]It is shown in Theorem 3.4.1 and Remark 3.4.2 that such a sequence exists.

A change of sign in the last component yields

$$\|\widehat{w}\|^p_{L^p(\mathbb{R}^d)} = \int_{\mathbb{R}^d_+} |w(\boldsymbol{x})|^p \, \mathrm{d}\boldsymbol{x} + \int_{\mathbb{R}^d_-} \left|w(\boldsymbol{x}', -x_d)\right|^p \, \mathrm{d}\boldsymbol{x} = 2\|w\|^p_{L^p(\mathbb{R}^d_+)}.$$

To show the (weak) differentiability of $\widehat{w}$, let $\varphi \in \mathcal{D}(\mathbb{R}^d)$ be given. Together with φ also $(\boldsymbol{x}', x_d) \mapsto \varphi(\boldsymbol{x}', x_d) + \varphi(\boldsymbol{x}', -x_d)$ is in $\mathcal{D}(\mathbb{R}^d)$ and Eq. (3.1) yields

$$\begin{aligned}
\int_{\mathbb{R}^d} \widehat{w}(\boldsymbol{x}) \frac{\partial \varphi}{\partial x_j}(\boldsymbol{x}) \, \mathrm{d}\boldsymbol{x} &= \int_{\mathbb{R}^d_+} w(\boldsymbol{x}) \frac{\partial}{\partial x_j} \big(\varphi(\boldsymbol{x}) + \varphi(\boldsymbol{x}', -x_d)\big) \, \mathrm{d}\boldsymbol{x} \\
&= - \int_{\mathbb{R}^d_+} \frac{\partial w}{\partial x_j}(\boldsymbol{x}) \big(\varphi(\boldsymbol{x}) + \varphi(\boldsymbol{x}', -x_d)\big) \, \mathrm{d}\boldsymbol{x} \\
&= - \int_{\mathbb{R}^d} \left(\frac{\partial w}{\partial x_j}(\boldsymbol{x}) \chi_{\mathbb{R}^d_+}(\boldsymbol{x}) + \frac{\partial w}{\partial x_j}(\boldsymbol{x}, -x_d) \chi_{\mathbb{R}^d_-}(\boldsymbol{x}) \right) \varphi(\boldsymbol{x}) \, \mathrm{d}\boldsymbol{x},
\end{aligned}$$

where χ_M is the indicator function of the set M. Hence, the weak derivative in x_j-direction of $\widehat{w}$ is given by the function in the parentheses in the above integrand and is an element of $L^p(\mathbb{R}^d)$. Finally, consider the derivative in x_d-direction. The function $(\boldsymbol{x}', x_d) \mapsto \varphi(\boldsymbol{x}', x_d) - \varphi(\boldsymbol{x}', -x_d)$ is in $\mathcal{D}(\mathbb{R}^d)$ and vanishes on $\{x_d = 0\}$. Therefore, Eq. (3.1) shows

$$\begin{aligned}
\int_{\mathbb{R}^d} \widehat{w}(\boldsymbol{x}) \frac{\partial \varphi}{\partial x_d}(\boldsymbol{x}) \, \mathrm{d}\boldsymbol{x} &= \int_{\mathbb{R}^d_+} w(\boldsymbol{x}) \frac{\partial}{\partial x_d} \big(\varphi(\boldsymbol{x}', x_d) - \varphi(\boldsymbol{x}', -x_d)\big) \, \mathrm{d}\boldsymbol{x} \\
&= - \int_{\mathbb{R}^d_+} \frac{\partial w}{\partial x_d}(\boldsymbol{x}) \big(\varphi(\boldsymbol{x}) - \varphi(\boldsymbol{x}', -x_d)\big) \, \mathrm{d}\boldsymbol{x} \\
&= - \int_{\mathbb{R}^d} \left(\frac{\partial w}{\partial x_d}(\boldsymbol{x}) \chi_{\mathbb{R}^d_+}(\boldsymbol{x}) - \frac{\partial w}{\partial x_d}(\boldsymbol{x}, -x_d) \chi_{\mathbb{R}^d_-}(\boldsymbol{x}) \right) \varphi(\boldsymbol{x}) \, \mathrm{d}\boldsymbol{x}.
\end{aligned}$$

Again the term in the parentheses in the above integrand is in $L^p(\mathbb{R}^d)$ and therefore the weak derivative of $\widehat{w}$. The explicit derivatives also show the required estimate $\|\widehat{w}\|_{W^{1,p}(\mathbb{R}^d)} \leq 2\|w\|_{W^{1,p}(\mathbb{R}^d_+)}$. □

Remark 3.3.2 In the previous Proposition 3.3.1, the support of the extension Eu of u can be chosen: Let $V \subset \mathbb{R}^d$ be a bounded set such that $\overline{\Omega} \subset V$. Then define a cut-off function $\phi \in \mathcal{D}(\mathbb{R}^d)$ with $\phi(\boldsymbol{x}) = 1$ for $\boldsymbol{x} \in \Omega$, $\operatorname{supp}(\phi) \subset V$. Then the product $v = \phi E u$ is an element of $W^{s,p}(\mathbb{R}^d)$ due to Proposition 3.2.1 and its support is in V. Furthermore, it is an extension of u and it holds

$$\|v\|_{W^{s,p}(V)} \leq C\|u\|_{W^{s,p}(\Omega)},$$

where C additionally depends on V, in particular on $\operatorname{dist}(V, \Omega) > 0$.

Remark 3.3.3 In [AF03], Definition 5.17, extension operators on $W^{k,p}(\Omega)$, $k \in \mathbb{N}$ are further distinguished: They are called simple-k, p-extension operators if the properties in the above Proposition 3.3.1 are fulfilled for given k and p. Furthermore, a strong-m-extension operator is a simple-k, p-extension operator for all $p \geq 1$ and $k \leq m$. Moreover, a total extension operator is a strong-m-extension operator for each $m \in \mathbb{N}$. In the proof of Proposition 3.3.1 the extension is constructed independently of p, therefore it is even a strong-1-extension-operator, but not a total extension operator. In [AF03] the authors refer to [Ste70] for a proof of the existence of a total extension operator for Ω possessing a "strong local Lipschitz condition".

In applications one faces boundary conditions on a part Γ of the boundary $\partial\Omega$. In order to proof existence and uniqueness of solutions of some partial differential equations, it is necessary to extend functions on Γ to all of $\partial\Omega$. This in turn poses a condition on the boundary of Γ because in general it is not possible to extend a $W^{s,p}(\Gamma)$-function continuously to $W^{s,p}(\partial\Omega)$. The next theorem therefore requires the boundary $\partial\Gamma$ of Γ to be Lipschitz as well. This means that locally after transforming $\Omega_i \cap \partial\Omega$ to $\mathbb{R}^{d-1}$, the set originating from Γ is Lipschitz in $\mathbb{R}^{d-1}$. This allows for an extension of a function in $W^{s,p}(\Gamma)$ to $W^{s,p}(\partial\Omega)$:

Proposition 3.3.4 ($W^{s,p}(\partial\Omega)$-Extensions) *Let $\Omega \subset \mathbb{R}^d$ be a Lipschitz domain and $\Gamma \subset \partial\Omega$ a relatively open subset of its boundary. Furthermore, let Γ itself be of Lipschitz type and $s \in [0, 1)$. Then there exists a linear and continuous extension operator $E : W^{s,p}(\Gamma) \to W^{s,p}(\partial\Omega)$, i.e., for all $u \in W^{s,p}(\Gamma)$ and almost every $\boldsymbol{x} \in \Gamma$ it is $Eu(\boldsymbol{x}) = u(\boldsymbol{x})$.*

Proof Let $u \in W^{s,p}(\Gamma)$ be given and let (V_i, h_i, Φ_i), $i = 1, \ldots, N$, be a localization of Ω as in Definition 2.2.1. The extension is composed of local contributions with respect to the sets $\Omega_i \cap \partial\Omega$. If Γ and $\partial\Omega$ coincide locally, i.e., if $\Gamma \cap \Omega_i = \partial\Omega \cap \Omega_i$, the function $\Phi_i u$ has compact support in $\partial\Omega \cap \Omega_i$ and belongs to $W^{s,p}(\partial\Omega)$. In case the boundary of Γ intersects Ω_i, define the local version of u on $\mathbb{R}^{d-1}$ as $v_i(\boldsymbol{x}') := \Phi_i u(\boldsymbol{x}', h_i(\boldsymbol{x}'))$ whenever $\boldsymbol{x}' \in \tilde{\Gamma}_i := \{\boldsymbol{y}' \in V_i \mid (\boldsymbol{y}', h_i(\boldsymbol{y}')) \in \Gamma\}$. Due to Propositions 3.2.1 and 3.2.4 together with Remark 3.2.2 the function v_i is in $W^{s,p}(\tilde{\Gamma}_i)$. According to the assumption, the set $\tilde{\Gamma}_i \subset \mathbb{R}^{d-1}$ is itself of Lipschitz type. Proposition 3.3.1 therefore gives an extension $v_i \in W^{s,p}(\mathbb{R}^{d-1})$ of the function v_i, where for simplicity the same name is used. It is furthermore assumed that $\operatorname{supp}(v_i) \subset V_i$, see Remark 3.3.2. Transforming this back to the boundary of Ω, $u_i(\boldsymbol{x}', h_i(\boldsymbol{x}')) = v_i(\boldsymbol{x}')$, yields an element in $W^{s,p}(\Omega_i \cap \partial\Omega)$ which extends $\Phi_i u$. Next, define the desired extension operator as the sum $Eu = \sum_{i=1}^{N} u_i$. Since the transformations to and from $\mathbb{R}^{d-1}$ as well as the extension operator on $\mathbb{R}^{d-1}$ are linear and continuous, the global extension E is, too:

$$\|Eu\|_{W^{s,p}(\partial\Omega)} \leq C \sum_{i=1}^{N} \|\Phi_i u\|_{W^{s,p}(\Gamma)} \leq CN\|u\|_{W^{s,p}(\Gamma)}.$$

□

Remark 3.3.5 A subset Γ of the boundary of a domain Ω which itself has a Lipschitz boundary is also called *admissible patch*. In this context the so-called *creased domains* are studied. These are Lipschitz domains together with two disjoint subsets of its boundary which are separated by a Lipschitz continuous interface, i.e., which are both admissable patches, see [MM07, AF03].

3.4 Density of Smooth Functions

The elements in Lebesgue and Sobolev spaces are only defined almost everywhere, i.e., unlike continuous functions, cannot be evaluated at points or other sets of zero measure. However in this section it is shown that certain smooth functions are dense in $W^{s,p}$-spaces. The main property of Sobolev spaces shown here is the density of the test functions $\mathcal{D}(\overline{\Omega})$ if the domain Ω is Lipschitz.

3.4.1 Density of Smooth Functions

In the proof of Proposition 3.1.3, *iii)*, it is shown that $\|u_\varepsilon\|_{L^p(U)} \leq \|u\|_{L^p(V)}$ where $U \subsetneq V$. Extending u by zero outside of Ω allows a meaningful definition of u_ε on all of Ω and gives $\|u_\varepsilon\|_{L^p(\Omega)} \leq \|u\|_{L^p(\Omega)}$. Furthermore the rest of that proof also extends to $L^p(\Omega)$, i.e., $u_\varepsilon \to u$ in $L^p(\Omega)$. However for $u \in W^{s,p}(\Omega)$, $s > 0$, an extension by zero is in general not in $W^{s,p}(\mathbb{R}^d)$, and the result above cannot be extended this way. However the local approximations enable also global ones, that means not only for every compact subset, i.e., in the spaces with index $_{\text{loc}}$, but also on all of Ω. The proof is taken from [Eva98], Theorem 2 in 5.3.2, where only the case $s \in \mathbb{N} \cup \{0\}$ is treated, see also Proposition 2.12 in [DDE12].

Theorem 3.4.1 **($C^\infty(\Omega)$ is Dense in $W^{s,p}(\Omega)$)** *Let $u \in W^{s,p}(\Omega)$, $p \geq 1$, $s \in [0,1]$ be given. Then there is a sequence of functions $u_i \in C^\infty(\Omega) \cap W^{s,p}(\Omega)$ which converges to u in $W^{s,p}(\Omega)$.*

Proof The idea is to partition the domain Ω into suitable subsets on which the local result from Proposition 3.1.3 can be applied. Let $\{U_i\}_{i\in\mathbb{N}}$ be a sequence of subsets of Ω which approach Ω from the inside, namely

$$U_i := \{\boldsymbol{x} \in \Omega \mid \operatorname{dist}(\boldsymbol{x}, \partial\Omega) > 1/i\}.$$

Then the union of all U_i is Ω and $U_i \subset U_j$ whenever $i < j$. For convenience, denote $U_{-1} = U_0 = \emptyset$. Next, define the differences $V_i := U_{i+2} \setminus \overline{U_i}$ for $i \in \mathbb{N} \cup \{0\}$. Then also the union of the V_i contains Ω, i.e., $\bigcup_{i=0}^\infty V_i = \Omega$. Let $\{\zeta_i\}_{i=0}^\infty \subset C^\infty(\Omega)$

be a partition of unity adapted to the sets V_i, so it holds

$$\zeta_i \in C_0^\infty(V_i), \qquad 0 \le \zeta_i \le 1, \qquad \sum_{i=0}^{\infty} \zeta_i = 1.$$

As $u \in W^{s,p}(\Omega)$ also the products $\zeta_i u$ are elements of $W^{s,p}(\Omega)$, see Proposition 3.2.1, and their support is contained in V_i, that is $\operatorname{supp}(\zeta_i u) \subset V_i$. Let $\delta > 0$ be given. For each i, due to Proposition 3.1.3, one can choose $\varepsilon_i > 0$ small enough so that the mollification $u_i := \eta_{\varepsilon_i} * (\zeta_i u)$ approximates $\zeta_i u$ with

$$\|u_i - \zeta_i u\|_{W^{s,p}(\Omega)} \le \frac{\delta}{2^{i+1}} \qquad \text{and}$$

$$\operatorname{supp}(u_i) \subset W_i := U_{i+3} \setminus \overline{U}_{i-1} \supset V_i.$$

Next, define $v = \sum_{i=0}^{\infty} u_i$ and note that at any point $\boldsymbol{x} \in \Omega$ only finitely many summands are nonzero due to the choice of the partition of Ω. Hence, $v \in C^\infty(\Omega)$ and it is

$$\begin{aligned} \|v - u\|_{W^{s,p}(\Omega)} &= \left\| \sum_{i=0}^{\infty} (u_i - \zeta_i u) \right\|_{W^{s,p}(\Omega)} \le \sum_{i=0}^{\infty} \|u_i - \zeta_i u\|_{W^{s,p}(\Omega)} \\ &\le \delta \sum_{i=0}^{\infty} \frac{1}{2^{i+1}} = \delta. \end{aligned}$$

□

Prior to a famous paper by Meyers and Serrin, another definition of $W^{k,p}$, $k \in \mathbb{N}$, existed, namely as the closure of $\{f \in C^\infty(\Omega) \mid \|f\|_{W^{k,p}} < \infty\}$ which is usually denoted by $H^{k,p}$. The previous Theorem shows that these two definitions are equivalent, even if Ω is not Lipschitz continuous. Both notations are still in use, however in this monograph the letter H is only used in the case of $p = 2$.

Remark 3.4.2 For unbounded domains Ω, define the subsets $U_i = \{\boldsymbol{x} \in \Omega \mid |\boldsymbol{x}| < i,\ \operatorname{dist}(\boldsymbol{x}, \partial\Omega) > 1/i\}$ in the previous proof and possibly choose a subsequence such that $U_1 \neq \emptyset$. This is how one can show that Theorem 3.4.1 holds true even for non-Lipschitz domains.

3.4.2 Density of Smooth Functions with Compact Support

Using mollifications it is shown in Theorem 3.4.1 that in bounded Lipschitz domains Ω smooth functions are dense in $W^{s,p}(\Omega)$. An improved result is the density of test functions on $\overline{\Omega}$ in $W^{s,p}(\Omega)$. For this purpose it is shown that the space of smooth functions with compact support are dense in $W^{s,p}(\mathbb{R}^d)$. Proofs can be found in [DDE12], Theorem 1.91, Proposition 2.29 ($s = 1$), and Proposition 4.27 ($0 <$

$s < 1$), as well as in [AF03], Corollary 2.30 ($s = 0$) and [DNPV12], Theorem 5.4 ($0 < s < 1$).

Proposition 3.4.3 ($\mathcal{D}(\mathbb{R}^d)$ is Dense in $W^{s,p}(\mathbb{R}^d)$) *Let $p \geq 1$ and $s \in [0, 1]$. Then the test functions $\mathcal{D}(\mathbb{R}^d)$ are dense in $W^{s,p}(\mathbb{R}^d)$.*

Proof Let $u \in W^{s,p}(\mathbb{R}^d)$ be given. A sequence of functions $u_n \in \mathcal{D}(\mathbb{R}^d)$ is constructed which converges to u in $W^{s,p}(\mathbb{R}^d)$.

Let $\phi \in \mathcal{D}(\mathbb{R}^d)$ be such that $\phi = 1$ in the ball around 0 with radius 1, $B(0,1) = \{\boldsymbol{x} \in \mathbb{R}^d \mid |\boldsymbol{x}| < 1\}$, ϕ equals zero outside of $B(0, 2)$, and $0 \leq \phi \leq 1$. Then its gradient $\nabla\phi$ is nonzero only in $B(0, 2) \setminus B(0, 1)$. In fact, this is a mollifier except that its integral is not one. Define a scaled version of ϕ as $\phi_n(\boldsymbol{x}) := \phi(\boldsymbol{x}/n)$ which is then 1 on $B(0, n)$ while its gradient is nonzero in $B(0, 2n) \setminus B(0, n)$. Next, define the product $u_n = \phi_n u \in W^{s,p}(\mathbb{R}^d)$ (Proposition 3.2.1) and assume that $\|u - u_n\|_{W^{s,p}(\mathbb{R}^d)} \to 0$ (which is shown below), then there exists $n_0 \in \mathbb{N}$ such that $\left\|u - u_{n_0}\right\|_{W^{s,p}(\mathbb{R}^d)} < \varepsilon/2$. Furthermore, according to Proposition 3.1.3, choose ε_0 small enough such that $\left\|u_{n_0} - (u_{n_0})_{\varepsilon_0}\right\|_{W^{s,p}(\mathbb{R}^d)} < \varepsilon/2$. Then it is

$$\left\|u - (u_{n_0})_{\varepsilon_0}\right\|_{W^{s,p}(\mathbb{R}^d)} \leq \left\|u - u_{n_0}\right\|_{W^{s,p}(\mathbb{R}^d)} + \left\|u_n - (u_{n_0})_{\varepsilon_0}\right\|_{W^{s,p}(\mathbb{R}^d)} \leq \varepsilon,$$

which proves the result. Hence, it remains to show that $\|u - u_n\|_{W^{s,p}(\mathbb{R}^d)} \to 0$, which is done separately for $s = 0$, $s = 1$, and $s \in (0, 1)$.

$s = 0$: It is

$$\begin{aligned} \|u - u_n\|^p_{L^p(\mathbb{R}^d)} = \|(1 - \phi_n)u\|^p_{L^p(\mathbb{R}^d)} &= \int_{\mathbb{R}^d \setminus B(0,n)} |(1 - \phi_n(\boldsymbol{x}))u(\boldsymbol{x})|^p \, \mathrm{d}\boldsymbol{x} \\ &\leq \int_{\mathbb{R}^d \setminus B(0,n)} |u(\boldsymbol{x})|^p \, \mathrm{d}\boldsymbol{x} \xrightarrow{n\to\infty} 0, \end{aligned}$$

because $|u|^p$ is in $L^p(\mathbb{R}^d)$.

$s = 1$: With the same reasoning it is also $\|(1 - \phi_n)\nabla u\|_{L^p(\mathbb{R}^d)} \to 0$ for $n \to \infty$. The gradient of u_n is $\nabla u_n = \nabla\phi_n u + \phi_n \nabla u$. The L^p-norm of the first term approaches zero:

$$\begin{aligned} \int_{\mathbb{R}^d} |(\nabla\phi_n u)(\boldsymbol{x})|^p \, \mathrm{d}\boldsymbol{x} &= \int_{B(0,2n)\setminus B(0,n)} |(\nabla\phi_n u)(\boldsymbol{x})|^p \, \mathrm{d}\boldsymbol{x} \\ &\leq \|\nabla\phi_n\|^p_{L^\infty(\mathbb{R}^d)} \int_{\mathbb{R}^d \setminus B(0,n)} |u(\boldsymbol{x})|^p \, \mathrm{d}\boldsymbol{x} \xrightarrow{n\to\infty} 0. \end{aligned}$$

Then u_n approaches u in $W^{1,p}(\Omega)$ because

$$\|\nabla(u - u_n)\|_{L^p(\mathbb{R}^d)} \leq \|(1 - \phi_n)\nabla u\|_{L^p(\mathbb{R}^d)} + \|\nabla\phi_n u\|_{L^p(\mathbb{R}^d)},$$

and both terms above tend to zero with $n \to \infty$.

$s \in (0, 1)$: Define

$$v_n(\boldsymbol{x}, \boldsymbol{y}) := \frac{|(u_n - u)(\boldsymbol{x}) - (u_n - u)(\boldsymbol{y})|}{|\boldsymbol{x} - \boldsymbol{y}|^{s+d/p}}$$

such that v_n^p is the integrand in the Sobolev–Slobodeckij semi-norm of $u_n - u$ and it remains to show that v_n converges to zero in $L^p(\mathbb{R}^d \times \mathbb{R}^d)$. Note that $v_n(\boldsymbol{x}, \boldsymbol{y})$ is zero whenever both its arguments $\boldsymbol{x}$ and $\boldsymbol{y}$ are in the support of ϕ_n, i.e., if $|\boldsymbol{x}| \leq n$ and $|\boldsymbol{y}| \leq n$. Furthermore, v_n is symmetric in the sense of $v_n(\boldsymbol{x}, \boldsymbol{y}) = v_n(\boldsymbol{y}, \boldsymbol{x})$ for all $\boldsymbol{x}, \boldsymbol{y} \in \mathbb{R}^d$. Hence, denoting $B_n = B(0, n)$ and $B_n^c = \mathbb{R}^d \setminus B_n$, the integrals in the L^p-norm of v_n can be split as follows

$$\begin{aligned}\int_{\mathbb{R}^d}\int_{\mathbb{R}^d}(v_n(\boldsymbol{x}, \boldsymbol{y}))^p \,\mathrm{d}\boldsymbol{x}\,\mathrm{d}\boldsymbol{y} &= \int_{B_n^c}\int_{B_n^c}(v_n(\boldsymbol{x}, \boldsymbol{y}))^p \,\mathrm{d}\boldsymbol{x}\,\mathrm{d}\boldsymbol{y} + 2\int_{B_n^c}\int_{B_n}(v_n(\boldsymbol{x}, \boldsymbol{y}))^p \,\mathrm{d}\boldsymbol{x}\,\mathrm{d}\boldsymbol{y} \\ &=: J_n + 2I_n \end{aligned} \tag{3.2}$$

where Fubini's theorem is applied to interchange the order of integration to obtain a second integral I_n, i.e., the factor 2 above. Since for $\boldsymbol{x} \in B_n$ it is $u_n(\boldsymbol{x}) = u(\boldsymbol{x})$ the integral I_n can be rewritten as

$$\begin{aligned} I_n &= \int_{B_n^c}\int_{B_n}\frac{|(\phi_n(\boldsymbol{y}) - 1)u(\boldsymbol{y})|^p}{|\boldsymbol{x} - \boldsymbol{y}|^{sp+d}}\,\mathrm{d}\boldsymbol{x}\,\mathrm{d}\boldsymbol{y} \\ &= \int_{B_n^c}|\phi_n(\boldsymbol{y}) - 1|^p|u(\boldsymbol{y})|^p\int_{B_n}\frac{1}{|\boldsymbol{x} - \boldsymbol{y}|^{sp+d}}\,\mathrm{d}\boldsymbol{x}\,\mathrm{d}\boldsymbol{y}. \end{aligned}$$

For $y \in B_n^c$ it is $\varepsilon := \mathrm{dist}(y, B_n) = |\boldsymbol{y} - n\boldsymbol{y}/|\boldsymbol{y}|| = ||\boldsymbol{y}| - n| > 0$ and the inner integral over B_n can be bounded by

$$\int_{B_n}\frac{1}{|\boldsymbol{x} - \boldsymbol{y}|^{sp+d}}\,\mathrm{d}\boldsymbol{x} \leq \int_{\mathbb{R}^d\setminus B(0,\varepsilon)}|z|^{-sp-d}\,\mathrm{d}z = \omega_d\int_{\varepsilon}^{\infty}r^{-sp-1}\,\mathrm{d}r = \frac{\omega_d}{sp}\varepsilon^{-sp}, \tag{3.3}$$

where polar coordinates are used and ω_d is the surface area of ∂B_1. In consequence it is

$$I_n \leq \frac{\omega_d}{sp}\int_{B_n^c}|u(\boldsymbol{y})|^p\frac{|\phi_n(\boldsymbol{y}) - 1|^p}{||\boldsymbol{y}| - n|^{sp}}\,\mathrm{d}\boldsymbol{y}.$$

The fraction in the integrand above is bounded on B_n^c which can be seen as follows:

$$\sup_{y\in B_n^c}\frac{|\phi_n(\boldsymbol{y}) - 1|^p}{||\boldsymbol{y}| - n|^{sp}} = \frac{1}{n^{sp}}\sup_{z\in\mathbb{R}^d\setminus B(0,1)}\underbrace{\frac{|\phi(z) - 1|^p}{||z| - 1|^{sp}}}_{=:f(z)}$$

and the function $f : \mathbb{R}^d \setminus B(0,1) \to \mathbb{R}$ is bounded by 1 on $\mathbb{R}^d \setminus B(0,2)$ and by $\|\nabla\phi\|^p_{L^\infty(\mathbb{R}^d)}$ otherwise, i.e.,

$$\begin{aligned}
|z| \geq 2 \quad &\implies f(z) = \frac{1}{||z|-1|^{sp}} \leq 1, \\
1 \leq |z| \leq 2 \quad &\implies f(z) = \frac{|\phi(z) - \phi(z/|z|)|^p}{||z|-1|^{sp}} \leq \|\nabla\phi\|^p_{L^\infty(\mathbb{R}^d)} ||z|-1|^{(1-s)p} \\
&\qquad \leq \|\nabla\phi\|^p_{L^\infty(\mathbb{R}^d)},
\end{aligned}$$

where in the second case the mean value theorem is applied. Hence, it is

$$I_n \leq \frac{1}{n^{sp}} \frac{\omega_d}{sp} \max\left\{1, \|\nabla\phi\|^p_{L^\infty(\mathbb{R}^d)}\right\} \|u\|^p_{L^p(\mathbb{R}^d)} \xrightarrow{n\to\infty} 0.$$

The other integral J_n in Eq. (3.2) can be decomposed using the inequality $(a+b)^p \leq 2^{p-1}(a^p + b^p)$:

$$\begin{aligned}
J_n &= \int_{B_n^c} \int_{B_n^c} \frac{|(u_n - u)(\boldsymbol{x}) - (u_n - u)(\boldsymbol{y})|^p}{|\boldsymbol{x} - \boldsymbol{y}|^{sp+d}} \,\mathrm{d}\boldsymbol{x}\,\mathrm{d}\boldsymbol{y} \\
&\leq 2^{p-1} \int_{B_n^c} \int_{B_n^c} \frac{|u_n(\boldsymbol{x}) - u_n(\boldsymbol{y})|^p}{|\boldsymbol{x} - \boldsymbol{y}|^{sp+d}} \,\mathrm{d}\boldsymbol{x}\,\mathrm{d}\boldsymbol{y} + 2^{p-1} \int_{B_n^c} \int_{B_n^c} \frac{|u(\boldsymbol{x}) - u(\boldsymbol{y})|^p}{|\boldsymbol{x} - \boldsymbol{y}|^{sp+d}} \,\mathrm{d}\boldsymbol{x}\,\mathrm{d}\boldsymbol{y}.
\end{aligned}$$

The second term on the right-hand side converges to zero because $u \in W^{s,p}(\mathbb{R}^d)$. Denote the first term by $2^{p-1}K_n$ and decompose it into

$$\begin{aligned}
K_n^{(1)} &= \int_{A_n} \int_{A_n} \frac{|u_n(\boldsymbol{x}) - u_n(\boldsymbol{y})|^p}{|\boldsymbol{x} - \boldsymbol{y}|^{sp+d}} \,\mathrm{d}\boldsymbol{x}\,\mathrm{d}\boldsymbol{y}, \\
K_n^{(2)} &= \int_{A_n} \int_{B_{2n}^c} \frac{|u_n(\boldsymbol{x}) - u_n(\boldsymbol{y})|^p}{|\boldsymbol{x} - \boldsymbol{y}|^{sp+d}} \,\mathrm{d}\boldsymbol{x}\,\mathrm{d}\boldsymbol{y},
\end{aligned}$$

where $A_n = B_{2n} \setminus B_n = B_n^c \setminus B_{2n}^c$. Note that u_n is zero in $\mathbb{R}^d \setminus B(0, 2n) = B_{2n}^c$ and hence $K_n = K_n^{(1)} + 2K_n^{(2)}$, again using Fubini's Theorem to get a second $K_n^{(2)}$. Since $u_n(\boldsymbol{x}) = 0$ for $\boldsymbol{x} \in B_{2n}^c$, $K_n^{(2)}$ is bounded by

$$K_n^{(2)} = \int_{A_n} \int_{B_{2n}^c} \frac{|u_n(\boldsymbol{y})|^p}{|\boldsymbol{x} - \boldsymbol{y}|^{sp+d}} \,\mathrm{d}\boldsymbol{x}\,\mathrm{d}\boldsymbol{y} \leq \frac{\omega_d}{sp} \int_{A_n} \frac{|u_n(\boldsymbol{y})|^p}{||\boldsymbol{y}| - 2n|^{sp}} \,\mathrm{d}\boldsymbol{y},$$

where the inner integral is handled very similarly to the inner integral of I_n in Eq. (3.3). As $\phi(2\boldsymbol{y}/|\boldsymbol{y}|) = 0$ and $u_n = \phi_n u$ it is

$$\begin{aligned} K_n^{(2)} &\le \frac{\omega_d}{sp} \int_{A_n} \frac{|(\phi(\boldsymbol{y}/n) - \phi(2\boldsymbol{y}/|\boldsymbol{y}|))u(\boldsymbol{y})|^p}{||\boldsymbol{y}| - 2n|^{sp}} \,\mathrm{d}\boldsymbol{y} \\ &\le \frac{\omega_d}{sp} \sup_{\boldsymbol{y}\in A_n} \left(\frac{|(\phi(\boldsymbol{y}/n) - \phi(2\boldsymbol{y}/|\boldsymbol{y}|))|^p}{||\boldsymbol{y}| - 2n|^{sp}} \right) \int_{A_n} |u(\boldsymbol{y})| \,\mathrm{d}\boldsymbol{y} \\ &\le \frac{1}{n^{sp}} \frac{\omega_d}{sp} \sup_{\boldsymbol{z}\in A_1} \left(\frac{|(\phi(\boldsymbol{z}) - \phi(2\boldsymbol{z}/|\boldsymbol{z}|))|^p}{||\boldsymbol{z}| - 2|^{sp}} \right) \|u\|_{L^p(A_n)}^p \end{aligned}$$

As before with f the supremum above is bounded by $\|\nabla\phi\|_{L^\infty(\mathbb{R}^d)}^p$ and, accordingly, $K_n^{(2)}$ converges to zero with $n \to \infty$. The term $K_n^{(1)}$ also converges to 0 which is shown using the same techniques as above, see also the proof of Proposition 3.2.1. In the numerator of its integrand one adds and subtracts $u(\boldsymbol{y})\phi_n(\boldsymbol{x})$, uses the inequality $(a+b)^p \le 2^{p-1}(a^p + b^p)$, and applies the mean value theorem to yield

$$\begin{aligned} \frac{K_n^{(1)}}{2^{p-1}} &\le \iint_{A_n^2} \frac{|\phi(\boldsymbol{y}/n) - \phi(\boldsymbol{x}/n)|^p |u(\boldsymbol{y})|^p}{|\boldsymbol{x} - \boldsymbol{y}|^{sp+d}} \,\mathrm{d}\boldsymbol{x}\,\mathrm{d}\boldsymbol{y} \\ &\quad + \iint_{A_n^2} \frac{|u(\boldsymbol{y}) - u(\boldsymbol{x})|^p |\phi(\boldsymbol{x}/n)|^p}{|\boldsymbol{x} - \boldsymbol{y}|^{sp+d}} \,\mathrm{d}\boldsymbol{x}\,\mathrm{d}\boldsymbol{y} \\ &\le \frac{\|\nabla\phi\|_{L^\infty(\mathbb{R}^d)}^p}{n^p} \int_{A_n} |u(\boldsymbol{y})|^p \int_{A_n} |\boldsymbol{x} - \boldsymbol{y}|^{(1-s)p-d} \,\mathrm{d}\boldsymbol{x}\,\mathrm{d}\boldsymbol{y} \\ &\quad + \underbrace{\|\phi\|_{L^\infty(\mathbb{R}^d)}^p}_{=1} |u|_{W^{s,p}(A_n)}^p. \end{aligned}$$

The inner integral is finite for all $\boldsymbol{y} \in A_n$, in fact using polar coordinates it is

$$\int_{A_n} |\boldsymbol{x} - \boldsymbol{y}|^{(1-s)p-d} \,\mathrm{d}\boldsymbol{x} \le \omega_d \frac{(2n)^{p(1-s)}}{p(1-s)}$$

and consequently

$$\begin{aligned} K_n^{(1)} &\le 2^{p-1} \left(\frac{\|\nabla\phi\|_{L^\infty(\mathbb{R}^d)}^p}{n^{ps}} \frac{\omega_d 2^{p(1-s)}}{p(1-s)} \|u\|_{L^p(A_n)}^p + |u|_{W^{s,p}(A_n)}^p \right) \\ &\le \frac{c}{n^{ps}} \|u\|_{W^{s,p}(A_n)}. \end{aligned}$$

Hence, $K_n = K_n^{(1)} + K_n^{(2)}$ converges to zero with $n \to \infty$ and therefore $u_n \to u$ in $W^{s,p}(\mathbb{R}^d)$. □

As an example, the previous result is used to show that small shifts in the argument of an L^p function only have a small effect.

Corollary 3.4.4 (Shift Operator) *Let* $\mathbf{h} \in \mathbb{R}^d$ *be given. Then the shift operator* $\tau_{\mathbf{h}} : L^p(\mathbb{R}^d) \to L^p(\mathbb{R}^d)$*, defined for all* $\boldsymbol{x} \in \mathbb{R}^d$ *as*

$$\tau_{\mathbf{h}} u(\boldsymbol{x}) = u(\boldsymbol{x} + \mathbf{h}),$$

preserves the norm, i.e., $\|\tau_{\mathbf{h}} u\|_{L^p(\mathbb{R}^d)} = \|u\|_{L^p(\mathbb{R}^d)}$*, and it holds* $\tau_{\mathbf{h}} u \to u$ *in* $L^p(\mathbb{R}^d)$ *for* $|\mathbf{h}| \to 0$.

Proof A change of variables $\boldsymbol{y} = \boldsymbol{x} - \mathbf{h}$ yields

$$\|\tau_{\mathbf{h}} u\|^p_{L^p(\mathbb{R}^d)} = \int_{\mathbb{R}^d} |u(\boldsymbol{x} - \mathbf{h})|^p \, \mathrm{d}\boldsymbol{x} = \int_{\mathbb{R}^d} |u(\boldsymbol{y})|^p \, \mathrm{d}\boldsymbol{y} = \|u\|^p_{L^p(\mathbb{R}^d)}.$$

To show the convergence, let $\varepsilon > 0$ be given. According to Proposition 3.4.3 there exists a function $v \in \mathcal{D}(\mathbb{R}^d)$ such that $\|u - v\|_{L^p(\mathbb{R}^d)} < \varepsilon/3$. Since v is continuous and has compact support, it is even uniformly continuous, hence there exists a $\mathbf{h}_0 \in \mathbb{R}^d$ such that for all $\mathbf{h} \in \mathbb{R}^d$ with $|\mathbf{h}| < |\mathbf{h}_0|$ it is

$$\|\tau_{\mathbf{h}} v - v\|_{L^\infty(\mathbb{R}^d)} \le \frac{\varepsilon}{3} (2\,|\mathrm{supp}(v)|)^{-1/p}.$$

This implies $\|\tau_{\mathbf{h}} v - v\|_{L^p(\mathbb{R}^d)} \le \|\tau_{\mathbf{h}} v - v\|_{L^\infty(\mathbb{R}^d)} (2|\mathrm{supp}(v)|)^{1/p} \le \varepsilon/3$. Finally, using $\|\tau_{\mathbf{h}} u - \tau_{\mathbf{h}} v\|_{L^p(\mathbb{R}^d)} = \|\tau_{\mathbf{h}}(u - v)\|_{L^p(\mathbb{R}^d)} = \|u - v\|_{L^p(\mathbb{R}^d)} \le \varepsilon/3$ and the triangle inequality, it is

$$\|\tau_{\mathbf{h}} u - u\|_{L^p(\mathbb{R}^d)} \le \|\tau_{\mathbf{h}} u - \tau_{\mathbf{h}} v\|_{L^p(\mathbb{R}^d)} + \|\tau_{\mathbf{h}} v - v\|_{L^p(\mathbb{R}^d)} + \|v - u\|_{L^p(\mathbb{R}^d)} \le \varepsilon.$$

□

3.4.3 Density of Smooth Functions up to the Boundary

With the density of test functions in $W^{s,p}(\mathbb{R}^d)$, Proposition 3.4.3, it is possible to proof density of smooth functions in $W^{s,p}(\Omega)$, see [DDE12], Corollary 2.71 ($s = 1$) and Proposition 4.52 ($s \in (0, 1)$). This is Proposition 3.4.3 above but $\mathbb{R}^d$ replaced by $\overline{\Omega}$. Another proof can be found in [Tar07], Lemma 12.3.

Theorem 3.4.5 ($\mathcal{D}(\overline{\Omega})$ is Dense in $W^{s,p}(\Omega)$) *Let* $\Omega \subset \mathbb{R}^d$ *be a bounded Lipschitz domain,* $p \ge 1$ *and* $s \in [0, 1]$*. Then* $\mathcal{D}(\overline{\Omega})$ *is dense in* $W^{s,p}(\Omega)$.

Proof Let $u \in W^{s,p}(\Omega)$ be given and according to Proposition 3.3.1 it has an extension $Eu \in W^{s,p}(\mathbb{R}^d)$. Since $\mathcal{D}(\mathbb{R}^d)$ is dense in $W^{s,p}(\mathbb{R}^d)$, according to Proposition 3.4.3, there is a sequence $u_n \in \mathcal{D}(\mathbb{R}^d)$ which converges in $W^{s,p}(\mathbb{R}^d)$ to

Eu. Then restricting each u_n to Ω yields a sequence in $\mathcal{D}(\overline{\Omega})$ which converges to u in $W^{s,p}(\Omega)$:

$$\|u - u_n\|_{W^{s,p}(\Omega)} = \|Eu - u_n\|_{W^{s,p}(\Omega)} \leq \|Eu - u_n\|_{W^{s,p}(\mathbb{R}^d)} \xrightarrow{n\to\infty} 0. \qquad \square$$

This subsection finishes with two interesting remarks which however have no particular impact on the rest of the monograph.

Remark 3.4.6 The spaces $W^{s,p}(\Omega)$ are nested. In [DNPV12] (Propositions 2.1 and 2.2) it is shown that for a Lipschitz domain Ω the space $W^{s',p}(\Omega)$ is continuously embedded into $W^{s,p}(\Omega)$ if $0 < s \leq s' \leq 1$. However unless $u \in W^{1,p}(\Omega)$ is constant the limit of $|u|_{W^{s,p}(\Omega)}$ with $s \to 1$ does not exist and in fact approaches infinity. However if one considers $(1-s)|u|_{W^{s,p}(\Omega)}$ instead, the respective limit can be bounded by the $W^{1,p}(\Omega)$ semi-norm:

$$\lim_{s\nearrow 1}(1-s)|u|^p_{W^{s,p}(\Omega)} = \frac{K_{p,d}}{p}|u|^p_{W^{1,p}(\Omega)}$$

with a constant $K_{p,d}$ which depends on p and d, see [Bre02].

Remark 3.4.7 So far s is in the unit interval. For larger s the space $W^{s,p}(\Omega)$ is not defined by simply inserting $s \geq 1$ into the definition of the Sobolev–Slobodeckij semi-norm. One reason is that

$$\iint_{\Omega\times\Omega} \frac{|u(\boldsymbol{x}) - u(\boldsymbol{y})|^p}{|\boldsymbol{x} - \boldsymbol{y}|^{d+1}} \,\mathrm{d}\boldsymbol{x}\,\mathrm{d}\boldsymbol{y} < \infty$$

for a measurable function $u : \Omega \to \mathbb{R}$ implies that u must be constant, see [Bre02], Proposition 1. Instead let $s = m + \sigma$ with $m \in \mathbb{N}$ and $\sigma \in (0, 1)$, then $W^{s,p}(\Omega)$ is defined to be the subspace of $W^{m,p}(\Omega)$ whose elements and all their derivatives up to order m are in $W^{\sigma,p}(\Omega)$.

3.5 Equivalent Norms

Later proofs use a localization of the domain Ω, see Definition 2.2.1. Then on $W^{1,p}(\Omega)$, one can define an equivalent norm using the sum of the local contributions as follows, see [DDE12], Proposition 2.68.

Proposition 3.5.1 (Equivalent Norm on $W^{1,p}(\Omega)$) *Let Ω be a Lipschitz domain which is covered by a finite set of domains Ω_i, $i \in \{0,\dots,N\}$, and $u \in L^p(\Omega)$. Then u is in $W^{1,p}(\Omega)$ if and only if $u \in W^{1,p}(\Omega_i \cap \Omega)$ for each i. The map*

$$u \mapsto \sum_{i=0}^{N} \|u\|_{W^{1,p}(\Omega_i\cap\Omega)}$$

is an equivalent norm to $\|u\|_{W^{1,p}(\Omega)}$.

Proof Let $u \in W^{1,p}(\Omega)$ and $i \in \{0, \dots, N\}$ be given. Clearly $u \in L^p(\Omega_i \cap \Omega)$ and any given $\varphi \in \mathcal{D}(\Omega_i \cap \Omega)$ can be extended by zero to Ω so that the weak derivative of u exists also on $\Omega_i \cap \Omega$, i.e., $u \in W^{1,p}(\Omega_i \cap \Omega)$. Furthermore it is $\sum_{i=0}^N \|u\|_{W^{1,p}(\Omega_i \cap \Omega)} \leq C\|u\|_{W^{1,p}(\Omega)}$, where C is bounded by the maximum number of subdomains Ω_i having a nonempty intersection (at most N).

On the other hand let $u \in W^{1,p}(\Omega_i \cap \Omega)$ for all $i \in \{0, \dots, N\}$ be given. In order to show that u is also in $W^{1,p}(\Omega)$ let $\varphi \in \mathcal{D}(\Omega)$ be given. Writing $\phi = \sum_{i=0}^N \Phi_i \varphi$ for a partition of unity $\{\Phi_i\}$ subordinate to the Ω_i gives

$$\begin{aligned}
\int_\Omega \left(u \frac{\partial \varphi}{\partial x_j} \right)(\boldsymbol{x})\, \mathrm{d}\boldsymbol{x} &= \sum_{i=0}^N \int_{\Omega_i \cap \Omega} u(\boldsymbol{x}) \frac{\partial (\Phi_i \varphi)}{\partial x_j}(\boldsymbol{x})\, \mathrm{d}\boldsymbol{x} \\
&= -\sum_{i=0}^N \int_{\Omega_i \cap \Omega} \frac{\partial u(\boldsymbol{x})}{\partial x_j} (\Phi_i \varphi)(\boldsymbol{x})\, \mathrm{d}\boldsymbol{x} \\
&= -\int_\Omega \left(\sum_{i=0}^N \Phi_i \frac{\partial u}{\partial x_j} \right)(\boldsymbol{x}) \varphi(\boldsymbol{x})\, \mathrm{d}\boldsymbol{x},
\end{aligned}$$

where the second equality uses the weak differentiability on $\Omega_i \cap \Omega$ ($\Phi_i \varphi \in \mathcal{D}(\Omega_i \cap \Omega)$). Hence, u is also weakly differentiable on Ω with its derivative given by $\sum_{i=0}^N \Phi_i \frac{\partial u}{\partial x_j}$. Finally, using $u = \sum_{i=0}^N \Phi_i u$, it is $\|u\|_{W^{1,p}(\Omega)} \leq C \sum_{i=0}^N \|u\|_{W^{1,p}(\Omega_i \cap \Omega)}$, where C depends on the Φ_i and its derivatives, i.e., on the cover $\{\Omega_i\}$. □

A similar approach is possible for $s < 1$ as well, however another equivalent norm on $W^{s,p}(\mathbb{R}^K)$ is useful with $K = d - 1$. The proof is mainly that of [DDE12], Lemma 3.27, where the case $s = 1 - 1/p$ is shown. However a shorter proof for general $s < 1$ can be found in the same book, Lemma 4.33, [DDE12].

Proposition 3.5.2 (Equivalent Norm on $W^{s,p}(\mathbb{R}^K)$) *Let $u \in L^p(\mathbb{R}^K)$, $p > 1$. Then the following two statements are equivalent:*

(i) $u \in W^{s,p}(\mathbb{R}^K)$,
(ii) for all $i \in \{1, \dots, K\}$ it is

$$|u|_{i,s,p}^p := \int_{\mathbb{R}^K} \int_{\mathbb{R}} \frac{|u(\boldsymbol{x} + t\mathbf{e}_i) - u(\boldsymbol{x})|^p}{t^{sp+1}}\, \mathrm{d}t\, \mathrm{d}\boldsymbol{x} < \infty.$$

Moreover, an equivalent norm to $\|\cdot\|_{W^{s,p}(\mathbb{R}^K)}$ is given by $\|u\|' := \|u\|_{L^p(\mathbb{R}^K)} + \sum_{i=1}^K |u|_{i,s,p}$.

Proof Note that in the case of $K = 1$ this is essentially a change of variables and the statement of this proposition does not need further proof. Therefore, assume $K > 1$.

(ii) $\Longrightarrow$ *(i)*: Let $u \in L^p(\mathbb{R}^K)$ and assume $|u|_{i,s,p} < \infty$ for all $i \in \{1, \dots, K\}$. Furthermore, for all $\boldsymbol{x} = (x_1, \dots, x_K)$, $\boldsymbol{y} = (y_1, \dots, y_K) \in \mathbb{R}^K$ define $\widehat{\boldsymbol{x}\boldsymbol{y}}_i :=$

$\sum_{j=1}^{i} x_j \mathbf{e}_j + \sum_{j=i+1}^{K} y_j \mathbf{e}_j = (x_1, \ldots, x_i, y_{i+1}, \ldots, y_K)$. In particular it is $\widehat{\boldsymbol{x}\boldsymbol{y}}_0 = \boldsymbol{y}$ and $\widehat{\boldsymbol{x}\boldsymbol{y}}_K = \boldsymbol{x}$. Then using a telescoping sum, it is

$$u(\boldsymbol{y}) - u(\boldsymbol{x}) = \sum_{i=0}^{K-1} \big(u(\widehat{\boldsymbol{x}\boldsymbol{y}}_i) - u(\widehat{\boldsymbol{x}\boldsymbol{y}}_{i+1})\big),$$

hence the Sobolev–Slobodeckij semi-norm can be bounded as follows:

$$\int_{\mathbb{R}^K} \int_{\mathbb{R}^K} \frac{|u(\boldsymbol{x}) - u(\boldsymbol{y})|^p}{|\boldsymbol{x} - \boldsymbol{y}|^{sp+K}} \, \mathrm{d}\boldsymbol{x} \, \mathrm{d}\boldsymbol{y} \le \sum_{i=0}^{K-1} I_i,$$

with

$$I_i := \int_{\mathbb{R}^K} \int_{\mathbb{R}^K} \frac{\big|u(\widehat{\boldsymbol{x}\boldsymbol{y}}_i) - u(\widehat{\boldsymbol{x}\boldsymbol{y}}_{i+1})\big|^p}{\Big(\sum_{j=1}^{K} |x_j - y_j|^2\Big)^q} \, \mathrm{d}\boldsymbol{x} \, \mathrm{d}\boldsymbol{y},$$

where $q = (sp + K)/2 \ge 1$. In order to estimate the denominators above consider the constants $C_1, \ldots, C_4$, depending only on q and K, such that

$$C_1 \Bigg(\sum_{j=1}^{K} |\eta_i|\Bigg)^{2q} \le C_2 \Bigg(\sum_{j=1}^{K} |\eta_i|^2\Bigg)^{q} \le \sum_{j=1}^{K} |\eta_i|^{2q} \le C_3 \Bigg(\sum_{j=1}^{K} |\eta_i|^2\Bigg)^{q} \le C_4 \Bigg(\sum_{j=1}^{K} |\eta_i|\Bigg)^{2q}.$$

The existence of the constants C_j is assured because these are all norms in the finite dimensional space $\mathbb{R}^K$. In fact, one can choose $C_3 = C_4 = 1$. Next, consider I_{K-1} and in particular the integration with respect to y_1:

$$\begin{aligned}
\int_{\mathbb{R}} \frac{\big|u(\widehat{\boldsymbol{x}\boldsymbol{y}}_{K-1}) - u(\boldsymbol{x})\big|^p}{\Big(\sum_{j=1}^{K} |x_j - y_j|^2\Big)^q} \, \mathrm{d}y_1 &\le \int_{\mathbb{R}} \frac{\big|u(\widehat{\boldsymbol{x}\boldsymbol{y}}_{K-1}) - u(\boldsymbol{x})\big|^p}{\frac{1}{C_3}|x_1 - y_1|^{2q} + \frac{1}{C_3} \sum_{j=2}^{K} |x_j - y_j|^{2q}} \, \mathrm{d}y_1 \\
&= 2C_3 \int_{x_1}^{\infty} \frac{\big|u(\widehat{\boldsymbol{x}\boldsymbol{y}}_{K-1}) - u(\boldsymbol{x})\big|^p}{|x_1 - y_1|^{2q} + \sum_{j=2}^{K} |x_j - y_j|^{2q}} \, \mathrm{d}y_1 \\
&\le 2C_3 \int_{x_1}^{\infty} \frac{\big|u(\widehat{\boldsymbol{x}\boldsymbol{y}}_{K-1}) - u(\boldsymbol{x})\big|^p}{|x_1 - y_1|^{2q} + C_1 \Big(\sum_{j=2}^{K} |x_j - y_j|\Big)^{2q}} \, \mathrm{d}y_1,
\end{aligned}$$

and applying the change of variables $z = (x_1 - y_1)\Big(\sum_{j=2}^{K} |x_j - y_j|\Big)^{-1}$ yields

$$= 2C_3 \frac{\big|u(\widehat{\boldsymbol{x}\boldsymbol{y}}_{K-1}) - u(\boldsymbol{x})\big|^p}{\Big(\sum_{j=2}^{K} |x_j - y_j|\Big)^{2q-1}} \int_0^{\infty} \frac{1}{z^{2q} + C_1} \, \mathrm{d}z.$$

The integral over z above is finite because $2q > 1$, hence combining all the constants in M_1, it is

$$\int_{\mathbb{R}} \frac{\left|u(\widehat{\boldsymbol{x}\boldsymbol{y}}_{K-1}) - u(\boldsymbol{x})\right|^p}{\left(\sum_{j=1}^K |x_j - y_j|^2\right)^q} \,\mathrm{d}y_1 \le M_1 \frac{\left|u(\widehat{\boldsymbol{x}\boldsymbol{y}}_{K-1}) - u(\boldsymbol{x})\right|^p}{\left(\sum_{j=2}^K |x_j - y_j|\right)^{sp+K-1}}.$$

Next, integrate this estimate over $\mathbb{R}$ with respect to y_2 and the same computations yield

$$\int_{\mathbb{R}} \int_{\mathbb{R}} \frac{\left|u(\widehat{\boldsymbol{x}\boldsymbol{y}}_{K-1}) - u(\boldsymbol{x})\right|^p}{\left(\sum_{j=1}^K |x_j - y_j|^2\right)^q} \,\mathrm{d}y_1 \,\mathrm{d}y_2 \le M_1 M_2 \frac{\left|u(\widehat{\boldsymbol{x}\boldsymbol{y}}_{K-1}) - u(\boldsymbol{x})\right|^p}{\left(\sum_{j=3}^K |x_j - y_j|\right)^{p+K-2}}.$$

In these computations it is $q = (sp + K - 1)/2$ such that $2q > 1$ assures the finiteness of the above integral over z.[5] By induction it is

$$\begin{aligned} I_{K-1} &\le \prod_{j=1}^{K-1} M_j \int_{\mathbb{R}^K} \int_{\mathbb{R}} \frac{\left|u(\widehat{\boldsymbol{x}\boldsymbol{y}}_{K-1}) - u(\boldsymbol{x})\right|^p}{|x_K - y_K|^{sp+1}} \,\mathrm{d}y_K \,\mathrm{d}\boldsymbol{x} \\ &= \prod_{j=1}^{K-1} M_j \int_{\mathbb{R}^K} \int_{\mathbb{R}} \frac{|u(\boldsymbol{x} + t\mathbf{e}_K) - u(\boldsymbol{x})|^p}{|t|^{sp+1}} \,\mathrm{d}t \,\mathrm{d}\boldsymbol{x}, \end{aligned}$$

which is finite due to the assumption *(ii)*. The other integrals I_i, $i \in \{1, \dots, K-2\}$, can be handled the same way proving *(i)*. Furthermore, it is shown that there exists a constant C which does not depend on u such that

$$\|u\|^p_{W^{1-1/p,p}(\mathbb{R}^K)} \le \|u\|^p_{L^p(\mathbb{R}^K)} + C \sum_{i=1}^K \int_{\mathbb{R}^K} \int_{\mathbb{R}} \frac{|u(\boldsymbol{x}) - u(\boldsymbol{x} + t\mathbf{e}_i)|^p}{|t|^{sp+1}} \,\mathrm{d}t \,\mathrm{d}\boldsymbol{x}.$$

(i) $\Longrightarrow$ *(ii)*: Let $u \in W^{s,p}(\mathbb{R}^K)$. The statement *(ii)* is shown for $i = K$, for other i the same steps work as well, with a slightly more inconvenient notation. Set

$$J_1 := \int_{\mathbb{R}^K} \int_0^\infty \frac{|u(\boldsymbol{x} + t_K\mathbf{e}_K) - u(\boldsymbol{x})|^p}{t_K^{sp+1}} \,\mathrm{d}t_K \,\mathrm{d}\boldsymbol{x}.$$

[5] Of course this is only true for $K \ge 2$, for $K = 1$ there is no integration over a second component y_2. Similarly in the following steps.

Note that it is $J_1 = \frac{1}{2}|u|^p_{K,1-1/p,p}$ because

$$\int_{\mathbb{R}^K}\int_{-\infty}^{0} \frac{|u(\boldsymbol{x}+t_K\mathbf{e}_K)-u(\boldsymbol{x})|^p}{|t_K|^{sp+1}}\,\mathrm{d}t_K\,\mathrm{d}\boldsymbol{x} = \int_{\mathbb{R}^K}\int_{0}^{\infty} \frac{|u(\boldsymbol{y})-u(\boldsymbol{y}+s_K\mathbf{e}_K)|^p}{s_K^{sp+1}}\,\mathrm{d}s_K\,\mathrm{d}\boldsymbol{y},$$

where $s_K = -t_K$ and $\boldsymbol{y} = \boldsymbol{x} + t_K\mathbf{e}_K$. Therefore, statement *(ii)* is true if J_1 is finite. Generalize J_1 to J_k, $2 \le k \le K$ as follows:

$$J_2 := \int_{\mathbb{R}^K}\int_{(0,\infty)^2} \frac{\left|u(\boldsymbol{x}+t_{K-1}\mathbf{e}_{\mathbf{e}_{K-1}}+t_K\mathbf{e}_K)-u(\boldsymbol{x})\right|^p}{(t_{K-1}+t_K)^{sp+2}}\,\mathrm{d}t_{K-1}\,\mathrm{d}t_K\,\mathrm{d}\boldsymbol{x},$$

$$J_k := \int_{\mathbb{R}^K}\int_{(0,\infty)^k} \frac{\left|u\left(\boldsymbol{x}+\sum_{j=K-k+1}^K t_j\mathbf{e}_j\right)-u(\boldsymbol{x})\right|^p}{\left(\sum_{j=K-k+1}^K t_j\right)^{sp+k}}\,\mathrm{d}t_{K-k+1}\ldots\mathrm{d}t_K\,\mathrm{d}\boldsymbol{x},$$

$$J_K := \int_{\mathbb{R}^K}\int_{(0,\infty)^K} \frac{\left|u\left(\boldsymbol{x}+\sum_{j=1}^K t_j\mathbf{e}_j\right)-u(\boldsymbol{x})\right|^p}{\left(\sum_{j=1}^K t_j\right)^{sp+K}}\,\mathrm{d}t_1\ldots\mathrm{d}t_K\,\mathrm{d}\boldsymbol{x}.$$

The sum in the denominator of J_K can be bounded from below by $\sqrt{\sum_{j=1}^K t_j^2}$ and with $\boldsymbol{y} = \boldsymbol{x}$ and $\boldsymbol{z} = \boldsymbol{x} + \sum_{j=1}^K t_j\mathbf{e}_j$ it is[6]

$$\begin{aligned}
J_K &\le \int_{\mathbb{R}^K}\int_{(0,\infty)^K+\{\boldsymbol{y}\}} \frac{|u(\boldsymbol{z})-u(\boldsymbol{y})|^p}{|\boldsymbol{z}-\boldsymbol{y}|^{sp+K}}\,\mathrm{d}\boldsymbol{z}\,\mathrm{d}\boldsymbol{y}\\
&\le \int_{\mathbb{R}^K}\int_{\mathbb{R}^K} \frac{|u(\boldsymbol{z})-u(\boldsymbol{y})|^p}{|\boldsymbol{z}-\boldsymbol{y}|^{sp+K}}\,\mathrm{d}\boldsymbol{z}\,\mathrm{d}\boldsymbol{y}\\
&\le \|u\|^p_{W^{s,p}(\mathbb{R}^K)}.
\end{aligned}$$

Due to the assumption *(i)* this is finite. In order to prove $J_1 \le C_1\|u\|^p_{W^{s,p}(\mathbb{R}^K)}$, it is shown that $J_{k+1} \le C_{k+1}\|u\|^p_{W^{s,p}(\mathbb{R}^K)} \implies J_k \le C_k\|u\|^p_{W^{s,p}(\mathbb{R}^K)}$ for all $k \in \{1,\ldots,K-1\}$ with suitable constants C_k and C_{k+1}, which do not depend on u. Therefore, assume $J_{k+1} \le C_{k+1}\|u\|^p_{W^{s,p}(\mathbb{R}^K)}$. The fundamental theorem of calculus

[6] The set $(0,\infty)^K + \{\boldsymbol{y}\}$ has to be understood clement-wise, i.e., it is defined as $(0,\infty)^K + \{\boldsymbol{y}\} = \left\{\boldsymbol{x}+\boldsymbol{y} \mid \boldsymbol{x} \in (0,\infty)^K\right\}$, see also Footnote 3 on page 15.

applied to the function $t' \mapsto \left(t' + \sum_{j=K-k+1}^{K} t_j\right)^{-sp-k}$ shows

$$\frac{1}{\left(\sum_{j=K-k+1}^{K} t_j\right)^{sp+k}} = (sp+k) \int_0^\infty \frac{1}{\left(t' + \sum_{j=K-k+1}^{K} t_j\right)^{sp+k+1}} \, \mathrm{d}t'.$$

Hence, J_k can be reformulated as

$$J_k = (sp+k) \int_{\mathbb{R}^K} \int_{(0,\infty)^{k+1}} \frac{\left|u\left(\boldsymbol{x} + \sum_{j=K-k+1}^{K} t_j \mathbf{e}_j\right) - u(\boldsymbol{x})\right|^p}{\left(t' + \sum_{j=K-k+1}^{K} t_j\right)^{sp+k+1}} \, \mathrm{d}t' \, \mathrm{d}t_{K-k+1} \ldots \mathrm{d}t_K \, \mathrm{d}\boldsymbol{x}.$$

Using the inequality $|a+b|^p \le 2^{p-1}(|a|^p + |b|^p)$ and a point $\boldsymbol{x}^*$, which is to be chosen, yields

$$J_k \le (sp+k)2^{p-1} \int_{\mathbb{R}^K} \int_{(0,\infty)^{k+1}} \frac{\left|u\left(\boldsymbol{x} + \sum_{j=K-k+1}^{K} t_j \mathbf{e}_j\right) - u(\boldsymbol{x}^*)\right|^p + |u(\boldsymbol{x}^*) - u(\boldsymbol{x})|^p}{\left(t' + \sum_{j=K-k+1}^{K} t_j\right)^{sp+k+1}} \, \mathrm{d}t' \, \mathrm{d}t_{K-k+1} \ldots \mathrm{d}t_K \, \mathrm{d}\boldsymbol{x}.$$

Write the last term as $(sp+k)2^{p-1}(B+A)$ where A and B are treated separately. Choose $\boldsymbol{x}^* = \boldsymbol{x} + \frac{1}{2}t'\mathbf{e}_{K-k} + \frac{3}{2}\sum_{j=K-k+1}^{K} t_j \mathbf{e}_j$, then using $t' + \sum t_j \ge \frac{2}{3}(\frac{1}{2}t' + \frac{3}{2}\sum t_j)$ leads to

$$A \le \left(\frac{3}{2}\right)^{sp+k+1} \int_{\mathbb{R}^K} \int_{(0,\infty)^{k+1}} \frac{\left|u\left(\boldsymbol{x} + \frac{1}{2}t'\mathbf{e}_{K-k} + \frac{3}{2}\sum_{j=K-k+1}^{K} t_j \mathbf{e}_j\right) - u(\boldsymbol{x})\right|^p}{\left(\frac{1}{2}t' + \frac{3}{2}\sum_{j=K-k+1}^{K} t_j\right)^{sp+k+1}} \, \mathrm{d}t' \, \mathrm{d}t_{K=k+1} \ldots \mathrm{d}t_K \, \mathrm{d}\boldsymbol{x}.$$

Changing the variables to $\tau_j = \frac{3}{2}t_j$, $j \in \{K-k+1, \ldots, K\}$, and $\tau_{K-k} = \frac{1}{2}t'$ the integral above reads

$$A \le \left(\frac{3}{2}\right)^{sp+k+1} 2 \left(\frac{2}{3}\right)^{k-1} \int_{\mathbb{R}^K} \int_{(0,\infty)^{k+1}} \frac{\left|u\left(\boldsymbol{x} + \sum_{j=K-k}^{K} \tau_j \mathbf{e}_j\right) - u(\boldsymbol{x})\right|^p}{\left(\sum_{j=K-k}^{K} \tau_j\right)^{sp+k+1}} \, \mathrm{d}\tau_{K-k} \ldots \mathrm{d}\tau_K \, \mathrm{d}\boldsymbol{x}$$

$$
\begin{aligned}
&= 2\left(\frac{3}{2}\right)^{sp+2} J_{k+1} \\
&\le 2\left(\frac{3}{2}\right)^{sp+2} C_{k+1}\|u\|^p_{W^{s,p}(\mathbb{R}^K)}.
\end{aligned}
$$

To bound B, a change of variables to $\boldsymbol{y} = \boldsymbol{x} + \sum_{j=K-k+1}^{K} t_j \mathbf{e}_j$, $\tau_j = \frac{1}{2}t_j$, $j \in \{K-k+1, \ldots, K\}$, and $\tau_{K-k} = \frac{1}{2}t'$ yields

$$
\begin{aligned}
B &= 2^{-sp-k-1}\int_{\mathbb{R}^K}\int_{(0,\infty)^{k+1}} \\
&\quad \frac{\left|u\left(\boldsymbol{x} + \sum_{j=K-k+1}^{K} t_j\mathbf{e}_j\right) - u\left(\boldsymbol{x} + \frac{1}{2}t' + \frac{3}{2}\sum_{j=K-k+1}^{K} t_j\mathbf{e}_j\right)\right|^p}{\left(\frac{1}{2}t' + \frac{1}{2}\sum_{j=K-k+1}^{K} t_j\right)^{sp+k+1}} \\
&\quad \mathrm{d}t'\,\mathrm{d}t_{K-k+1}\ldots\mathrm{d}t_K\,\mathrm{d}\boldsymbol{x} \\
&= 2^{-sp-k-1}2^k\int_{\mathbb{R}^K}\int_{(0,\infty)^{k+1}} \frac{\left|u(\boldsymbol{y}) - u\left(\boldsymbol{y} + \sum_{j=K-k}^{K} \tau_j\mathbf{e}_j\right)\right|^p}{\left(\sum_{j=K-k}^{K}\tau_j\right)^{sp+k+1}}\,\mathrm{d}\tau_{K-k}\ldots\mathrm{d}\tau_K\,\mathrm{d}\boldsymbol{x} \\
&= 2^{-sp-1}J_{k+1} \\
&\le 2^{-sp-1}C_{k+1}\|u\|^p_{W^{s,p}(\mathbb{R}^K)}.
\end{aligned}
$$

Together this shows that $J_k \le C_k\|u\|^p_{W^{s,p}(\mathbb{R}^K)}$ with $C_k = (sp+k)2^{(1-s)p-2}(1 + 3^{sp-2})C_{k+1}$. By induction over k it follows that $\|u\|^p_{K,s,p} = 2J_1$ is bounded by $C\|u\|^p_{W^{s,p}(\mathbb{R}^K)}$, where C only depends on p and K, but not on u. Proceeding the same way to bound $\|u\|^p_{i,s,p}$, $i \ne K$, results in the estimate $\sum_{i=1}^{K}|u|_{i,s,p} \le C\|u\|_{W^{s,p}(\mathbb{R}^K)}$ with a different C. □

3.5.1 Further Properties

The following important inequality bounds the gradient of a vector by its symmetric part, also called the deformation tensor,

$$
\mathsf{D}(\boldsymbol{v}) := \frac{1}{2}\left(\nabla\boldsymbol{v} + \nabla\boldsymbol{v}^\top\right).
$$

Lemma 3.5.3 (Korn Inequality) *For all* $\boldsymbol{v} \in \mathbf{H}^1(\Omega)$ *it is*

$$\|\nabla \boldsymbol{v}\|_0 \leq C(\|\mathsf{D}(\boldsymbol{v})\|_0 + \|\boldsymbol{v}\|_0) \tag{3.4}$$

where C *does not depend on* $\boldsymbol{v}$.

This is remarkable because while the gradient $\nabla \boldsymbol{v}$ has in general d^2 different entries, the deformation tensor only has $(d^2 + d)/2$ due to its symmetry. The proof can be found for example in [KO88, Theorem 5.13]. Furthermore, there is also a version for $p \neq 2$, an entire chapter in [DDE12] considers Korn's inequality in L^p.

Some important and well known properties of Sobolev spaces are not shown in this monograph, most notably the Sobolev embedding theorems, which are stated here for completeness and can be found for example in [DDE12],

Theorem 3.5.4 ([DDE12], Theorem 2.72 and Corollary 4.53) *Let* $s \in (0, 1]$ *and* $p > 1$ *be given. Further let* $\Omega \subset \mathbb{R}^d$ *be an open bounded Lipschitz set. Then it holds:*

- *If* $sp < d$, *then* $W^{s,p}(\Omega)$ *is continuously embedded in* $L^q(\Omega)$ *for all* $q \leq dp/(d - sp)$.
- *If* $sp = d$, *then* $W^{s,p}(\Omega)$ *is continuously embedded in* $L^q(\Omega)$ *for all* $q \leq \infty$,
- *If* $sp > d$, *then* $W^{s,p}(\Omega)$ *is continuously embedded in* $L^\infty(\Omega)$ *and even in a suitable Hölder space.*

Chapter 4
Traces

One important property of certain Sobolev spaces is the fact that there is a well defined restriction onto the boundary, even though the boundary has measure zero. Such restrictions are known as traces and allow for prescribed boundary data of solutions of partial differential equations.

A notion of trace extends that of the restriction of continuous functions to Sobolev spaces. A first result shows that such an operator exists and that it continuously maps into a Lebesgue space on the boundary of a Lipschitz domain. This is what many textbooks also cover, especially in the finite element community. A second theorem is more involved and states that the trace operator in fact maps into a proper subspace, namely a suitable Sobolev–Slobodeckij space, on the boundary. This is shown in steps; at first for the half space, then on all of $\partial\Omega$, and finally on a part of the boundary.

Of special interest are the kernels of these trace operators, i.e., the sets which have zero boundary values in a sense which is to be made precise. This kind of analysis leads to the so-called Lions–Magenes spaces which are studied in Sect. 4.3.1.

Together with traces, the questions of inverses arises. Every trace operator in this chapter is introduced with a suitable right inverse which is of importance to define weak formulations of some partial differential equations in Chap. 5. Of course to define right inverses the trace operator needs to be surjective which is the main motivation in studying the image spaces of the trace operators.

4.1 The Trace Operator

Theorem 4.1.1 (Trace Operator) *Let Ω be a bounded Lipschitz domain and $p > 1$. Then there exists a linear and continuous map $T : W^{1,p}(\Omega) \to L^p(\partial\Omega)$ which extends the restriction of continuous functions, i.e., for all $u \in C(\overline{\Omega}) \cap W^{1,p}(\Omega)$ the function Tu is the restriction onto the boundary: $Tu(\boldsymbol{x}) = u(\boldsymbol{x})$ for all $\boldsymbol{x} \in \partial\Omega$.*

U. Wilbrandt, *Stokes–Darcy Equations*, Advances in Mathematical Fluid Mechanics,
https://doi.org/10.1007/978-3-030-02904-3_4

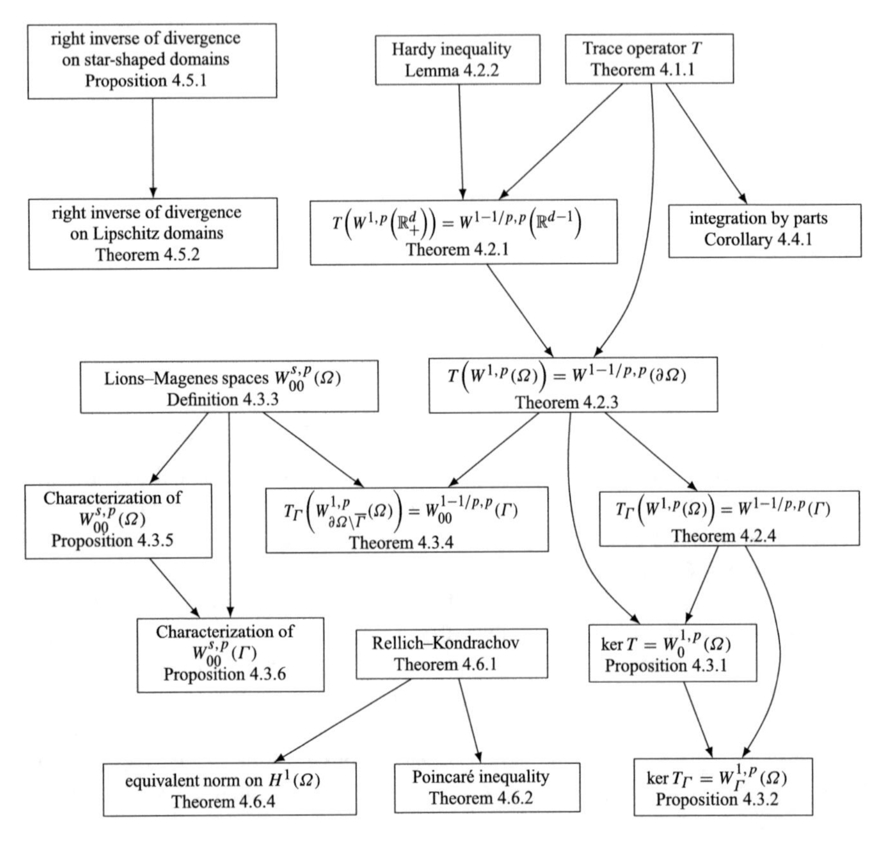

Fig. 4.1 A graph showing the dependencies among all theorems, definitions, corollaries, lemmas as well as propositions in Chap. 4. Note that implicit dependencies are not always shown

Proof This proof mainly follows [Eva10], Theorem 1 in Chapter 5.5, however there it is only shown for $\mathcal{C}^1$ domains. See also [DDE12], Theorem 2.86, for the Lipschitz case.

Consider a localization (V_i, h_i, Φ_i), $i \in [0, N]$, of Ω as in Definition 2.2.1 and, at first, assume $u \in \mathcal{C}^\infty(\overline{\Omega})$. Let v_i be the local representation of u, i.e., for all $(\boldsymbol{x}', x_d) \in \mathbb{R}^{d-1} \times \mathbb{R}_+$ define $v_i(\boldsymbol{x}', x_d) = (\Phi_i u)(\boldsymbol{x}', x_d + h_i(\boldsymbol{x}'))$. The function v_i is an element of $W^{1,p}(\mathbb{R}^d_+)$, see Propositions 3.2.1 and 3.2.3, and is Lipschitz continuous because h_i is. According to the fundamental theorem of calculus it is

$$\begin{aligned}\int_{\mathbb{R}^{d-1}} \left|v_i(\boldsymbol{x}', 0)\right|^p \,\mathrm{d}\boldsymbol{x}' &= -\int_{\mathbb{R}^d_+} \frac{\partial |v_i|^p}{\partial x_d}(\boldsymbol{x}) \,\mathrm{d}\boldsymbol{x} \\ &= -\int_{\mathbb{R}^d_+} p|v_i(\boldsymbol{x})|^{p-1} \operatorname{sgn}(v_i(\boldsymbol{x})) \frac{\partial v_i}{\partial x_d}(\boldsymbol{x}) \,\mathrm{d}\boldsymbol{x}\end{aligned}$$

$$\leq (p-1)\int_{\mathbb{R}^d_+} |v_i(\boldsymbol{x})|^p \,\mathrm{d}\boldsymbol{x} + \int_{\mathbb{R}^d_+} \left|\frac{\partial v_i}{\partial x_d}(\boldsymbol{x})\right|^p \,\mathrm{d}\boldsymbol{x}$$

$$\leq (p-1)\|v_i\|^p_{L^p(\mathbb{R}^d_+)} + \|\nabla v_i\|^p_{L^p(\mathbb{R}^d_+)},$$

where the first estimate follows from the Young inequality, Eq. (2.8), with $a = |\frac{\partial v_i}{\partial x_d}(\boldsymbol{x})|$ and $b = |v_i(\boldsymbol{x})|^{p-1}$. It therefore is $\|v_i(\cdot,0)\|_{L^p(\mathbb{R}^{d-1})} \leq \max\{(p-1)^{1/p}, 1\}\|v_i\|_{W^{1,p}(\mathbb{R}^d_+)}$. Since v_i is a Lipschitz transformation of $\Phi_i u$, Proposition 3.2.3 extends this estimate to $\Phi_i u$ on $\partial\Omega \cap \Omega_i$:

$$\|\Phi_i u\|_{L^p(\partial\Omega\cap\Omega_i)} \leq C\|\Phi_i u\|_{W^{1,p}(\Omega_i\cap\Omega)},$$

where C depends on Ω through h_i and can be chosen independently of i. Combining all the local contributions and Proposition 3.5.1 yields

$$\|u\|_{L^p(\partial\Omega)} \leq \sum_{i=1}^{N}\|\Phi_i u\|_{L^p(\partial\Omega\cap\Omega_i)} \leq C\sum_{i=1}^{N}\|\Phi_i u\|_{W^{1,p}(\Omega_i\cap\Omega)} \leq C\|u\|_{W^{1,p}(\Omega)}. \tag{4.1}$$

Thus the L^p norm of the restriction to the boundary of the continuous function u is bounded by its $W^{1,p}$ norm inside the domain. Now assume $u \in W^{1,p}(\Omega)$ and let $u_n \in \mathcal{D}(\overline{\Omega})$ be a sequence of smooth function converging to u in $W^{1,p}(\Omega)$, according to Theorem 3.4.5. Then, due to Eq. (4.1) applied to $u_n - u_m$, $u_n\big|_{\partial\Omega}$ is a Cauchy sequence in $L^p(\partial\Omega)$ and hence it converges. Define the trace of u to be its limit:

$$Tu := \lim_{n\to\infty} u_n\big|_{\partial\Omega}.$$

In case $u \in \mathcal{C}(\overline{\Omega}) \cap W^{1,p}(\Omega)$ note that the u_n above converge uniformly to u, because the domain is bounded, i.e., $Tu = u\big|_{\partial\Omega}$ as claimed. □

Remark 4.1.2 In case one only considers a nonzero relatively open part Γ of $\partial\Omega$, a trace $T_\Gamma : W^{1,p}(\Omega) \to L^p(\Gamma)$ can be defined restricting the trace $T = T_{\partial\Omega}$ to Γ. Hence, such a trace operator is also continuous and identical to a restriction on the set $\mathcal{C}(\overline{\Omega}) \cap W^{1,p}(\Omega)$.

Remark 4.1.3 In the case of vector valued spaces such as $\mathbf{H}^1(\Omega)$ a trace operator is defined by the vector of traces for each component. The same name T with possible subscripts is used here.

4.2 Continuity of the Trace Operator on $W^{1,p}(\Omega)$

In this section it is shown that the trace operator T from Theorem 4.1.1 maps not into the entire $L^p(\Gamma)$, but only into the Sobolev–Slobodeckij space $W^{1-1/p,p}(\Gamma)$. A rather lengthy proof is taken from the book [DDE12]. First, the half space

$$\mathbb{R}^d_+ = \mathbb{R}^{d-1} \times (0, \infty) = \left\{ \boldsymbol{x} \in \mathbb{R}^d \;\middle|\; \boldsymbol{x} = (x_1, \ldots, x_d), x_d > 0 \right\}$$

is considered. Later a more general open set Ω with Lipschitz continuous boundary is studied.

Theorem 4.2.1 (Theorem 3.9, [DDE12]) *For $d \geq 2$ the image of the trace operator $T : W^{1,p}(\mathbb{R}^d_+) \to L^p(\mathbb{R}^{d-1})$ from Theorem 4.1.1 is the Sobolev–Slobodeckij space on the boundary, i.e.,*

$$T\left(W^{1,p}(\mathbb{R}^d_+)\right) = W^{1-1/p,p}\left(\mathbb{R}^{d-1}\right).$$

Furthermore, as a map from $W^{1,p}(\mathbb{R}^d_+)$ to $W^{1-1/p,p}\left(\mathbb{R}^{d-1}\right)$ it is continuous and possesses a linear and continuous right inverse $E : W^{1-1/p,p}\left(\mathbb{R}^{d-1}\right) \to W^{1,p}(\mathbb{R}^d_+)$.

The proof uses another result which is introduced first, namely an inequality due to Hardy:

Lemma 4.2.2 (Hardy's Inequality) *Let $f \in L^p(\mathbb{R}_+)$, $p > 1$, $\mathbb{R}_+ = (0, \infty)$, and define $g : \mathbb{R}_+ \to \mathbb{R}$ as $g(t) = \frac{1}{t}\int_0^t f(s)\,\mathrm{d}s$. Then g is in $L^p(\mathbb{R}_+)$ and it holds*

$$\|g\|_{L^p(\mathbb{R}_+)} \leq \frac{p}{p-1}\|f\|_{L^p(\mathbb{R}_+)}.$$

Proof This proof is from Lemma 13.4 in [Tar07]. In [DDE12], Lemma 3.14, a more general version is proved but with a larger constant.

Using the Hölder inequality, Theorem 2.3.1, it is

$$\begin{aligned} |g(t)| &\leq \frac{1}{t}\left(\int_0^t |f(s)|^p\,\mathrm{d}s\right)^{1/p} t^{1-1/p} = t^{-1/p}\|f\|_{L^p((0,t))} \\ \Longrightarrow \qquad t|g(t)|^p &\leq \|f\|^p_{L^p((0,t))}. \end{aligned}$$

In particular $t|g(t)|^p \to 0$ for $t \to 0$. Because $p > 1$ the definition of g shows $t|g(t)|^p \to 0$ for $t \to \infty$. Furthermore, the derivative of g is given by

$$g'(t) = -\frac{1}{t^2}\int_0^t f(s)\,\mathrm{d}s + \frac{1}{t}f(t) = -\frac{1}{t}(g(t) - f(t)).$$

Next, g is shown to be in $L^p((0, M))$ for any $M \in \mathbb{R}_+$. Using the fact $t|g(t)|^p \to 0$ for $t \to 0$ from above and integration by parts gives

$$\begin{aligned}\int_0^M |g(t)|^p \, dt &= -p \int_0^M t|g(t)|^{p-2} g(t) g'(t) \, dt + M|g(M)|^p \\ &= p \int_0^M |g(t)|^{p-2} g(t)(g(t) - f(t)) \, dt + M|g(M)|^p \\ &= p \int_0^M |g(t)|^p - p \int_0^M |g(t)|^{p-2} g(t) f(t) \, dt + M|g(M)|^p \\ \implies (1-p) \int_0^M |g(t)|^p \, dt &= -p \int_0^M |g(t)|^{p-2} g(t) f(t) \, dt + M|g(M)|^p.\end{aligned}$$

Taking the absolute value followed by an application of the Hölder inequality, Theorem 2.3.1, gives

$$\begin{aligned}(p-1)\|g\|^p_{L^p((0,M))} &\le p \int_0^M |g(t)|^{p-1} |f(t)| \, dt + M|g(M)|^p \\ &\le p \Big(\int_0^M |g(t)|^{(p-1)p/(p-1)} \, dt \Big)^{1-1/p} \|f\|_{L^p(0,M)} + M|g(M)|^p \\ &= p\|g\|^{p-1}_{L^p(0,M)} \|f\|_{L^p(0,M)} + M|g(M)|^p \\ \overset{M\to\infty}{\Longrightarrow} \|g\|_{L^p(\mathbb{R}_+)} &\le \frac{p}{p-1} \|f\|_{L^p(\mathbb{R}_+)}\end{aligned}$$

as claimed. Note that in the last step it is used that $t|g(t)|^p \to 0$ for $t \to \infty$. □

Having proved Hardy's inequality, Theorem 4.2.1 can be shown:

Proof of Theorem 4.2.1 In this proof it is assumed that $d \ge 2$. In [DDE12] the case $d = 2$ is treated separately.

The first step in this proof is to show that $W^{1-1/p,p}(\mathbb{R}^{d-1})$ is continuously embedded in $T(W^{1,p}(\mathbb{R}^d_+))$. To this end, let $u \in W^{1-1/p,p}(\mathbb{R}^{d-1})$ be given. The aim is to construct a function $Eu = v \in W^{1,p}(\mathbb{R}^d_+)$ such that $Tv = u$ and $\|v\|_{W^{1,p}(\mathbb{R}^d_+)} \le c\|u\|_{W^{1-1/p,p}(\mathbb{R}^{d-1})}$. Then the extension operator has the claimed properties.

Let $\varphi \in \mathcal{D}(\mathbb{R})$ be a test function with $\varphi(0) = 1$ and $\|\varphi\|_{L^\infty(\mathbb{R})} = 1$, then define $v : \mathbb{R}^d_+ \to \mathbb{R}$ as

$$v(\boldsymbol{x}, t) := \frac{\varphi(t)}{t^{d-1}} \int_{(0,t)^{d-1}} u(\boldsymbol{x} + \boldsymbol{y}) \, d\boldsymbol{y} = \varphi(t) \int_{(0,1)^{d-1}} u(\boldsymbol{x} + t\boldsymbol{y}) \, d\boldsymbol{y}.$$

Using Hölder's inequality, Theorem 2.3.1, and Fubini's Theorem to exchange the order of integration shows that v is in $L^p(\mathbb{R}^d_+)$:

$$\begin{aligned}\|v\|^p_{L^p(\mathbb{R}^d_+)} &= \int_{\mathbb{R}^{d-1}}\int_{(0,\infty)}\left|\varphi(t)\int_{(0,1)^{d-1}} u(\boldsymbol{x}+t\boldsymbol{y})\,\mathrm{d}\boldsymbol{y}\right|^p \mathrm{d}t\,\mathrm{d}\boldsymbol{x}\\ &\le \int_{\mathbb{R}^{d-1}}\int_{(0,\infty)}|\varphi(t)|^p\int_{(0,1)^{d-1}}|u(\boldsymbol{x}+t\boldsymbol{y})|^p\,\mathrm{d}\boldsymbol{y}\,\mathrm{d}t\,\mathrm{d}\boldsymbol{x}\\ &= \int_{(0,1)^{d-1}}\int_{(0,\infty)}|\varphi(t)|^p\int_{\mathbb{R}^{d-1}}|u(\boldsymbol{x}+t\boldsymbol{y})|^p\,\mathrm{d}\boldsymbol{x}\,\mathrm{d}t\,\mathrm{d}\boldsymbol{y}\\ &\le \|u\|^p_{L^p(\mathbb{R}^{d-1})}\|\varphi\|_{L^\infty(\mathbb{R})}|\mathrm{supp}(\varphi)|.\end{aligned}$$

For almost all $\boldsymbol{x}\in\mathbb{R}^{d-1}$ and all $t\in\mathbb{R}$ it is

$$\begin{aligned}|v(\boldsymbol{x},t)-u(\boldsymbol{x})| &= \left|\int_{(0,1)^{d-1}}\varphi(t)u(\boldsymbol{x}+t\boldsymbol{y})-u(\boldsymbol{x})\,\mathrm{d}\boldsymbol{y}\right|\\ &\le \|\varphi\|_{L^\infty(\mathbb{R})}\left|\int_{(0,1)^{d-1}}u(\boldsymbol{x}+t\boldsymbol{y})-u(\boldsymbol{x})\,\mathrm{d}\boldsymbol{y}\right| + |u(\boldsymbol{x})|\,|\varphi(t)-1|.\end{aligned}$$

In the limit $t\to 0$ both terms above approach zero, the first due to the continuity of the shift operator on $L^p(\mathbb{R}^{d-1})$, see Corollary 3.4.4, the second because φ is continuous and $\varphi(0)$ is chosen to be 1. Taking the p-th power and integrating with respect to $\boldsymbol{x}$ over $\mathbb{R}^{d-1}$ therefore shows $v(\cdot,t)\to u(\cdot)$ in $L^p(\mathbb{R}^{d-1})$ for $t\to 0$, which implies $Tv=u$. It remains to show that v is in fact an element in $W^{1,p}(\mathbb{R}^d_+)$. Consider the derivative of v in x_i direction. The integral is split into two where the inner one integrates over the i-th variable z_i while the outer one integrates over the remaining $d-2$ variables. For $i\in[1,d-1]$ and $\boldsymbol{y}=(y_1,\dots,y_{d-1})\in(0,1)^{d-1}$ define $\check{\boldsymbol{y}}_i=(y_1,\dots,y_{i-1},0,y_{i+1},\dots,y_{d-1})$, i.e., $\boldsymbol{y}=\check{\boldsymbol{y}}_i+y_i\mathbf{e}_i$. Furthermore, the inner integral is transformed such that x_i no longer enters the argument of u. This avoids having to differentiate u in the application of the Leibniz integral rule, which would be impossible because in general it is only in $W^{1-1/p,p}(\mathbb{R}^{d-1})$ rather than $W^{1,p}(\mathbb{R}^{d-1})$.

$$\begin{aligned}\frac{\partial v}{\partial x_i}(\boldsymbol{x},t) &= \frac{\partial}{\partial x_i}\left(\frac{\varphi(t)}{t}\int_{(0,1)^{d-2}}\int_{(x_i,x_i+t)}u(\check{\boldsymbol{x}}_i+t\check{\boldsymbol{z}}_i+z_i\mathbf{e}_i)\,\mathrm{d}z_i\,\mathrm{d}\check{\boldsymbol{z}}_i\right)\\ &= \frac{\varphi(t)}{t}\int_{(0,1)^{d-2}}\frac{\partial}{\partial x_i}\left(\int_{(x_i,x_i+t)}u(\check{\boldsymbol{x}}_i+t\check{\boldsymbol{z}}_i+z_i\mathbf{e}_i)\,\mathrm{d}z_i\right)\mathrm{d}\check{\boldsymbol{z}}_i\\ &= \frac{\varphi(t)}{t}\int_{(0,1)^{d-2}}u(\check{\boldsymbol{x}}_i+t\check{\boldsymbol{z}}_i+(x_i+t)\mathbf{e}_i)-u(\check{\boldsymbol{x}}_i+t\check{\boldsymbol{z}}_i+x_i\mathbf{e}_i)\,\mathrm{d}\check{\boldsymbol{z}}_i\\ &= \varphi(t)\int_{(0,1)^{d-2}}\frac{u(\boldsymbol{x}+t\check{\boldsymbol{z}}_i+t\mathbf{e}_i)-u(\boldsymbol{x}+t\check{\boldsymbol{z}}_i)}{t}\,\mathrm{d}\check{\boldsymbol{z}}_i.\end{aligned}$$

Integrating with respect to $\boldsymbol{x}$ and t over $\mathbb{R}^{d-1}$ and $\mathbb{R}_+$ followed by a change of variables $\boldsymbol{y} = \boldsymbol{x} + t\breve{\boldsymbol{z}}_i$ leads to (note that $\|\varphi\|_{L^\infty(\mathbb{R})} = 1$ is assumed)

$$\int_{\mathbb{R}^{d-1}} \int_{\mathbb{R}} \left| \frac{\partial v}{\partial x_i}(\boldsymbol{x}, t) \right|^p \mathrm{d}t\, \mathrm{d}\boldsymbol{x} \leq \int_{\mathbb{R}^{d-1}} \int_{\mathbb{R}_+} \int_{(0,1)^{d-2}} \left| \frac{u(\boldsymbol{x} + t\breve{\boldsymbol{z}}_i + t\mathbf{e}_i) - u(\boldsymbol{x} + t\breve{\boldsymbol{z}}_i)}{t} \right|^p \mathrm{d}\breve{\boldsymbol{z}}_i\, \mathrm{d}t\, \mathrm{d}\boldsymbol{x}$$

$$= \int_{\mathbb{R}^{d-1}} \int_{\mathbb{R}_+} \left| \frac{u(\boldsymbol{y} + t\mathbf{e}_i) - u(\boldsymbol{y})}{t} \right|^p \mathrm{d}t\, \mathrm{d}\boldsymbol{y}$$

$$= \frac{1}{2} |u|^p_{i,1-1/p,p},$$

where the last semi-norm is defined in Proposition 3.5.2. The derivative of v with respect to t can be bounded in a similar fashion:

$$\frac{\partial v}{\partial t}(\boldsymbol{x}, t) = \frac{\varphi'(t)}{t^{d-1}} \int_{(0,t)^{d-1}} u(\boldsymbol{x} + \boldsymbol{z})\, \mathrm{d}\boldsymbol{z} + \frac{\varphi(t)}{t^d}(1-d) \int_{(0,t)^{d-1}} u(\boldsymbol{x} + \boldsymbol{z})\, \mathrm{d}\boldsymbol{z} + \frac{\varphi(t)}{t^{d-1}} \int_{\partial((0,t)^{d-1})} u(\boldsymbol{x} + \boldsymbol{z}) \frac{\partial \boldsymbol{z}}{\partial t} \cdot \boldsymbol{n}\, \mathrm{d}\boldsymbol{z}.$$

In the last integral, for almost every $\boldsymbol{z} \in \partial\big((0,t)^{d-1}\big)$, the normal $\boldsymbol{n}$ is either $-\mathbf{e}_i$ and $z_i = 0$ or $\boldsymbol{n} = \mathbf{e}_i$ and $z_i = t$ for some i. Therefore, the term $\frac{\partial \boldsymbol{z}}{\partial t} \cdot \boldsymbol{n}$ vanishes whenever $z_i = 0$ and reduces to 1 otherwise, hence, leaving a sum of integrals over all boundary faces with $\boldsymbol{z} = \breve{\boldsymbol{z}}_i + t\mathbf{e}_i$:

$$\begin{aligned}
\frac{\partial v}{\partial t}(\boldsymbol{x}, t) &= \frac{\varphi'(t)}{t^{d-1}} \int_{(0,t)^{d-1}} u(\boldsymbol{x} + \boldsymbol{z})\, \mathrm{d}\boldsymbol{z} - \frac{\varphi(t)}{t^{d-1}} \sum_{i=1}^{d-1} \int_{(0,t)^{d-1}} \frac{u(\boldsymbol{x} + \boldsymbol{z})}{t}\, \mathrm{d}\boldsymbol{z} \\
&\quad + \frac{\varphi(t)}{t^{d-1}} \sum_{i=1}^{d-1} \int_{(0,t)^{d-2}} u(\boldsymbol{x} + \breve{\boldsymbol{z}}_i + t\mathbf{e}_i)\, \mathrm{d}\breve{\boldsymbol{z}}_i \\
&= \frac{\varphi'(t)}{t^{d-1}} \int_{(0,t)^{d-1}} u(\boldsymbol{x} + \boldsymbol{z})\, \mathrm{d}\boldsymbol{z} \\
&\quad + \frac{\varphi(t)}{t^{d-1}} \sum_{i=1}^{d-1} \int_{(0,t)^{d-1}} \frac{u(\boldsymbol{x} + \breve{\boldsymbol{z}}_i + t\mathbf{e}_i) - u(\boldsymbol{x} + \boldsymbol{z})}{t}\, \mathrm{d}\boldsymbol{z} \\
&= \frac{\varphi'(t)}{t^{d-1}} \int_{(0,t)^{d-1}} u(\boldsymbol{x} + \boldsymbol{z})\, \mathrm{d}\boldsymbol{z} \\
&\quad + \varphi(t) \sum_{i=1}^{d-1} \int_{(0,1)^{d-1}} \frac{u(\boldsymbol{x} + t\breve{\boldsymbol{y}}_i + t\mathbf{e}_i) - u(\boldsymbol{x} + t\boldsymbol{y})}{t}\, \mathrm{d}\boldsymbol{y},
\end{aligned}$$

where in the last step it is $t\boldsymbol{y} = \boldsymbol{z}$. The first term on the right-hand side above clearly is a function in $L^p(\mathbb{R}^d_+)$. Taking the p-th power and integrating over $\mathbb{R}^d_+$ followed by a change of variables $\boldsymbol{x}' = \boldsymbol{x} + t\boldsymbol{y}$ and $s = t(1 - y_i)$ in the second term leads to

$$\begin{aligned}
&\sum_{i=1}^{d-1} \int_{\mathbb{R}^{d-1}} \int_{\mathbb{R}_+} \int_{(0,1)^{d-1}} \left| \frac{u(\boldsymbol{x}' + s\mathbf{e}_i) - u(\boldsymbol{x}')}{s} \right|^p (1 - y_i)^{p-1} \,\mathrm{d}\boldsymbol{y}\,\mathrm{d}s\,\mathrm{d}\boldsymbol{x}' \\
&= \frac{1}{p} \sum_{i=1}^{d-1} \int_{\mathbb{R}^{d-1}} \int_{\mathbb{R}_+} \left| \frac{u(\boldsymbol{x}' + s\mathbf{e}_i) - u(\boldsymbol{x}')}{s} \right|^p \mathrm{d}s\,\mathrm{d}\boldsymbol{x}' \\
&= \frac{1}{2p} |u|^p_{i,1-1/p,p}.
\end{aligned}$$

Due to Proposition 3.5.2 all partial derivatives of u can be bounded by the Sobolev–Slobodeckij norm of u with a constant c which depends on p, d, and φ but not on u. It therefore is

$$\|v\|_{W^{1,p}(\mathbb{R}^d_+)} \le c\|u\|_{W^{1-1/p,p}(\mathbb{R}^{d-1})},$$

as claimed. With this construction the extension operator E is a linear and continuous right inverse of the trace T.

Conversely, now let $v \in W^{1,p}(\mathbb{R}^d_+)$ be given and it must be shown that the trace $u = Tv$, defined as $u(\boldsymbol{x}) = v(\boldsymbol{x}, 0)$ for almost all $\boldsymbol{x}$, is a function in $W^{1-1/p,p}(\mathbb{R}^{d-1})$ whose norm can be bounded by that of v. According to Proposition 3.5.2 it suffices to prove $|u|_{i,1-1/p,p} \le c\|v\|_{W^{1,p}(\mathbb{R}^d_+)}$ for all $i \in \{1, \ldots, d-1\}$. Using $(a + b)^p \le 2^{p-1}(a^p + b^p)$ it is

$$\begin{aligned}
|u|^p_{i,1-1/p,p} &= \int_{\mathbb{R}^{d-1}} \int_{\mathbb{R}} \left| \frac{v(\boldsymbol{x} + t\mathbf{e}_i, 0) - v(\boldsymbol{x}, 0)}{t} \right|^p \mathrm{d}t\,\mathrm{d}\boldsymbol{x} \\
&\le 2^{p-1} \int_{\mathbb{R}^{d-1}} \int_{\mathbb{R}} \left| \frac{v(\boldsymbol{x} + t\mathbf{e}_i, 0) - v(\boldsymbol{x} + \frac{t}{2}\mathbf{e}_i, \frac{t}{2})}{t} \right|^p \mathrm{d}t\,\mathrm{d}\boldsymbol{x} \\
&\quad + 2^{p-1} \int_{\mathbb{R}^{d-1}} \int_{\mathbb{R}} \left| \frac{v(\boldsymbol{x} + \frac{t}{2}\mathbf{e}_i, \frac{t}{2}) - v(\boldsymbol{x}, 0)}{t} \right|^p \mathrm{d}t\,\mathrm{d}\boldsymbol{x}.
\end{aligned}$$

Consider the first of the last two terms, the second can be handled in the same way. First, a change of variables ($\boldsymbol{y} = \boldsymbol{x} + t\mathbf{e}_i$, $\tau = \frac{1}{2}t$) is followed by an application of the fundamental theorem of calculus:

$$
\begin{aligned}
&2^{p-1} \int_{\mathbb{R}^{d-1}} \int_{\mathbb{R}} \left| \frac{v(\boldsymbol{x}+t\mathbf{e}_i, 0) - v(\boldsymbol{x}+\frac{t}{2}\mathbf{e}_i, \frac{t}{2})}{t} \right|^p \mathrm{d}t\, \mathrm{d}\boldsymbol{x} \\
&= 2^p \int_{\mathbb{R}^{d-1}} \int_{\mathbb{R}} \left| \frac{v(\boldsymbol{y}, 0) - v(\boldsymbol{y}-\tau\mathbf{e}_i, \tau)}{2\tau} \right|^p \mathrm{d}\tau\, \mathrm{d}\boldsymbol{y} \\
&= \int_{\mathbb{R}^{d-1}} \int_{\mathbb{R}} \left| \frac{1}{\tau} \int_{(0,\tau)} \frac{\partial}{\partial s} v(\boldsymbol{y}-s\mathbf{e}_i, s)\, \mathrm{d}s \right|^p \mathrm{d}\tau\, \mathrm{d}\boldsymbol{y} \\
&\leq \int_{\mathbb{R}^{d-1}} \int_{\mathbb{R}} \left| \frac{1}{\tau} \int_{(0,\tau)} |\nabla v(\boldsymbol{y}-s\mathbf{e}_i, s)|\, \mathrm{d}s \right|^p \mathrm{d}\tau\, \mathrm{d}\boldsymbol{y} \\
&\leq c(p) \int_{\mathbb{R}^{d-1}} \int_{\mathbb{R}} |\nabla v(\boldsymbol{y}-\tau\mathbf{e}_i, \tau)|^p\, \mathrm{d}\tau\, \mathrm{d}\boldsymbol{y} \qquad \text{Hardy's inequality, Lemma 4.2.2,} \\
&= c(p) \int_{\mathbb{R}^d} |\nabla v(\boldsymbol{z})|^p\, \mathrm{d}\boldsymbol{z} \\
&\leq c(p) \|v\|^p_{W^{1,p}(\mathbb{R}^d_+)}.
\end{aligned}
$$

Combining these results yields the bound $|u|_{i,1-1/p,p} \leq 2c(p)^{1/p}\|v\|_{W^{1,p}(\mathbb{R}^d_+)}$ with $c(p)$ being the constant from Hardy's inequality, Lemma 4.2.2. Therefore, with an application of Proposition 3.5.2 it is $u \in W^{1-1/p,p}(\mathbb{R}^{d-1})$, i.e., the continuity of the trace $T : W^{1,p}(\mathbb{R}^d_+) \to W^{1-1/p,p}(\mathbb{R}^{d-1})$ holds as claimed. □

To prove such a result for a trace on Lipschitz domains, a partition of unity is used and the local results are suitably combined. The proof of the following theorem is mainly from [DDE12], Proposition 3.31.

Theorem 4.2.3 (Trace on Lipschitz Domains) *Let $\Omega \subset \mathbb{R}^d$ be a Lipschitz domain. Then the image of the trace map $T : W^{1,p}(\Omega) \to L^p(\partial\Omega)$ introduced in Theorem 4.1.1 is the Sobolev–Slobodeckij space on the boundary, i.e.,*

$$T\Big(W^{1,p}(\Omega)\Big) = W^{1-1/p,p}(\partial\Omega).$$

Furthermore, as a map from $W^{1,p}(\Omega)$ to $W^{1-1/p,p}(\partial\Omega)$ it is continuous and possesses a linear and continuous right inverse $E : W^{1-1/p,p}(\partial\Omega) \to W^{1,p}(\Omega)$.

Proof Let $u \in W^{1,p}(\Omega)$ be given and let (V_i, h_i, Φ_i), $i = 0, \ldots, N$, be a (Lipschitz) localization of Ω, see Definition 2.2.1. Let $v_i : \mathbb{R}^d_+ \to \mathbb{R}$ be the local part of u transformed by h_i:

$$v_i(\boldsymbol{x}', x_d) := (\Phi_i u)\big(\boldsymbol{x}', h_i(\boldsymbol{x}') + x_d\big),$$

where Φ_i is extended by zero outside of its domain. Note that v_i has compact support in $\mathbb{R}^d_+$ and is an element of $W^{1,p}(\mathbb{R}^d_+)$ due to Propositions 3.2.3 and 3.2.1 applied to the product $\Phi_i u$ and the transformation $(\boldsymbol{x}', x_d) \mapsto (\boldsymbol{x}', h_i(\boldsymbol{x}') + x_d)$, respectively. Additionally, the estimate

$$\|v_i\|_{W^{1,p}(\mathbb{R}^d_+)} \leq C\|\Phi_i u\|_{W^{1,p}(\Omega_i \cap \Omega)} \leq C\|u\|_{W^{1,p}(\Omega_i \cap \Omega)}$$

holds. Theorem 4.2.1 implies that its trace is in the respective Sobolev–Slobodeckij space, i.e., $T v_i \in W^{1-1/p,p}(\mathbb{R}^{d-1})$, with the bound

$$\|T v_i\|_{W^{1-1/p,p}(\mathbb{R}^{d-1})} \leq \|v\|_{W^{1,p}(\mathbb{R}^d_+)}.$$

According to Proposition 3.2.4 the product $\Phi_i u$ is in $W^{1-1/p,p}(\Omega_i \cap \partial\Omega)$ because it is a Lipschitz transformation of $T v_i$ and it holds

$$\|\Phi_i u\|_{W^{1-1/p,p}(\Omega_i \cap \partial\Omega)} \leq \|T v_i\|_{W^{1-1/p,p}(\mathbb{R}^{d-1})}.$$

Combining all these partial results leads to

$$\begin{aligned}
\|Tu\|_{W^{1-1/p,p}(\Gamma)} &\leq \sum_{i=1}^{N} \|\Phi_i u\|_{W^{1-1/p,p}(\Omega_i \cap \partial\Omega)} \\
&\leq C' \sum_{i=1}^{N} \|u\|_{W^{1,p}(\Omega_i \cap \Omega)} \\
&\leq C\|u\|_{W^{1,p}(\Omega)},
\end{aligned}$$

where the last inequality is due to Proposition 3.5.1, showing the continuity of the trace operator.

Conversely, let $u \in W^{1-1/p,p}(\partial\Omega)$ be given. The aim is to construct an extension $Eu \in W^{1,p}(\Omega)$ such that $T \circ E = \mathrm{id}$ on $W^{1-1/p,p}(\partial\Omega)$. First, a local extension is constructed. Let $v_i : \mathbb{R}^{d-1} \to \mathbb{R}$ be defined by $v_i(\boldsymbol{x}) = (\Phi_i u)(\boldsymbol{x}, h_i(\boldsymbol{x}))$. According to Propositions 3.2.4 and 3.2.1, the function v_i belongs to $W^{1-1/p,p}(\mathbb{R}^{d-1})$ with the estimate

$$\|v_i\|_{W^{1-1/p,p}(\mathbb{R}^{d-1})} \leq c\|\Phi_i u\|_{W^{1-1/p,p}(\partial\Omega \cap \Omega_i)},$$

where c depends on Ω. Next, in the setting of the half space, there is a continuous extension $E v_i \in W^{1,p}(\mathbb{R}^d_+)$ such that

$$\|E v_i\|_{W^{1,p}(\mathbb{R}^d_+)} \leq C\|v_i\|_{W^{1-1/p,p}(\mathbb{R}^{d-1})},$$

see Theorem 4.2.1. This extension can be transformed to $W^{1,p}(\Omega_i \cap \Omega)$ applying Proposition 3.2.3 which yields $V_i \in W^{1,p}(\Omega_i \cap \Omega)$ with $V_i(\boldsymbol{x}', x_d) := Ev_i(\boldsymbol{x}', x_d - h_i(\boldsymbol{x}'))$ for almost all $(\boldsymbol{x}', x_d) \in \Omega_i \cap \Omega$, and

$$\|V_i\|_{W^{1,p}(\Omega_i \cap \Omega)} \leq \|Ev_i\|_{W^{1,p}(\mathbb{R}^d_+)}.$$

Due to the construction it is $TV_i = v_i$. The desired extension operator is then the sum of the local ones, $Eu := \sum_{i=1}^N V_i$. Combining these results yields

$$\begin{aligned}
\|Eu\|_{W^{1,p}(\Omega)} &\leq \sum_{i=1}^N \|V_i\|_{W^{1,p}(\Omega_i \cap \Omega)} \\
&\leq C \sum_{i=1}^N \|\Phi_i u\|_{W^{1-1/p,p}(\partial\Omega \cap \Omega_i)} \\
&\leq C \|u\|_{W^{1-1/p,p}(\partial\Omega)},
\end{aligned}$$

hence the continuity of this extension operator. □

In practice, solutions of partial differential equations have prescribed boundary values often only on a subset Γ of the boundary $\partial\Omega$. While the trace can be further restricted to Γ without difficulties, the extension operator in general does not exist. Additionally to the Lipschitz continuity of the boundary $\partial\Omega$, another such condition on the boundary $\partial\Gamma$ of Γ is needed.

Theorem 4.2.4 *Let $\Omega \subset \mathbb{R}^d$ be a Lipschitz domain and $\Gamma \subset \partial\Omega$ a relatively open subset of its boundary. Moreover, assume the boundary $\partial\Gamma$ of Γ to be Lipschitz. Then the image of the trace map $T_\Gamma : W^{1,p}(\Omega) \to L^p(\Gamma)$ introduced in Remark 4.1.2 is the Sobolev–Slobodeckij space on that boundary part, i.e.,*

$$T_\Gamma\Big(W^{1,p}(\Omega)\Big) = W^{1-1/p,p}(\Gamma).$$

Furthermore, as a map from $W^{1,p}(\Omega)$ to $W^{1-1/p,p}(\Gamma)$ it is continuous and possesses a linear and continuous right inverse $E : W^{1-1/p,p}(\Gamma) \to W^{1,p}(\Omega)$.

Proof Let $u \in W^{1,p}(\Omega)$. Then, using Theorem 4.2.3, its trace Tu onto $\partial\Omega$ is in $W^{1-1/p,p}(\partial\Omega)$. Restricting Tu onto Γ to yield $T_\Gamma u$ is then in $W^{1-1/p,p}(\Gamma)$ and it holds

$$\|T_\Gamma u\|_{W^{1-1/p,p}(\Gamma)} \leq \|Tu\|_{W^{1-1/p,p}(\partial\Omega)} \leq C\|u\|_{W^{1,p}(\Omega)}.$$

Conversely, let $u \in W^{1-1/p,p}(\Gamma)$ be given. Due to Proposition 3.3.4 there is an extension $u \in W^{1-1/p,p}(\partial\Omega)$ which in turn can be extended to $W^{1,p}(\Omega)$, according to Theorem 4.2.3, where the extensions keep the same name Eu for simplicity.

These extensions are linear and continuous, i.e., there exists a constant C which does not depend on u such that

$$\|Eu\|_{W^{1,p}(\Omega)} \leq C\|u\|_{W^{1-1/p,p}(\Gamma)}. \qquad \square$$

4.3 Characterization of the Kernel of the Trace Operator

In Sect. 3.4 it is shown that certain smooth functions are dense in some Sobolev spaces. In particular $\mathcal{D}(\mathbb{R}^d)$ is dense in $W^{s,p}(\mathbb{R}^d)$ and so is $\mathcal{D}(\overline{\Omega})$ in $W^{s,p}(\Omega)$, see Proposition 3.4.3 and Theorem 3.4.5, respectively. However $\mathcal{D}(\Omega)$ is dense in $W^{s,p}(\Omega)$ if, and only if, $s \leq 1/2$, see [LM72], Section 11.1. Yet, for bounded Lipschitz domains Ω, the case which is most important for this monograph, and $s = 1$ the two aforementioned spaces do not coincide. In fact the closure

$$W_0^{1,p}(\Omega) := \overline{\mathcal{D}(\Omega)}^{W^{1,p}(\Omega)} \tag{4.2}$$

is a particular subspace of $W^{1,p}(\Omega)$ which can be characterized as the kernel of the trace operator:

Proposition 4.3.1 *The kernel of the trace operator from Theorem 4.2.3, $T : W^{1,p}(\Omega) \to W^{1-1/p,p}(\partial\Omega)$, is $W_0^{1,p}(\Omega)$.*

Proof This proof can be found in [Eva98], Theorem 2 in Section 5.5, as well as in [Tar07], Lemma 13.7, and [Tri92], Theorem 1, Section 1.5.5.

Let $u \in W_0^{1,p}(\Omega)$ be given together with an approximating sequence $u_n \in \mathcal{D}(\Omega)$, according to Eq. (4.2). Because all u_n have compact support, their restrictions onto the boundary are zero, hence so are their traces. The continuity of the trace operator T now ensures that $Tu = 0$, i.e., $u \in \ker(T)$.

Now let $u \in \ker(T) \subset W^{1,p}(\Omega)$ be given and let (V_i, h_i, Φ_i), $i = 0, \ldots, N$, be a (Lipschitz) localization of Ω, see Definition 2.2.1. As in previous proofs let v_i be the local part of u transformed by h_i:

$$v_i(\boldsymbol{x}', x_d) := (\Phi_i u)(\boldsymbol{x}', h_i(\boldsymbol{x}') + x_d).$$

Due to Propositions 3.2.1 and 3.2.3 it is $v_i \in W^{1,p}(\mathbb{R}^d_+)$ with compact support in $\overline{\mathbb{R}^d_+}$ and $Tv_i = 0$ on $\mathbb{R}^{d-1}$. Next, consider a smooth function $\zeta \in C^\infty(\mathbb{R})$ such that

$$\zeta = 1 \text{ on } [0, 1], \quad \zeta = 0 \text{ on } [2, \infty), \quad \text{and} \quad 0 \leq \zeta \leq 1.$$

Let ζ_n be a scaled version of ζ, $\zeta_n(x) = \zeta(nx)$, and define[1] $w_n \in W^{1,p}(\mathbb{R}^d_+)$ as $w_n(\boldsymbol{x}) = v_i(\boldsymbol{x})(1 - \zeta_n(x_d))$. As $1 - \zeta_n$ converges to 1 point-wise in $\mathbb{R}^d_+$ the w_n

[1]The index $_i$ in the definition of w_n is omitted for better readability.

converge to v_i in $L^p(\mathbb{R}^d_+)$. To show that the derivatives also converge in $L^p(\mathbb{R}^d_+)$ note that

$$\int_{\mathbb{R}^d_+} |\nabla(v_i - w_n)(\boldsymbol{x})|^p \, \mathrm{d}\boldsymbol{x} \leq 2^{p-1} \int_{\mathbb{R}^d_+} |\nabla v_i(\boldsymbol{x}) \zeta_n(x_d)|^p + \left| n v_i(\boldsymbol{x}) \zeta'(n x_d) \right|^p \mathrm{d}\boldsymbol{x}.$$

The first integral above approaches 0 with $n \to \infty$ because the support of ζ_n (i.e., $[0, 2/n]$) intersected with that of v_i vanishes. The second can be bounded by

$$B := C n^p \int_0^{\frac{2}{n}} \int_{\mathbb{R}^{d-1}} \left| v_i(\boldsymbol{x}', x_d) \right|^p \mathrm{d}\boldsymbol{x}' \, \mathrm{d}x_d,$$

where C depends on the derivative of ζ and p. To bound this, further note that according to Theorem 3.4.5 for any $f \in W^{1,p}(\mathbb{R}^d_+)$ there is a sequence $f_n \in \mathcal{D}(\overline{\mathbb{R}^d_+})$ which converges to f in $W^{1,p}(\mathbb{R}^d_+)$. The continuity of the trace operator yields

$$T f_n \xrightarrow{n \to \infty} T f \qquad \text{in } W^{1-1/p,p}(\mathbb{R}^{d-1}).$$

For f_n and $(\boldsymbol{x}', x_d) \in \mathbb{R}^d_+$ the fundamental theorem of calculus yields

$$\left| f_n(\boldsymbol{x}', x_d) \right| \leq \left| f_n(\boldsymbol{x}', 0) \right| + \int_0^{x_d} \left| \frac{\partial f_n}{\partial x_d}(\boldsymbol{x}', t) \right| \mathrm{d}t.$$

The inequality $(a + b)^p \leq 2^{p-1}(a^p + b^p)$ together with the Hölder inequality, Theorem 2.3.1, imply

$$\left| f_n(\boldsymbol{x}', x_d) \right|^p \leq 2^{p-1} \left(\left| f_n(\boldsymbol{x}', 0) \right|^p + x_d^{p-1} \int_0^{x_d} \left| \frac{\partial f_n}{\partial x_d}(\boldsymbol{x}', t) \right|^p \mathrm{d}t \right).$$

Integrating this equation over $\mathbb{R}^{d-1}$ gives

$$\int_{\mathbb{R}^{d-1}} \left| f_n(\boldsymbol{x}', x_d) \right|^p \mathrm{d}\boldsymbol{x}' \leq 2^{p-1} \left(\| T f_n \|^p_{L^p(\mathbb{R}^{d-1})} + x_d^{p-1} \| \nabla f_n \|^p_{L^p(\mathbb{R}^{d-1} \times [0, x_d])} \right),$$

which also holds for f because f_n as well as its trace converge appropriately. Hence, this result can be inserted for the inner integral of B with $f = v_i$. Noting that $T v_i = 0$ it is

$$\begin{aligned} B &\leq C n^p 2^{p-1} \int_0^{\frac{2}{n}} x_d^{p-1} \| \nabla v_i \|^p_{L^p(\mathbb{R}^{d-1} \times [0, x_d])} \, \mathrm{d}x_d \\ &\leq C \frac{2^{2p-1}}{p} \| \nabla v_i \|^p_{L^p(\mathbb{R}^{d-1} \times [0, 2/n])}. \end{aligned}$$

The norm above converges to zero for $n \to \infty$, therefore so does B. Thus, it is shown that the w_n converge to v_i in $W^{1,p}(\mathbb{R}^d_+)$. By definition $w_n(\boldsymbol{x})$ vanishes whenever $x_d < 1/n$ and therefore has compact support. Then mollifications $(w_n)_\varepsilon$ with $\varepsilon < 1/n$ converge to v_i in $W^{1,p}(\mathbb{R}^d_+)$, have compact support and are smooth, see Proposition 3.1.3. This shows $v_i \in W_0^{1,p}(\mathbb{R}^d_+)$ and also $u\Phi_i \in W_0^{1,p}(\Omega)$ for each $i = 1, \ldots, N$. Since $u = \sum_{i=0}^N u\Phi_i$ it is $u \in W_0^{1,p}(\Omega)$. □

In case one only considers a part Γ of the boundary $\partial\Omega$ and the corresponding trace operator, there is a very similar result to the previous one. First, similarly to Eq. (4.2) define

$$W_\Gamma^{1,p}(\Omega) := \overline{\mathcal{D}(\Omega \cup \partial\Omega \setminus \overline{\Gamma})}^{W^{1,p}(\Omega)}. \tag{4.3}$$

The support of a function in $\mathcal{D}(\Omega \cup \partial\Omega \setminus \overline{\Gamma})$ does not intersect with Γ but it may do so with the rest of the boundary. The proof of the following proposition is not given here. Basically, the same considerations as in the previous proof suffice.

Proposition 4.3.2 *The kernel of the trace operator* $T_\Gamma : W^{1,p}(\Omega) \to W^{1-1/p,p}(\Gamma)$ *from Theorem 4.2.4 is* $W_\Gamma^{1,p}(\Omega)$.

4.3.1 The Lions–Magenes Spaces $W_{00}^{s,p}$

Together with traces on Γ one can as well consider traces on the rest $\partial\Omega \setminus \overline{\Gamma}$. In the following the range of the trace operator T_Γ restricted to the kernel of $T_{\partial\Omega\setminus\overline{\Gamma}}$ is studied. The resulting spaces are strictly smaller than $W^{1-1/p,p}(\Gamma)$ and typically associated with Lions and Magenes, [LM72] (only $p = 2$), and denoted by $W_{00}^{1-1/p,p}(\Gamma)$ (and $H_{00}^{1/2}(\Gamma)$ for $p = 2$).

Definition 4.3.3 Let $\Omega \subset \mathbb{R}^d$ be a Lipschitz domain, $0 < s < 1$, and $p > 1$. For a given $u \in W^{s,p}(\Omega)$ denote its extension by zero to all of $\mathbb{R}^d$ by $\tilde{u}$. Then define

$$W_{00}^{s,p}(\Omega) := \left\{ u \in W^{s,p}(\Omega) \;\middle|\; \tilde{u} \in W^{s,p}\big(\mathbb{R}^d\big) \right\}$$

with norm $\|u\|_{W_{00}^{s,p}(\Omega)} := \|\tilde{u}\|_{W^{s,p}(\mathbb{R}^d)}$. In case $u \in W^{s,p}(\Gamma)$ for some $\Gamma \subset \partial\Omega$ and $\tilde{u}$ is its extension by zero to all of $\partial\Omega$ define

$$W_{00}^{s,p}(\Gamma) := \left\{ u \in W^{s,p}(\Gamma) \;\middle|\; \tilde{u} \in W^{s,p}(\partial\Omega) \right\}$$

again with norm $\|u\|_{W_{00}^{s,p}(\Gamma)} := \|\tilde{u}\|_{W^{s,p}(\partial\Omega)}$.

Next, a connection to an appropriate range of traces can be established:

Theorem 4.3.4 *Let $\Omega \subset \mathbb{R}^d$ be a Lipschitz domain and $\Gamma \subset \partial\Omega$ be a part of its boundary which is itself Lipschitz continuous. Then it is*

$$T_\Gamma\left(W^{1,p}_{\partial\Omega\setminus\overline{\Gamma}}(\Omega)\right) = W^{1-1/p,p}_{00}(\Gamma).$$

Furthermore, as a map from $W^{1,p}_{\partial\Omega\setminus\overline{\Gamma}}(\Omega)$ to $W^{1-1/p,p}_{00}(\Gamma)$ it is continuous and there exists a continuous right inverse $E_\Gamma : W^{1-1/p,p}_{00}(\Gamma) \to W^{1,p}_{\partial\Omega\setminus\overline{\Gamma}}(\Omega)$.

Proof Let $u \in W^{1,p}_{\partial\Omega\setminus\overline{\Gamma}}(\Omega)$ be given. Note that with $v = T_\Gamma u$ it is $\tilde{v} = T_{\partial\Omega} u \in W^{1-1/p,p}(\partial\Omega)$. The continuity of the trace map $T_{\partial\Omega}$ shows

$$\|T_\Gamma u\|_{W^{1-1/p,p}_{00}(\Gamma)} = \|T_{\partial\Omega} u\|_{W^{1-1/p,p}(\partial\Omega)} \leq C\|u\|_{W^{1,p}(\Omega)},$$

i.e., the continuity of $T_\Gamma : W^{1,p}_{\partial\Omega\setminus\overline{\Gamma}}(\Omega) \to W^{1-1/p,p}_{00}(\Gamma)$.

Conversely, let $u \in W^{1-1/p,p}_{00}(\Gamma)$ be given. Then, by Definition 4.3.3, its extension $\tilde{u}$ by zero is an element of $W^{1-1/p,p}(\partial\Omega)$ and can in turn be extended to an element $E\tilde{u}$ of $W^{1,p}(\Omega)$ due to Theorem 4.2.3. According to the construction, $\tilde{u}$ vanishes on $\partial\Omega\setminus\overline{\Gamma}$, i.e., $\tilde{u} \in W^{1,p}_{\partial\Omega\setminus\overline{\Gamma}}(\Omega)$ and $T_\Gamma\tilde{u} = u$. Hence, E_Γ can be defined as the composition of $\tilde{\cdot}$ (extension by zero to $\partial\Omega$) and E from Theorem 4.2.3. Then it is continuous:

$$\|E_\Gamma u\|_{W^{1,p}(\Omega)} = \|E\tilde{u}\|_{W^{1,p}(\Omega)} \leq C\|\tilde{u}\|_{W^{1-1/p,p}(\partial\Omega)} = C\|u\|_{W^{1-1/p,p}_{00}(\Gamma)}. \qquad \square$$

A slight generalization of these concepts is discussed now. Consider a domain Ω with three disjoint (connected) parts Γ_1, Γ_2, and Γ_3, such that $\partial\Omega = \bigcup_{i=1}^{3} \overline{\Gamma_i}$. Then the space

$$T_{\Gamma_1}\left(W^{1,p}_{\Gamma_2}(\Omega)\right)$$

is strictly between $W^{1-1/p,p}_{00}(\Gamma_1)$ and $W^{1-1/p,p}(\Gamma_1)$ if and only if Γ_1 and Γ_3 have a common boundary. In that case, the extension by zero of a function in $T_{\Gamma_1}(W^{1,p}_{\Gamma_2}(\Omega))$ is an element of $W^{1-1/p,p}(\overline{\Gamma_1 \cup \Gamma_2})$. The respective trace and extension operators remain continuous. A characterization of the Lions–Magenes spaces that does not consider extensions at all is proved next. It provides an equivalent norm and shows that functions in these spaces vanish at the boundary at a certain rate. A version for a domain Ω is followed by one for its boundary.

Proposition 4.3.5 *Let $\Omega \subset \mathbb{R}^d$ be a Lipschitz domain, $0 < s < 1$, $p > 1$, and $u \in W^{s,p}(\Omega)$. Furthermore, define $\varrho : \Omega \to \mathbb{R}$ to be the distance function to the boundary: $\varrho(\boldsymbol{x}) := \operatorname{dist}(\boldsymbol{x}, \partial\Omega)$. Then it holds*

$$u \in W^{s,p}_{00}(\Omega) \iff u/\varrho^s \in L^p(\Omega).$$

Additionally, there exist positive constants C_1 and C_2 such that

$$C_1\|u\|_{W^{s,p}_{00}(\Omega)} \le \left(\|u\|^p_{W^{s,p}(\Omega)} + \|u/\varrho^s\|^p_{L^p(\Omega)}\right)^{1/p} \le C_2\|u\|_{W^{s,p}_{00}(\Omega)}.$$

Proof This proof can be found in [Tar07], Lemma 37.1. Again, denote the extension of u to all of $\mathbb{R}^d$ by $\tilde{u}$ such that $\|u\|_{W^{s,p}_{00}(\Omega)} = \|\tilde{u}\|_{W^{s,p}(\mathbb{R}^d)}$.

"$\Longleftarrow$": The $L^p(\mathbb{R}^d)$-norm of $\tilde{u}$ is finite, in fact it is equal to the $L^p(\Omega)$-norm of u. The Sobolev–Slobodeckij semi-norm is split as follows:

$$|\tilde{u}|^p_{W^{s,p}(\mathbb{R}^d)} = \iint_{\Omega\times\Omega} \frac{|u(\boldsymbol{x}) - u(\boldsymbol{y})|^p}{|\boldsymbol{x} - \boldsymbol{y}|^{sp+d}}\,\mathrm{d}\boldsymbol{x}\,\mathrm{d}\boldsymbol{y} + 2\iint_{\Omega\times(\mathbb{R}^d\setminus\Omega)} \frac{|u(\boldsymbol{x})|^p}{|\boldsymbol{x} - \boldsymbol{y}|^{sp+d}}\,\mathrm{d}\boldsymbol{y}\,\mathrm{d}\boldsymbol{x}.$$

The first integral above is $|u|^p_{W^{s,p}(\Omega)}$ and the second can be written as $\|u\varphi\|^p_{L^p(\Omega)}$ where φ is defined for all $\boldsymbol{x} \in \Omega$ such that

$$|\varphi(\boldsymbol{x})|^p = \int_{\mathbb{R}^d\setminus\Omega} \frac{1}{|\boldsymbol{x} - \boldsymbol{y}|^{sp+d}}\,\mathrm{d}\boldsymbol{y}.$$

Note that a ball around $\boldsymbol{x}$ with radius $\varrho(\boldsymbol{x})$ is a subset of Ω, hence

$$|\varphi(\boldsymbol{x})|^p \le \int_{\mathbb{R}^d\setminus B(\boldsymbol{x},\varrho(\boldsymbol{x}))} \frac{1}{|\boldsymbol{x} - \boldsymbol{y}|^{sp+d}}\,\mathrm{d}\boldsymbol{y} = \int_{\mathbb{R}^d\setminus B(0,\varrho(\boldsymbol{x}))} \frac{1}{|\boldsymbol{z}|^{sp+d}}\,\mathrm{d}\boldsymbol{z} = \omega_d\varrho(\boldsymbol{x})^{-sp},$$

where polar coordinates are used. A similar reasoning is employed in the proof of Proposition 3.4.3 as well. The norm $\|u\varphi\|^p_{L^p(\Omega)}$ therefore can be bounded by $\omega_d\|u/\varrho^s\|^p_{L^p(\Omega)}$ which is finite by assumption. Together, this shows that there is a constant C such that

$$\|u\|^p_{W^{s,p}_{00}(\mathbb{R}^d)} = \|\tilde{u}\|^p_{W^{s,p}(\mathbb{R}^d)} \le C\left(\|u\|^p_{W^{s,p}(\Omega)} + \|u/\varrho^s\|^p_{L^p(\Omega)}\right),$$

i.e., setting $C_1 = C^{-1}$ suffices.

"$\Longrightarrow$": Consider a localization (V_i, h_i, ϕ_i), $i = 0, \ldots, N$, of Ω, see Definition 2.2.1, then it is

$$\|u/\varrho^s\|^p_{L^p(\Omega)} = \int_\Omega \left|\frac{\sum_{i=0}^N(\phi_i u)(\boldsymbol{x})}{\varrho^s(\boldsymbol{x})}\right|^p \mathrm{d}\boldsymbol{x} \le (N+1)^{p-1}\sum_{i=0}^N \int_{\Omega_i\cap\Omega} \left|\frac{(\phi_i u)(\boldsymbol{x})}{\varrho^s(\boldsymbol{x})}\right|^p \mathrm{d}\boldsymbol{x}.$$

The term for $i = 0$ in the above sum is bounded by the $L^p(\Omega)$-norm of u, because ϕ_0 has compact support in Ω. For $i \geq 1$ one would like to straighten out the boundary. To find out how ϱ^s in the denominator can be handled, for $\boldsymbol{x} = (\boldsymbol{x}', x_d) \in \Omega_i \cap \Omega$, let $p(\boldsymbol{x}) = (\boldsymbol{y}', h_i(\boldsymbol{y}')) \subset \partial\Omega \cap \Omega_i$ be a point such that $\varrho(\boldsymbol{x}) = \operatorname{dist}(\boldsymbol{x}, p(\boldsymbol{x})) = |\boldsymbol{x} - p(\boldsymbol{x})|$. Then it is

$$\begin{aligned} \left|x_d - h_i(\boldsymbol{x}')\right| &\leq \left|x_d - h_i(\boldsymbol{y}')\right| + \left|h_i(\boldsymbol{y}') - h_i(\boldsymbol{x}')\right| \\ &\leq \left|x_d - h_i(\boldsymbol{y}')\right| + \|\nabla h_i\|_\infty \left|\boldsymbol{x}' - \boldsymbol{y}'\right| \\ &\leq C\varrho(\boldsymbol{x}), \end{aligned} \tag{4.4}$$

where C depends on the Lipschitz continuity of h_i and on the space dimension d. Additionally, assuming the 1-norm on $\mathbb{R}^d$ for simplicity, with each $\boldsymbol{y} \in \mathbb{R}^d_+$ and $\mathbb{R}^d_- = \mathbb{R}^{d-1} \times (-\infty, 0)$ it is

$$\begin{aligned} \int_{\mathbb{R}^d_-} \frac{1}{|\boldsymbol{x} - \boldsymbol{y}|^{sp+d}} \,\mathrm{d}\boldsymbol{x} &= \left(\int_{y_d}^\infty z_d^{-sp-1} \,\mathrm{d}z_d\right)\left(\int_{\mathbb{R}^{d-1}} \frac{1}{|1 + |z'||^{sp+d}} \,\mathrm{d}z'\right) \\ &= \frac{\omega_{d-1}}{sp} y_d^{-sp} \int_0^\infty \frac{r^{d-1}}{(1+r)^{sp+d}} \,\mathrm{d}r, \end{aligned}$$

where at first the transformation $z_d = x_d - y_d$, $z_i = (x_i - y_i)/(x_d - y_d)$, $i = 1, \ldots, d-1$, is applied, while in the second equation polar coordinates are employed. Combining all the constants above into K yields

$$\int_{\mathbb{R}^d_-} |\boldsymbol{x} - \boldsymbol{y}|^{-sp-d} \,\mathrm{d}\boldsymbol{x} = K y_d^{-sp}. \tag{4.5}$$

With these results the relevant norm ($i \geq 1$) can be bounded using the transformation $(\boldsymbol{y}', y_d) = (\boldsymbol{x}', x_d - h_i(\boldsymbol{x}'))$:

$$\begin{aligned} \int_{\Omega_i \cap \Omega} \left|\frac{(\phi_i u)(\boldsymbol{x})}{\varrho^s(\boldsymbol{x})}\right|^p \mathrm{d}\boldsymbol{x} &\overset{(4.4)}{\leq} C^{sp} \int_{\Omega_i \cap \Omega} \left|\frac{(\phi_i u)(\boldsymbol{x})}{|x_d - h_i(\boldsymbol{x}')|^s}\right|^p \mathrm{d}\boldsymbol{x} \\ &= C^{sp} \int_{\mathbb{R}^d_+} \frac{\left|(\phi_i u)(\boldsymbol{y}', y_d + h_i(\boldsymbol{y}'))\right|^p}{|y_d|^{sp}} \,\mathrm{d}\boldsymbol{y} \\ &\overset{(4.5)}{\leq} \frac{C^{sp}}{K} \int_{\mathbb{R}^d_+} \int_{\mathbb{R}^d_-} \frac{\left|(\phi_i u)(\boldsymbol{y}', y_d + h_i(\boldsymbol{y}'))\right|^p}{|\boldsymbol{x} - \boldsymbol{y}|^{sp+d}} \,\mathrm{d}\boldsymbol{x}\,\mathrm{d}\boldsymbol{y}. \end{aligned}$$

The function $v_i : \boldsymbol{y} \mapsto (\phi_i u)(\boldsymbol{y}', y_d + h_i(\boldsymbol{y}'))$ in the numerator is in $W^{s,p}(\mathbb{R}^d_+)$ due to Propositions 3.2.1 and 3.2.3 and its extension by zero $\tilde{v}$ is even in $W^{s,p}(\mathbb{R}^d)$ because it is the transformed version of $\tilde{u}$. Hence, it is

$$\int_{\Omega_i \cap \Omega} \left| \frac{(\phi_i u)(\boldsymbol{x})}{\varrho^s(\boldsymbol{x})} \right|^p \mathrm{d}\boldsymbol{x} \le \frac{1}{CK} \|\tilde{v}\|^p_{W^{s,p}(\mathbb{R}^d)}.$$

Combining all these results gives the bound $\|u/\varrho^s\|_{L^p(\Omega)} \le C\|\tilde{u}\|_{W^{s,p}(\mathbb{R}^d)}$ for some constant C and hence the claim with $C_2 = (1 + C^p)^{1/p}$. □

The previous proposition shows the equivalence of two norms on $W^{s,p}_{00}(\Omega)$. On the boundary $\partial\Omega$ essentially the same result holds:

Proposition 4.3.6 *Let $\Omega \subset \mathbb{R}^d$ be a Lipschitz domain, $\Gamma \subset \partial\Omega$ a part of its boundary which itself is Lipschitz continuous, $s \in (0, 1)$, $p > 1$. Then it is*

$$u \in W^{s,p}_{00}(\Gamma) \iff u/\varrho^s \in L^p(\Gamma),$$

where ϱ is the distance from $\partial\Gamma$ within Γ.

An explicit proof is not given here. Just note that transforming to local coordinates in $\mathbb{R}^{d-1}$ and extension by zero commute. Then, locally, Proposition 4.3.5 yields the desired result. Note that this representation is not affected by the existence of more than two boundary parts. Furthermore, it is completely local in the sense that no information of the rest $\partial\Omega \setminus \Gamma$ is needed.

4.4 Integration by Parts

Having established a well defined notion of trace for Sobolev spaces, the important integration by parts formula can be extended to these spaces.

Corollary 4.4.1 *For all $\varphi \in H^1(\Omega)$ and all $\boldsymbol{v} \in \boldsymbol{H}^1(\Omega)$ it holds*[2]

$$(\boldsymbol{v}, \nabla\varphi)_0 + (\nabla \cdot \boldsymbol{v}, \varphi)_0 = (\boldsymbol{v} \cdot \boldsymbol{n}, \varphi)_{0,\partial\Omega},$$

where on the right-hand side both $\boldsymbol{v}$ and φ have to be understood in the sense of traces as in Theorem 4.1.1.

Proof For all $\varphi \in \mathcal{D}(\overline{\Omega})$ and $\boldsymbol{v} \in \left(\mathcal{D}(\overline{\Omega})\right)^d$ integration by parts yields

$$(\boldsymbol{v}, \nabla\varphi)_{L^2(\Omega)} + (\nabla \cdot \boldsymbol{v}, \varphi)_{L^2(\Omega)} = \int_{\partial\Omega} \varphi(\boldsymbol{x})(\boldsymbol{v} \cdot \boldsymbol{n})(\boldsymbol{x})\, \mathrm{d}\sigma_{\boldsymbol{x}}. \tag{4.6}$$

[2] As in Sect. 2.3 it is $(\cdot,\cdot)_0 = (\cdot,\cdot)_{L^2(\Omega)}$ and $(\cdot,\cdot)_{0,M} = (\cdot,\cdot)_{L^2(M)}$ for sets M.

Both sides are linear in $\boldsymbol{v}$ as well as φ and can be bounded by the product $\|\boldsymbol{v}\|_{\boldsymbol{H}^1(\Omega)}\|\varphi\|_{H^1(\Omega)}$ with a suitable constant. Due to Theorem 3.4.5 this equation holds also for $\varphi \in H^1(\Omega)$, where on the right-hand side φ has to be understood in the sense of a trace. The same reasoning applies to $\boldsymbol{v} \in \boldsymbol{H}^1(\Omega)$ where its trace is multiplied by the normal $\boldsymbol{n}$ which is in $L^\infty(\partial\Omega)$. □

Remark 4.4.2 For $p \neq 2$ an integration by parts formula can be shown with the exact same steps. Let q be given such that $\frac{1}{p} + \frac{1}{q} = 1$. Then for all $\varphi \in W^{1,p}(\Omega)$ and all $\boldsymbol{v} \in \mathbf{W}^{1,q}(\Omega)$ it is

$$\int_\Omega \boldsymbol{v}(\boldsymbol{x}) \cdot \nabla\varphi(\boldsymbol{x})\,\mathrm{d}\boldsymbol{x} + \int_\Omega (\nabla \cdot \boldsymbol{v})(\boldsymbol{x})\varphi(\boldsymbol{x})\,\mathrm{d}\boldsymbol{x} = \int_{\partial\Omega} \varphi(\boldsymbol{x})\boldsymbol{v} \cdot \boldsymbol{n}(\boldsymbol{x})\,\mathrm{d}\sigma_{\boldsymbol{x}}.$$

Remark 4.4.3 The integration by parts formula can be extended to hold also for vector fields $\boldsymbol{v} \in \boldsymbol{H}(\mathrm{div}, \Omega) = \left\{\boldsymbol{v} \in \left(L^2(\Omega)\right)^d \,\middle|\, \nabla \cdot \boldsymbol{v} \in L^2(\Omega)\right\}$. Such functions also admit a trace but only in normal direction and the inner product on the boundary then turns into a duality product in $\left(H^{1/2}(\partial\Omega)\right)^* \times H^{1/2}(\partial\Omega)$. Special care then has to be taken if φ is in a space such as $H^1_\Gamma(\Omega)$ or if the normal trace $\boldsymbol{v} \cdot \boldsymbol{n}$ vanishes on a part of the boundary. See [GR86], Theorems 2.2 and 2.3 and their corollaries in Chapter 1 as well as Proposition 3.58 in [DDE12] (also for $p \neq 2$).

4.5 A Right Inverse of the Divergence Operator

In order to prove the existence and uniqueness of a solution to the Stokes equations in Sect. 5.3.2, a vector field $\mathbf{u}$ is used whose divergence is a given function $f \in L^p(\Omega)$. It turns out that if $f \in L^p_0(\Omega)$, there is a function $\mathbf{u} \in \mathbf{W}^{1,p}_0(\Omega)$ such that $\nabla \cdot \mathbf{u} = f$ and $\|\mathbf{u}\|_{W^{1,p}(\Omega)} \leq \|f\|_{L^p(\Omega)}$. Note that $L^p_0(\Omega)$ is the subspace of $L^p(\Omega)$ with functions whose integral over Ω vanishes, see Eq. (2.9). The Sobolev space $W^{1,p}_0(\Omega)$ is the completion of smooth functions with compact support in $W^{1,p}(\Omega)$, see Definition 2.4.2 and Eq. (4.2).

The typical proof of the existence of such a vector $\mathbf{u}$ for $p = 2$ uses results which are not (yet) shown in this monograph, namely higher regularity results and existence and uniqueness of solutions to Laplace problems. Such proofs can be found in [GR86], Lemma 3.2, and [Tem77], Lemma 2.4.

There is however another approach which explicitly constructs the vector field $\mathbf{u}$ for a given f. Its proof is more involved and not presented in full detail here. Another advantage is that this works not only for $p = 2$. The following proposition shows that the divergence has a right inverse when restricted to appropriate spaces.

Proposition 4.5.1 *Let $\Omega \subset \mathbb{R}^d$ be a bounded Lipschitz domain, which is star-shaped*[3] *with respect to a ball $B \subset \Omega$, and $p > 1$. For all $f \in L_0^p(\Omega)$, there exists a $\mathbf{u} \in \mathbf{W}_0^{1,p}(\Omega)$ such that*

$$\nabla \cdot \mathbf{u} = f \qquad \textit{and} \qquad \|\nabla \mathbf{u}\|_{(L^p(\Omega))^{d\times d}} \leq \|f\|_{L^p(\Omega)}.$$

Proof Only a part of the proof is given here. The remaining details can be found in [DM01], see also [Rus13], Theorem 1 and the references therein.

Let $\omega \in \mathcal{D}(\Omega)$ be a test function whose support is contained in B and whose integral is one, $\int_\Omega \omega(\boldsymbol{x})\,\mathrm{d}\boldsymbol{x} = 1$. Next, for any test function $\varphi \in \mathcal{D}(\Omega)$ denote

$$\overline{\varphi} := \int_\Omega \varphi(\boldsymbol{x})\omega(\boldsymbol{x})\,\mathrm{d}\boldsymbol{x}.$$

Then for any $\boldsymbol{x} \in \Omega$ it is

$$\begin{aligned}
\varphi(\boldsymbol{x}) - \overline{\varphi} &= \int_\Omega (\varphi(\boldsymbol{x}) - \varphi(\boldsymbol{y}))\omega(\boldsymbol{y})\,\mathrm{d}\boldsymbol{y} \\
&= \int_\Omega \omega(\boldsymbol{y}) \left(\int_0^1 (\boldsymbol{x} - \boldsymbol{y}) \cdot \nabla\varphi(\boldsymbol{y} + s(\boldsymbol{x} - \boldsymbol{y}))\,\mathrm{d}s \right) \mathrm{d}\boldsymbol{y} \\
&= \int_0^1 \int_\Omega \frac{1}{t^{d+1}} (\boldsymbol{x} - \boldsymbol{z}) \cdot \nabla\varphi(\boldsymbol{z})\, \omega\left(\boldsymbol{x} + \frac{\boldsymbol{z} - \boldsymbol{x}}{t} \right) \mathrm{d}\boldsymbol{z}\,\mathrm{d}t \\
&= - \int_\Omega \mathbf{G}(\boldsymbol{z}, \boldsymbol{x}) \cdot \nabla\varphi(\boldsymbol{z})\,\mathrm{d}\boldsymbol{z},
\end{aligned}$$

with

$$\mathbf{G}(\boldsymbol{z}, \boldsymbol{x}) := (\boldsymbol{z} - \boldsymbol{x}) \int_0^1 \frac{1}{t^{d+1}} \omega\left(\boldsymbol{x} + \frac{\boldsymbol{z} - \boldsymbol{x}}{t} \right) \mathrm{d}t.$$

In the second equality above, the fundamental theorem of calculus is applied, in the third the transformations $\boldsymbol{z} = (1 - s)\boldsymbol{y} + s\boldsymbol{x}$ and $t = 1 - s$. The vector $\mathbf{G}(\boldsymbol{z}, \boldsymbol{x})$ is bounded, because due to ω having compact support, the integrand is zero for $t < t_0$ for some positive t_0 which depends on $|\boldsymbol{z} - \boldsymbol{x}|$ and the diameter of Ω. Hence, changing the order of integration is possible (Fubini's theorem).

[3] A domain $\Omega \subset \mathbb{R}^d$ is called star-shaped with respect to $B \subset \Omega$, if for all $\boldsymbol{x} \in \Omega$ and all $\boldsymbol{y} \in B$ the connecting line segment $\{\boldsymbol{z} \in \mathbb{R}^d \mid \exists t \in [0, 1] : \boldsymbol{z} = \boldsymbol{y} + t(\boldsymbol{x} - \boldsymbol{y})\}$ is a subset of Ω.

Since $f \in L_0^p(\Omega)$ by assumption, it is

$$\begin{aligned}\int_\Omega f(\boldsymbol{x})\varphi(\boldsymbol{x})\,\mathrm{d}\boldsymbol{x} &= \int_\Omega f(\boldsymbol{x})(\varphi(\boldsymbol{x}) - \overline{\varphi})\,\mathrm{d}\boldsymbol{x} \\ &= -\int_\Omega \int_\Omega f(\boldsymbol{x})\mathbf{G}(z,\boldsymbol{x}) \cdot \nabla\varphi(z)\,\mathrm{d}z\,\mathrm{d}\boldsymbol{x} \\ &= -\int_\Omega \mathbf{u}(z) \cdot \nabla\varphi(z)\,\mathrm{d}z,\end{aligned}$$

where $\mathbf{u}$ is defined to be

$$\mathbf{u}(z) := \int_\Omega \mathbf{G}(z,\boldsymbol{x}) f(\boldsymbol{x})\,\mathrm{d}\boldsymbol{x}.$$

According to the definition of the weak derivative, the above calculations show that $\mathbf{u}$ is the divergence of f, $\nabla \cdot \mathbf{u} = f$, see Definition 2.4.1. Additionally, $G(z,\boldsymbol{x}) = \mathbf{0}$ for $z \in \partial\Omega$, because $\boldsymbol{y} = \boldsymbol{x} + (z - \boldsymbol{x})/t \notin B$ (otherwise it would be $z = t\boldsymbol{y} + (1 - t)\boldsymbol{x} \in \Omega$ because Ω is assumed to be star-shaped). Hence, it is $\mathbf{u} = \mathbf{0}$ on $\partial\Omega$. It remains to show that $\mathbf{u} \in \mathbf{W}^{1,p}(\Omega)$ and the bound on its gradient. Note that due to Hölder's inequality, Theorem 2.3.1, it is

$$|\mathbf{u}(z)| \le \|G(z,\cdot)\|_{(L^q(\Omega))^d} \|f\|_{L^p(\Omega)}$$

and therefore $\|\mathbf{u}\|_{(L^p(\Omega))^d} \le C\|f\|_{L^p(\Omega)}$, i.e., $\mathbf{u} \in (L^p(\Omega))^p$. In [DM01] it is shown also that $\mathbf{u} \in \mathbf{W}^{1,p}(\Omega)$ and $\|\nabla\mathbf{u}\|_{(L^p(\Omega))^{d\times d}} \le \|f\|_{L^p(\Omega)}$ as claimed. □

The restriction on the domain to be star-shaped can be dropped to yield:

Theorem 4.5.2 *Let $\Omega \subset \mathbb{R}^d$ be a bounded Lipschitz domain and $p > 1$. For all $f \in L_0^p(\Omega)$, there exists a $\mathbf{u} \in \mathbf{W}_0^{1,p}(\Omega)$ such that*

$$\nabla \cdot \mathbf{u} = f \qquad \textit{and} \qquad \|\nabla\mathbf{u}\|_{(L^p(\Omega))^{d\times d}} \le C\|f\|_{L^p(\Omega)},$$

with a constant C independent of f.

Proof Consider a covering of the domain Ω with subdomains $\Omega_i \subset \Omega$, $i = 1,\ldots,N$, which are star-shaped with respect to some open ball $B_i \subset \Omega_i$. Such a covering exists, because Ω is assumed to be Lipschitz continuous. Additionally, let $\phi_i \in \mathcal{D}(\Omega_i)$ be a partition of unity, i.e., $\sum_{i=1}^N \phi_i = 1$ on Ω. Due to Proposition 4.5.1 there exist functions $\mathbf{u}_i \in \mathbf{W}_0^{1,p}(\Omega_i)$ such that $\nabla \cdot \mathbf{u}_i = f\phi_i$ in $L^p(\Omega_i)$ and through extension by zero also in $L^p(\Omega)$. Furthermore, it is $\|\nabla\mathbf{u}_i\|_{(L^p(\Omega))^{d\times d}} \le \|f\phi_i\|_{L^p(\Omega)}$. Then the combined function $\mathbf{u} = \sum_{i=1}^N \mathbf{u}_i$ is an element of $\mathbf{W}^{1,p}(\Omega)$, see Proposition 3.5.1. Furthermore, $\mathbf{u}$ vanishes on the boundary because all the $\mathbf{u}_i$

do, hence $\mathbf{u} \in \mathbf{W}_0^{1,p}(\Omega)$. Finally, it is

$$\nabla \cdot \mathbf{u} = \sum_{i=1}^{N} \nabla \cdot \mathbf{u}_i = f \sum_{i=1}^{N} \phi_i = f$$

and

$$\|\nabla \mathbf{u}\|_{(L^p(\Omega))^{d\times d}} \le \sum_{i=1}^{N} \|\nabla \mathbf{u}_i\|_{(L^p(\Omega))^{d\times d}} \le \sum_{i=1}^{N} \|f\phi_i\|_{L^p(\Omega)} \le C\|f\|_{L^p(\Omega)}. \qquad \square$$

This result can be extended to more general, so-called, John domains, see [Rus13], Theorem 2. Furthermore, the above Theorem 4.5.2 is equivalent to other well known results such as the Jaques-Louis Lions lemma, the Neças inequality, or a "coarse" version of de Rhams's theorem, see [ACM14, ACM15] and the references therein.

4.6 Equivalent Norms on $H^1(\Omega)$

In the analysis of partial differential equations it is sometimes necessary to define norms on (sub spaces of) $H^1(\Omega)$ which are equivalent to the standard one. So-called Poincaré type inequalities often provide the proof of this equivalence and some of its variants are presented here. The proof of the Poincaré type inequalities use the Rellich–Kondrachov theorem which is shown for C^1 domains in [Eva98], Theorem 1 in Section 5.7, and in [AF03], Theorem 6.3, for the case of a domain satisfying a cone condition, which Lipschitz domains do. See also a proof using extensions in [Tar07], Lemma 14.5.

Theorem 4.6.1 (Rellich–Kondrachov) *Let Ω be a Lipschitz domain in $\mathbb{R}^d$, $p \ge 1$. Then $W^{1,p}(\Omega)$ is compactly embedded in $L^p(\Omega)$. In other words, any uniformly bounded sequence in $W^{1,p}(\Omega)$ has a subsequence which converges in $L^p(\Omega)$.*

With the help of the Rellich–Kondrachov theorem it is possible to prove the Poincaré inequality. It can be found in many textbooks which treat Sobolev spaces, see for example [Tri92], Theorem 3 in Section 6.1.5, or Section 10 in [Tar07].

Theorem 4.6.2 (Poincaré Inequality) *There is a constant c_P which only depends on Ω such that for all $u \in H^1(\Omega)$ it is*

$$\|u - \overline{u}\|_0 \le c_P \|\nabla u\|_0, \tag{4.7}$$

where $\overline{u}$ is the mean value of u, $\overline{u} = |\Omega|^{-1} \int_\Omega u(\boldsymbol{x})\,\mathrm{d}\boldsymbol{x}$.

Proof See § 5.8 in [Eva98]. Assume the inequality is not true, then there is a sequence $u_k \in H^1(\Omega)$ such that

$$\|u_k - \overline{u_k}\|_0 \geq k\|\nabla u_k\|_0.$$

Then define

$$v_k := \frac{u_k - \overline{u_k}}{\|u_k - \overline{u_k}\|_0}.$$

This sequence has the properties $\|v_k\|_0 = 1$, by construction $\overline{v_k} = 0$, and by assumption $\|\nabla v_k\|_0 \leq \frac{1}{k}$, i.e., it is a uniformly bounded sequence in $H^1(\Omega)$. Now the Rellich–Kondrachov theorem, Theorem 4.6.1, states there is a subsequence, still denoted by v_k, which converges in $L^2(\Omega)$ to, say, $v \in L^2(\Omega)$. Due to continuity of the norm it is $\|v\|_0 = 1$. Furthermore, for any $\phi \in \mathcal{D}(\Omega)$ it is

$$\int_\Omega v \frac{\partial \phi}{\partial \boldsymbol{x}_i} = \lim_{k\to\infty} \int_\Omega v_k \frac{\partial \phi}{\partial \boldsymbol{x}_i} = -\lim_{k\to\infty} \int_\Omega \frac{\partial v_k}{\partial \boldsymbol{x}_i}\phi = 0,$$

which means that v is in $H^1(\Omega)$ and has vanishing weak derivatives, implying v is constant, $v = |\Omega|^{-1/2}$. Since taking the mean value is continuous on $H^1(\Omega)$ there is a contradiction

$$0 = \lim_{k\to\infty} \overline{v_k} \neq \overline{v} = |\Omega|^{-1/2}. \tag{4.8}$$

□

Remark 4.6.3 In the proof of the previous theorem it is only used that taking the mean value, $v \mapsto \overline{v}$ is linear and continuous and that the constant function 1 is mapped to unity. In fact if one replaces the mean value by any such functional l the theorem is still valid, i.e., for a given linear and continuous $l : H^1(\Omega) \to \mathbb{R}$ with $l(1) = 1$ it is

$$\|u - l(u)\|_0 \leq c_P\|\nabla u\|_0. \tag{4.9}$$

However, c_P might now depend on l.

With the help of the Poincaré inequality one can easily prove that

$$|||v||| := \|\nabla v\|_0 + |\overline{v}|.$$

is an equivalent norm on $H^1(\Omega)$. Using the previous Remark 4.6.3, the mean value $\overline{v}$ can be replaced by $l(v)$. Another common approach to define equivalent norms on $H^1(\Omega)$ uses bilinear forms. For C^1-domains a proof can also be found in [Tri92], Theorem 1, Paragraph 6.1.5.

Theorem 4.6.4 *Let $l : H^1(\Omega) \to \mathbb{R}$ be linear and continuous and let $e : H^1(\Omega) \times H^1(\Omega) \to \mathbb{R}$ be a continuous, coercive, and symmetric bilinear form with the additional property $l(1)^2 + e(1, 1) > 0$. Then*

$$|||v||| := \sqrt{\|\nabla v\|_0^2 + |l(v)|^2 + e(v, v)}$$

defines an equivalent norm on $H^1(\Omega)$.

Proof To show that $|||\cdot|||$ in fact defines a norm, at first two properties of the bilinear form e are derived which are very similar to those of an inner product. Let $v, w \in H^1(\Omega)$ and $\lambda = e(v, v)/e(v, w)$. Then

$$\begin{aligned}
0 \leq e(v - \lambda w, v - \lambda w) &= e(v, v) - 2\lambda e(v, w) + \lambda^2 e(w, w) \\
&= e(v, v) - 2e(v, v) + \frac{e(v, v)^2 e(w, w)}{e(v, w)^2} \\
\Longrightarrow\ e(v, w) \leq \sqrt{e(v, v)}\sqrt{e(w, w)},
\end{aligned}$$

which gives

$$\begin{aligned}
e(v + w, v + w) &= e(v, v) + 2e(v, w) + e(w, w) \\
&\leq e(v, v) + 2\sqrt{e(v, v)}\sqrt{e(w, w)} + e(w, w) \\
&= \left(\sqrt{e(v, v)} + \sqrt{e(w, w)}\right)^2.
\end{aligned}$$

Next, the properties of a norm are checked: Let again $v, w \in H^1(\Omega)$ and $\eta \in \mathbb{R}$ be given, then

$$\begin{aligned}
|||v||| = 0 &\Longleftrightarrow \|\nabla v\|_0 = 0 \quad \text{and} \quad |l(v)|^2 + e(v, v) = 0 \\
&\Longleftrightarrow v(\boldsymbol{x}) = c \quad \text{and} \quad |l(v)|^2 + e(v, v) = c^2\left(|l(1)|^2 + e(1, 1)\right) = 0 \\
&\Longleftrightarrow v = 0, \\
|||\eta v||| &= |\eta|\, |||v|||\,, \\
|||v + w|||^2 &\leq (\|\nabla v\|_0 + \|\nabla w\|_0)^2 + (|l(v)| + |l(w)|)^2 \\
&\quad + \left(\sqrt{e(v, v)} + \sqrt{e(w, w)}\right)^2 \\
&= |||v|||^2 + |||w|||^2 \\
&\quad + 2\left(\|\nabla v\|_0 \|\nabla w\|_0 + |l(v)||l(w)| + \sqrt{e(v, v)}\sqrt{e(w, w)}\right) \\
&\leq |||v|||^2 + |||w|||^2 + 2\, |||v|||\, |||w||| \\
&= (|||v||| + |||w|||)^2.
\end{aligned}$$

The second estimate above uses the fact that for all nonnegative a_1, a_2, b_1, and b_2 it is $a_1a_2 + b_1b_2 \le \left(a_1^2 + b_1^2\right)^{1/2}\left(a_2^2 + b_2^2\right)^{1/2}$ and hence $a_1a_2 + b_1b_2 + c_1c_2 \le \left(a_1^2 + b_1^2 + c_1^2\right)^{1/2}\left(a_2^2 + b_2^2 + c_2^2\right)^{1/2}$ for positive c_1 and c_2. Due to the continuity of l and e the triple norm $|||\cdot|||$ is bounded by the standard norm on $H^1(\Omega)$, i.e., for all $v \in H^1(\Omega)$ it is

$$|||v||| \le \left(\|\nabla v\|_0^2 + \|l\|^2\|v\|_1^2 + \|e\|\|v\|_1^2\right)^{1/2} \le \left(1 + \|l\|^2 + \|e\|\right)^{1/2}\|v\|_1. \tag{4.10}$$

Next, the opposite is shown, namely: There is a constant $c > 0$ such that for all $u \in H^1(\Omega)$ it is $\|u\|_1 \le c\,|||u|||$. To this end, it is enough to show $\|u\|_0 \le c\,|||u|||$ (for some other constant c). Assume the opposite is true, i.e., there exists a sequence $\{u_k\}_{k=1}^{\infty} \subset H^1(\Omega)$ such that

$$1 = \|u_k\|_0 > k\,|||u_k||| \quad \Longrightarrow \quad \|\nabla u_k\|_0 \le |||u_k||| < \frac{1}{k}. \tag{4.11}$$

The sequence $\{u_k\}_{k=1}^{\mathbb{N}}$ is therefore uniformly bounded in $H^1(\Omega)$ and has, according to the theorem of Rellich–Kondrachov, Theorem 4.6.1, a subsequence which converges in $L^2(\Omega)$ and is denoted by the same name here. This sequence is also a Cauchy sequence in $H^1(\Omega)$ because

$$\begin{aligned}\|u_k - u_m\|_1 &\le \sqrt{(\underbrace{\|\nabla u_k\|_0}_{\le 1/k} + \underbrace{\|\nabla u_m\|_0}_{\le 1/m})^2 + \|u_k - u_m\|_0^2} \\ &\le \frac{1}{k} + \frac{1}{m} + \|u_k - u_m\|_0 \to 0,\end{aligned}$$

where k and m approach infinity independently. Therefore, u_k also converges in $H^1(\Omega)$ to, say, $u \in H^1(\Omega)$. Due to the continuity of the norm it is $\|u\|_0 = 1$ and $0 = \lim_{k\to\infty} \frac{1}{k} \ge \lim_{k\to\infty}\|\nabla u_k\|_0 = \|\nabla u\|_0$, so $u = a$ is constant. Furthermore, u_k also converges with respect to the triple norm, because of the estimate (4.10). Hence, it is $|||u||| = \lim_{k\to\infty} |||u_k||| = 0$ according to estimate (4.11) but $|||u||| = |||a||| = \sqrt{0 + a^2(l(1)^2 + e(1, 1))} > 0$ according to the assumption of the theorem, hence, there is a contradiction. □

Remark 4.6.5 If $l = 0$ in the previous theorem, the norm also corresponds to an inner product which can be shown with the parallelogram law (2.2) and Eq. (2.3)

$$|||v||| = \sqrt{((v, v))}, \qquad \text{with} \qquad ((v, w)) := (\nabla v, \nabla w)_0 + e(v, w).$$

Note, that Theorem 4.6.4 is a generalization of the Poincaré inequality, Theorem 4.6.2, setting l to be the mean value and e to zero. Next a few examples are provided.

Example 4.6.6 On $H^1(\Omega)$ the following norms are equivalent:

$$\|v\|_1 = \|v\|_0 + \|\nabla v\|_0,$$

$$|||v||| = \|\nabla v\|_0 + \left|\int_{\Omega_0} \eta v\right|,$$ where $\Omega_0 \subset \Omega$ has positive measure and $\eta \in L^2(\Omega_0)$ with $\int_{\Omega_0} \eta(\boldsymbol{x})\,\mathrm{d}\boldsymbol{x} \neq 0$,

$$|||v||| = \|\nabla v\|_0 + \left|\int_{\Gamma} \eta v\right|,$$ where $\Gamma \subset \partial\Omega$ has positive measure and $\eta \in L^2(\Gamma)$ with $\int_{\Gamma} \eta(\boldsymbol{x})\,\mathrm{d}\boldsymbol{x} \neq 0$,

$$|||v||| = \left(\|\nabla v\|_0^2 + \int_{\Omega_0} \eta v^2\right)^{1/2},$$ where $\Omega_0 \subset \Omega$ has positive measure and $\eta \in L^\infty(\Omega_0)$ is positive,

$$|||v||| = \left(\|\nabla v\|_0^2 + \int_{\Gamma} \eta v^2\right)^{1/2},$$ where $\Gamma \subset \partial\Omega$ has positive measure and $\eta \in L^\infty(\Gamma)$ is positive.

Often one is interested in the subspace $H^1_\Gamma(\Omega)$ where the last norm in the previous example reduces to $\|\nabla v\|_0$ which is therefore an equivalent norm on $H^1_\Gamma(\Omega)$, in particular it is

$$\|v\|_0 \leq C_P \|\nabla v\|_0, \tag{4.12}$$

with the Poincaré constant $C_P \geq 1$. This equation is often referred to as the Poincaré inequality in the literature and proved without the use of the Rellich–Kondrachov theorem 4.6.1. Then the constant C_P is given by s if Ω is contained in a d-dimensional cube of edge length s, see [Bra07a].

Remark 4.6.7 Due to the inequality of Korn (3.4) it is possible to define equivalent norms on $\mathbf{H}^1(\Omega)$ similar to the ones above where the part $\|\nabla \mathbf{u}\|_0$ is replaced by $\|\mathsf{D}(\mathbf{u})\|_0$. In case the space is restricted on the boundary, another Poincaré type inequality holds: Kikuchi and Oden prove in [KO88] that there is a constant $c > 0$ such that for all $\boldsymbol{v} \in \mathbf{H}^1_\Gamma(\Omega) = \left\{\boldsymbol{v} \in \mathbf{H}^1(\Omega) \;\middle|\; \boldsymbol{v} = \mathbf{0} \text{ on } \Gamma\right\}$ it is

$$\|\boldsymbol{v}\|_1 \leq c\,\|\mathsf{D}(\boldsymbol{v})\|_0. \tag{4.13}$$

Chapter 5
Subproblems Individually

In this section the existence and uniqueness of solutions of the involved subproblems are proved. Besides the Laplace equation this includes the so-called saddle point theory which is an abstract framework and is then applied to the Stokes equations as well as to the coupled Stokes–Darcy problem in Chap. 6.

In each case, let $\Omega \subset \mathbb{R}^d$ be an open domain with Lipschitz boundary. As before, the exterior unit normal vector on the boundary $\partial\Omega$ is denoted by $\boldsymbol{n}$, see Remark 2.5.1.

First, the Laplace equation is introduced followed by the general formulation of saddle point problems. The last section in this chapter applies the abstract theory to the Stokes problem. Furthermore, for each of these problem classes, operators which map data on part of the boundary to solutions in the domain are introduced. These are useful to understand the Stokes–Darcy coupling in Chap. 6.

Each equation is introduced in a similar manner. Starting from the partial differential equation a weak form is derived, for which existence and uniqueness of solutions is established. Finally, the solving can be viewed as an operator mapping data on a part Γ of the boundary to a solution inside the domain Ω. This data may be either of Dirichlet or Robin type resulting in operators K_{D} and K_{R}. Some interesting and helpful properties of both K_{D} and K_{R} are shown.

5.1 Laplace Equation

The Laplace equation and its extensions model a wide variety of applications. In this work it is the pressure of a fluid in a porous medium. This is also why the solution is denoted by p instead of the much more common notation u.

U. Wilbrandt, *Stokes–Darcy Equations*, Advances in Mathematical Fluid Mechanics,
https://doi.org/10.1007/978-3-030-02904-3_5

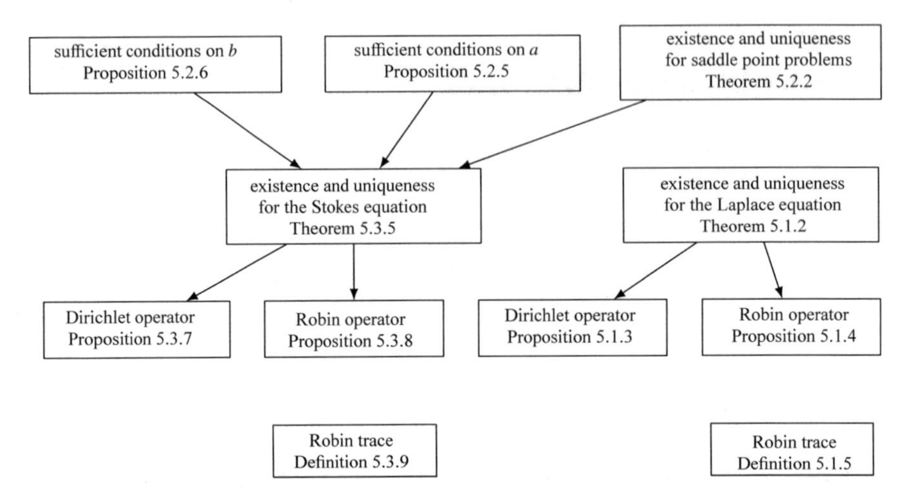

Fig. 5.1 A graph showing the dependencies among all theorems, definitions, as well as propositions in this chapter. Note that implicit dependencies are not always shown

The Laplace problem is as follows: Find $p : \Omega \to \mathbb{R}$ such that

$$-\nabla \cdot (\mathsf{K}\nabla p) = f \qquad \text{in } \Omega. \tag{5.1}$$

Here the source term $f : \Omega \to \mathbb{R}$ and the hydraulic conductivity tensor $\mathsf{K} : \Omega \to \mathbb{R}^{d \times d}$ are given data. Generally, it is assumed that K is symmetric and positive definite. Additionally, a solution p has to satisfy boundary conditions. Define two disjoint (relatively open) subsets Γ_D and Γ_R which in turn have a Lipschitz boundary, such that $\overline{\Gamma_\mathrm{D}} \cup \overline{\Gamma_\mathrm{R}} = \partial\Omega$. The boundary conditions read

$$\begin{cases} p = p_\mathrm{D} & \text{on } \Gamma_\mathrm{D}, \\ \mathsf{K}\nabla p \cdot \boldsymbol{n} + \gamma p = g_\mathrm{R} & \text{on } \Gamma_\mathrm{R}. \end{cases} \tag{5.2}$$

Here $p_\mathrm{D} : \Gamma_\mathrm{D} \to \mathbb{R}$, $g_\mathrm{R} : \Gamma_\mathrm{R} \to \mathbb{R}$ and $\gamma : \Gamma_\mathrm{R} \to \mathbb{R}$ are given functions. The coefficient function γ is not negative, $\gamma(\boldsymbol{x}) \geq 0$ for $\boldsymbol{x} \in \Gamma_\mathrm{R}$. The first boundary condition is of Dirichlet type, while the second is called Robin boundary condition. The case $\gamma = 0$ is referred to as of Neumann type. These different restrictions on the boundary are treated differently in the weak formulation. The Dirichlet condition is enforced in the definition of the solution space whereas the Robin (or Neumann respectively) condition becomes part of the formulation itself. Often the former are therefore called essential and the latter natural.

5.1.1 Weak Form

Assume the above Eq. (5.1) admits a smooth solution $p \in \mathcal{D}(\overline{\Omega})$ satisfying the boundary conditions (5.2). Next, multiply Eq. (5.1) with a test function $q \in C^\infty(\Omega)$ which vanishes in a neighborhood of the Dirichlet boundary part Γ_D,

$$q \in \mathcal{D}(\Omega \cup \partial\Omega \setminus \overline{\Gamma_\mathrm{D}}).$$

Integrating this equation over the domain Ω and integration by parts lead to

$$(\mathsf{K}\nabla p, \nabla q)_0 - (\mathsf{K}\nabla p \cdot \boldsymbol{n}, q)_{\Gamma_\mathrm{R},0} = (f, q)_0.$$

The integral over the Robin boundary can be replaced using the second condition in (5.2) yielding

$$a(p, q) = \ell(q) \tag{5.3}$$

with the bilinear form a and linear form ℓ defined by

$$\begin{aligned} a(p, q) &:= (\mathsf{K}\nabla p, \nabla q)_0 + (\gamma p, q)_{0,\Gamma_\mathrm{R}}, \\ \ell(q) &:= (f, q)_0 + (g_\mathrm{R}, q)_{0,\Gamma_\mathrm{R}}. \end{aligned}$$

Through approximation Eq. (5.3) can be meaningfully interpreted for $p \in H^1(\Omega)$ and $q \in H^1_{\Gamma_\mathrm{D}}(\Omega)$, see Theorem 3.4.5 and Eq. (4.3) on page 70. Therefore, the bilinear form a maps from $H^1(\Omega) \times H^1(\Omega)$ to $\mathbb{R}$ and the data must fulfill the following minimal smoothness properties

$$\begin{aligned} f &\in \left(H^1_{\Gamma_\mathrm{D}}(\Omega)\right)^*, \\ \mathsf{K} &\in \left(L^\infty(\Omega)\right)^{d\times d} \qquad \text{symmetric and positive definite,} \\ p_\mathrm{D} &\in H^{1/2}(\Gamma_\mathrm{D}), \\ \gamma &\in L^\infty(\Gamma_\mathrm{R}), \\ g_\mathrm{R} &\in \left(H^{1/2}_{00}(\Gamma_\mathrm{R})\right)^*. \end{aligned}$$

Note that in case f and g_R are not in more regular spaces (e.g. $L^2(\Omega)$ and $L^2(\Gamma_\mathrm{R})$) the definition of the functional ℓ must be understood as the respective duality pairings

$$\ell(q) = (f, q)_{\left(H^1_{\Gamma_\mathrm{D}}(\Omega)\right)^* \times H^1_{\Gamma_\mathrm{D}}(\Omega)} + (g_\mathrm{R}, q)_{\left(H^{1/2}_{00}(\Gamma_\mathrm{R})\right)^* \times H^{1/2}_{00}(\Gamma_\mathrm{R})}.$$

For the ease of presentation many authors still write it as an L^2 inner product. Summarizing, the weak formulation of the Laplace equation (5.1) together with the boundary condition (5.2) is: Find $p \in H^1(\Omega)$ such that $p - p_D \in H^1_{\Gamma_D}(\Omega)$ and for all $q \in H^1_{\Gamma_D}(\Omega)$ it is

$$a(p,q) = \ell(q). \tag{5.4}$$

Remark 5.1.1 The expression $p - p_D$ is at first not reasonable, because p and p_D do not belong to the same vector space. Instead, a continuous extension of p_D according to Theorem 4.2.4 is used. It is called p_D again to simplify notation, i.e., $\|p_D\|_{1,\Omega} \le c_{E_{\Gamma_D}} \|p_D\|_{1/2,\Gamma_D}$.

5.1.2 Existence and Uniqueness of a Solution

The theorem of Lax–Milgram, Theorem 2.1.13, is the key to establish existence and uniqueness of a solution to the weak formulation (5.4) of the Laplace equation. The Hilbert space in that theorem is $Q = H^1_{\Gamma_D}(\Omega)$. Since the solution p is in general not in Q (unless $p_D = 0$), an equivalent problem is considered: As described in the previous remark, let p_D denote not just the essential boundary data but also an extension of it into the domain Ω, $p_D \in H^1(\Omega)$ and let $P = p - p_D \in Q$. Then consider the modified problem

$$a(P,q) = \ell(q) - a(p_D, q). \tag{5.5}$$

Solving this problem is equivalent to the original weak formulation, p can be recovered as $P + p_D$. This transformation is often used to justify an assumption of homogeneous Dirichlet data p_D.

In order to apply the Lax–Milgram theorem, Theorem 2.1.13, to the bilinear form a restricted to Q and the right-hand side $L(q) = \ell(q) + a(p_D, q)$, one has to show continuity of a and L as well as the coercivity of a. This is done step-by-step in the following. Then, existence and uniqueness of the weak formulation (5.4) is established.

5.1.2.1 Continuity of *a*

Let $p, q \in Q$ be given, then, using the Hölder inequality, Theorem 2.3.1, to extract the L^∞-terms and a second time to split the products yields

$$\begin{aligned} a(p,q) &= (\mathsf{K}\nabla p, \nabla q)_0 + (\gamma p, q)_{0,\Gamma_R} \\ &\le \|\mathsf{K}\|_{(L^\infty(\Omega))^{d\times d}} \|\nabla p \cdot \nabla q\|_{L^1(\Omega)} + \|\gamma\|_{L^\infty(\Gamma_R)} \|p q\|_{L^1(\Gamma_R)} \end{aligned}$$

$$\leq \|\mathsf{K}\|_{(L^\infty(\Omega))^{d\times d}}\|p\|_1\|q\|_1 + \|\gamma\|_{L^\infty(\Gamma_\mathrm{R})}\|p\|_{0,\Gamma_\mathrm{R}}\|q\|_{0,\Gamma_\mathrm{R}}$$
$$\leq c_a\|p\|_1\|q\|_1.$$

The constant c_a depends on the hydraulic conductivity tensor K, the boundary data γ, and the domain Ω in form of the continuity constant of the trace operator mapping $Q \subset H^1(\Omega)$ to $L^2(\Gamma_\mathrm{R})$, Theorem 4.1.1.

5.1.2.2 Continuity of *L*

Let $q \in Q$ be given, then the continuity of the data f and g_R as well as of the bilinear form a yields

$$\begin{aligned}
L(q) &= (f,q)_{\left(H^1_{\Gamma_\mathrm{D}}(\Omega)\right)^*\times H^1_{\Gamma_\mathrm{D}}(\Omega)} + (g_\mathrm{R},q)_{\left(H^{1/2}_{00}(\Gamma_\mathrm{R})\right)^*\times H^{1/2}_{00}(\Gamma_\mathrm{R})} - a(p_\mathrm{D},q)\\
&\leq \|f\|_{\left(H^1_{\Gamma_\mathrm{D}}(\Omega)\right)^*}\|q\|_1 + \|g_\mathrm{R}\|_{\left(H^{1/2}_{00}(\Gamma_\mathrm{R})\right)^*}\|q\|_{H^{1/2}_{00}(\Gamma_\mathrm{R})} + c_a\|p_\mathrm{D}\|_1\|q\|_1\\
&\leq \left(\|f\|_{\left(H^1_{\Gamma_\mathrm{D}}(\Omega)\right)^*} + c_{T_\Gamma}\|g_\mathrm{R}\|_{\left(H^{1/2}_{00}(\Gamma_\mathrm{R})\right)^*} + c_a\|p_\mathrm{D}\|_1\right)\|q\|_1\\
&= c_L\|q\|_1.
\end{aligned} \tag{5.6}$$

The constant c_L depends on c_a and additionally on f, g_R, and the continuity constant C_{T_Γ} of the trace operator mapping Q to $H^{1/2}_{00}(\Gamma_\mathrm{R})$, see Theorem 4.3.4.

5.1.2.3 Coercivity of *a*

Let $q \in Q$ be given, using the lower bounds on the hydraulic conductivity tensor K and $\gamma \geq 0$ leads to

$$\begin{aligned}
a(q,q) &= (\mathsf{K}\nabla q, \nabla q)_0 + (\gamma q, q)_{0,\Gamma_\mathrm{R}}\\
&\geq \alpha_\mathsf{K}\|\nabla q\|_0^2 + \inf_{\boldsymbol{x}\in\Gamma_\mathrm{R}} \gamma(\boldsymbol{x})\ (q,q)_{0,\Gamma_\mathrm{R}}\\
&\geq \alpha_a\|q\|_1^2.
\end{aligned}$$

The constant α_a depends on α_K, the constant describing the positive definiteness of K, and on the inverse of the Poincaré constant C_P from Eq. (4.12). Note that α_a scales with K. If γ is not zero and q has a nonzero trace on the Robin boundary Γ_R the constant α_a could be chosen larger with an additional dependency on the boundary coefficient γ.

5.1.2.4 Application of the Lax–Milgram Theorem

According to the theorem of Lax–Milgram, Theorem 2.1.13, there exists a unique solution $P \in Q$ to (5.5) and consequently a unique solution $p = P + p_\mathrm{D}$ to the weak formulation (5.4) of the Laplace problem (5.1). For reference, this is repeated here in detail:

Theorem 5.1.2 *Let $\Omega \subset \mathbb{R}^d$ be a Lipschitz domain together with two disjoint subsets Γ_D and Γ_R of the boundary $\partial\Omega$, which are themselves Lipschitz, such that $\overline{\Gamma_\mathrm{D}} \cup \overline{\Gamma_\mathrm{R}} = \partial\Omega$. Furthermore, let*

$$\mathsf{K} \in \left(L^\infty(\Omega)\right)^{d\times d}, \qquad \gamma \in L^\infty(\Gamma_\mathrm{R}),$$

where K is symmetric and positive definite, γ nonnegative, and

$$f \in \left(H^1_{\Gamma_\mathrm{D}}(\Omega)\right)^*, \qquad p_\mathrm{D} \in H^{1/2}(\Gamma_\mathrm{D}), \qquad g_\mathrm{R} \in \left(H^{1/2}_{00}(\Gamma_\mathrm{R})\right)^*.$$

Then there exists a unique solution $p \in H^1(\Omega)$ such that $T_{\Gamma_\mathrm{D}} p = p_\mathrm{D}$, i.e., $p = p_\mathrm{D}$ on Γ_D, and for all $q \in H^1_{\Gamma_\mathrm{D}}(\Omega)$ it is

$$a(p, q) = \ell(q), \qquad \text{(5.4 revisited)}$$

with

$$\begin{aligned} a(p, q) &:= (\mathsf{K}\nabla p, \nabla q)_0 + (\gamma p, q)_{0,\Gamma_\mathrm{R}}, \\ \ell(q) &:= (f, q)_{\left(H^1_{\Gamma_\mathrm{D}}(\Omega)\right)^* \times H^1_{\Gamma_\mathrm{D}}(\Omega)} + \left(g_\mathrm{R}, T_{\Gamma_\mathrm{R}} q\right)_{\left(H^{1/2}_{00}(\Gamma_\mathrm{R})\right)^* \times H^{1/2}_{00}(\Gamma_\mathrm{R})}. \end{aligned}$$

An a priori estimate for the solution and right-hand side is (see also the end of the proof of the theorem of Lax–Milgram, Theorem 2.1.13) given for $P = p - p_\mathrm{D}$ from Eq. (5.5) as

$$\begin{aligned} \|P\|_1 &\le \frac{c_L}{\alpha_a}, \\ c_L &\le c_a \|P\|_1. \end{aligned}$$

These bounds lead to $\left|\|p\|_1 - \|p_\mathrm{D}\|_1\right| \le \frac{c_L}{\alpha_a}$ and $c_L \le c_a\|p\|_1 + c_a\|p_\mathrm{D}\|_1$ which in turn imply

$$\|p\|_1 \le \frac{c_L}{\alpha_a} + \|p_\mathrm{D}\|_1 = \frac{\tilde{c}_L}{\alpha_a} + \left(\frac{c_a}{\alpha_a} + 1\right)\|p_\mathrm{D}\|_1, \qquad (5.7)$$

$$\tilde{c}_L \le c_a \|p\|_1, \qquad (5.8)$$

with

$$\tilde{c}_L = c_L - c_a \|p_D\|_1 = \|f\|_{\left(H^1_{\Gamma_D}(\Omega)\right)^*} + c_{T_\Gamma} \|g_R\|_{\left(H^{1/2}_{00}(\Gamma_R)\right)^*},$$

see also the definition of c_L in Eq. (5.6). In the case of a small hydraulic conductivity K the constant on the right-hand side in the first estimate (5.7) becomes large.

5.1.3 View as an Operator on a Part of the Boundary

The existence and uniqueness result for the Laplace problem, Theorem 5.1.2, and the bounds above allow the definition of particular extension operators. For example, setting all data to be zero except p_D on the Dirichlet boundary and solving Eq. (5.4) yields a continuous extension operator, similar to the one defined in Theorem 4.2.4. Analogously, setting all data to zero except for the Robin data g_R, i.e., solving a Neumann problem, leads to a continuous extension operator, similar to that defined in Theorem 4.3.4. Such operators are introduced in detail in the following, which facilitates the coupling of Stokes–Darcy equations in Chap. 6.

Suppose the boundary $\partial\Omega$ is composed of three rather than two disjoint parts,

$$\partial\Omega = \overline{\Gamma_D} \cup \overline{\Gamma_R} \cup \overline{\Gamma}.$$

Then the weak solution of the Laplace equation with either Dirichlet or Robin data on Γ exists according to the previously introduced theory. Keeping other data fixed, one can define operators taking data on Γ and returning the weak solution. To make this more precise, define

$$\begin{aligned} K_D : H^{1/2}(\Gamma) &\to H^1(\Omega), \\ \lambda &\mapsto K_D(\lambda), \end{aligned}$$

where for all $v \in Q_D = H^1_{\Gamma_D \cup \Gamma}(\Omega)$ the function $\phi_\lambda := K_D(\lambda)$ solves

$$(\mathsf{K}\nabla\phi_\lambda, \nabla v)_0 + (\gamma\phi_\lambda, q)_{0,\Gamma_R} = (f, q)_{Q_D^* \times Q_D} + (g_R, q)_{T_{\Gamma_R}(Q_D)^* \times T_{\Gamma_R}(Q_D)}$$

and has the correct traces $T_{\Gamma_D}\phi_\lambda = p_D$ and $T_\Gamma\phi_\lambda = \lambda$. Note that the space $T_{\Gamma_R}(Q_D)$ is $H^{1/2}_{00}(\Gamma_R)$, see also Theorem 4.3.4. There are restrictions on the pre-image space of K_D because the combined data on the entire Dirichlet boundary $\Gamma_D \cup \Gamma$ has to be in $H^{1/2}$. This is made more precise in Proposition 5.1.3.

Instead of Dirichlet data on Γ consider $Q_{\mathrm{R}} = H^1_{\Gamma_{\mathrm{D}}}(\Omega)$ and Robin conditions to define a second operator

$$\begin{aligned} K_{\mathrm{R}} : (T_\Gamma Q_{\mathrm{R}})^* &\to H^1(\Omega), \\ \mu &\mapsto K_{\mathrm{R}}(\mu), \end{aligned}$$

where for all $q \in Q_{\mathrm{R}}$ the function $\phi_\mu := K_{\mathrm{R}}(\mu)$ solves

$$\begin{aligned} &\left(\mathsf{K}\nabla\phi_\mu, \nabla q\right)_0 + \left(\gamma\phi_\mu, q\right)_{0,\Gamma_{\mathrm{R}}\cup\Gamma} \\ &\quad = (f, q)_{Q_{\mathrm{R}}^*\times Q_{\mathrm{R}}} + (g_{\mathrm{R}}, q)_{(T_{\Gamma_{\mathrm{R}}}(Q_{\mathrm{R}}))^*\times T_{\Gamma_{\mathrm{R}}}(Q_{\mathrm{R}})} + (\mu, q)_{(T_\Gamma(Q_{\mathrm{R}}))^*\times T_\Gamma(Q_{\mathrm{R}})} \end{aligned}$$

and has the correct trace $T_{\Gamma_{\mathrm{D}}}\phi_\lambda = u_{\mathrm{D}}$.

Of special interest is the case where all data entering the right-hand side vanishes except the ones on Γ, that is f, p_{D}, and g_{R}. The coefficients K and γ remain as before. Then both K_{D} and K_{R} are linear operators which are continuous and bounded away from zero, which is shown in the following. Furthermore, they allow a deeper understanding of the Stokes–Darcy coupled problem in Chap. 6.

5.1.3.1 Dirichlet Operator on Γ

Consider the case where Dirichlet data is prescribed on the boundary Γ. Since the data on the other Dirichlet boundary Γ_{D} is now assumed to be zero, any function λ in the pre-image space of K_{D} must be in $H^{1/2}(\Gamma_{\mathrm{D}} \cup \Gamma)$, when extended by zero outside of Γ. This imposes no further restriction if Γ and Γ_{D} do not touch, i.e., $\lambda \in H^{1/2}(\Gamma)$. If however Γ_{D} and Γ do touch then λ must be in a Lions–Magenes-type space between $H^{1/2}(\Gamma)$ and $H^{1/2}_{00}(\Gamma)$. It is $H^{1/2}_{00}(\Gamma)$ if for example Γ_{R} is empty. In general, this space can be characterized as $T_\Gamma\big(H^1_{\Gamma_{\mathrm{D}}}(\Omega)\big)$ and is denoted by $H^{1/2}_{\Gamma_{\mathrm{D}}}(\Gamma)$ for brevity.

Proposition 5.1.3 *Define the operator* $K_{\mathrm{D}} : H^{1/2}_{\Gamma_{\mathrm{D}}}(\Gamma) \to H^1_{\Gamma_{\mathrm{D}}}(\Omega)$, *where for all* $q \in H^1_{\Gamma_{\mathrm{D}}\cup\Gamma}$ *the function* $\phi_\lambda := K_{\mathrm{D}}(\lambda)$ *solves*

$$(\mathsf{K}\nabla\phi_\lambda, \nabla q)_0 + (\gamma\phi_\lambda, q)_{0,\Gamma_{\mathrm{R}}} = 0 \tag{5.9}$$

and $T_\Gamma\phi_\lambda = \lambda$. *This operator is linear, continuous and positive, i.e., there exist constants* $c_{K_{\mathrm{D}}}$ *and* $\alpha_{K_{\mathrm{D}}}$ *such that*

$$\|K_{\mathrm{D}}(\lambda)\|_1 \le c_{K_{\mathrm{D}}} \|\lambda\|_{H^{1/2}_{\Gamma_{\mathrm{D}}}(\Gamma)},$$

$$\|K_{\mathrm{D}}(\lambda)\|_1 \ge \alpha_{K_{\mathrm{D}}} \|\lambda\|_{H^{1/2}_{\Gamma_{\mathrm{D}}}(\Gamma)}.$$

Proof The linearity of K_{D} follows directly from the linearity of Eq. (5.9) and of the extension operator E_{Γ}. Continuity is shown in the section on the existence and uniqueness of the Laplace weak form, Sect. 5.1.2, with a constant $c_{K_{\mathrm{D}}}$ depending on the continuity of the bilinear form, c_a, and the inverse of its coercivity constant α_a, in particular it is $c_{K_{\mathrm{D}}} = (c_a/\alpha_a + 1)c_{E_{\Gamma}}$, see Theorem 5.1.2 and the bound (5.7) where here it is $\tilde{c}_L = 0$. The continuity of the trace operator T_{Γ}, see Theorem 4.3.4 and the subsequent discussion, gives

$$\|\lambda\|_{H^{1/2}_{\Gamma_{\mathrm{D}}}(\Gamma)} = \|T_{\Gamma}(K_{\mathrm{D}}(\lambda))\|_{H^{1/2}_{\Gamma_{\mathrm{D}}}(\Gamma)} \leq c_{T_{\Gamma}} \|K_{\mathrm{D}}(\lambda)\|_1$$

and therefore the positivity of K_{D} with $\alpha_{K_{\mathrm{D}}} = c_{T_{\Gamma}}^{-1}$. □

The operator K_{D} defined above is therefore another linear and continuous right inverse of T_{Γ}, see Theorem 4.3.4 and the subsequent discussion.

5.1.3.2 Robin Operator on Γ

Consider the case where Robin data is prescribed on the boundary Γ. The pre-image space of the operator K_{R} consists of functionals on $H^{1/2}_{\Gamma_{\mathrm{D}}}(\Gamma) = T_{\Gamma}\left(H^1_{\Gamma_{\mathrm{D}}}(\Omega)\right)$.

Proposition 5.1.4 *Define the operator* $K_{\mathrm{R}} : \left(H^{1/2}_{\Gamma_{\mathrm{D}}}(\Gamma)\right)^* \to H^1_{\Gamma_{\mathrm{D}}}(\Omega)$, *where for all* $q \in H^1_{\Gamma_{\mathrm{D}}}$ *the function* $\phi_\mu := K_{\mathrm{R}}(\mu)$ *solves*

$$(\mathsf{K}\nabla\phi_\mu, \nabla q)_0 + (\gamma\phi_\mu, q)_{0,\Gamma_{\mathrm{R}}\cup\Gamma} = (\mu, q)_{\left(H^{1/2}_{\Gamma_{\mathrm{D}}}(\Gamma)\right)^* \times H^{1/2}_{\Gamma_{\mathrm{D}}}(\Gamma)}. \tag{5.10}$$

This operator is linear, continuous and positive, i.e., there exist constants $c_{K_{\mathrm{R}}}$ *and* $\alpha_{K_{\mathrm{R}}}$ *such that*

$$\|K_{\mathrm{R}}(\mu)\|_1 \leq c_{K_{\mathrm{R}}} \|\mu\|_{\left(H^{1/2}_{\Gamma_{\mathrm{D}}}(\Gamma)\right)^*},$$

$$\|K_{\mathrm{R}}(\mu)\|_1 \geq \alpha_{K_{\mathrm{R}}} \|\mu\|_{\left(H^{1/2}_{\Gamma_{\mathrm{D}}}(\Gamma)\right)^*}.$$

Proof Because Eq. (5.10) is linear in ϕ_μ, so is the operator K_{R}. As with K_{D} before the continuity is shown in the section on existence and uniqueness of the Laplace weak form, Sect. 5.1.2, specifically in Theorem 5.1.2. The constant $c_{K_{\mathrm{R}}}$ depends on the inverse of the coercivity constant α_a of the bilinear form a, in particular it is $c_{K_{\mathrm{R}}} = c_{T_{\Gamma}}/\alpha_a$, see Eq. (5.7) where here it is $p_{\mathrm{D}} = 0$ and $f = 0$. Moreover, Eq. (5.8) yields $\alpha_{K_{\mathrm{R}}} = c_{T_{\Gamma}}/c_a$. □

Note that the Robin operator above is a continuous extension operator which maps μ to a function $\phi_\mu = K_R\mu$ which, in a weak sense, satisfies $K\nabla\phi_\mu\cdot\boldsymbol{n}+\gamma\phi_\mu = \mu$ on the boundary part Γ. In fact, this operator has a left inverse T_Γ^R, i.e., it is $T_\Gamma^R\phi_\mu = \mu$:

Definition 5.1.5 Let $\phi \in H^1_{\Gamma_D}(\Omega)$ be given such that for all $q \in H^1_{\Gamma_D\cup\Gamma}(\Omega)$ it is

$$(\mathsf{K}\nabla\phi, \nabla q)_0 + (\gamma\phi, q)_{0,\Gamma_R} = 0.$$

Then its Robin data $T_\Gamma^R\phi \in \left(H^{1/2}_{\Gamma_D}(\Gamma)\right)^*$ on Γ is defined as

$$\left(T_\Gamma^R\phi, \xi\right)_{\left(H^{1/2}_{\Gamma_D}(\Gamma)\right)^*\times H^{1/2}_{\Gamma_D}(\Gamma)} := (\mathsf{K}\nabla\phi, \nabla(E_\Gamma\xi))_0 + (\gamma\phi, E_\Gamma\xi)_{0,\Gamma_R\cup\Gamma},$$

with the extension operator $E_\Gamma : H^{1/2}_{\Gamma_D}(\Gamma) \to H^1_{\Gamma_D}(\Omega)$ defined in Theorem 4.3.4 or in its subsequent discussion, respectively.

Due to the assumption on ϕ, the above definition does not depend on the particular extension operator E_Γ. Additionally, T_Γ^R is an extension of $p \mapsto \mathsf{K}\nabla p \cdot \boldsymbol{n} + \gamma p$ on Γ. Furthermore, just as the trace operator T_Γ is the left inverse of the Dirichlet operator K_D, the left inverse T_Γ^R of K_R may be called a Robin trace operator.

5.1.3.3 A Connection of the Dirichlet and Robin Operator on Γ

Suppose a $\mu \in \left(H^{1/2}_{\Gamma_D}(\Gamma)\right)^*$ is given. Then, for the function $\phi_\mu = K_R(\mu)$ and all $q \in H^1_{\Gamma_D\cup\Gamma}$ it is

$$\left(\mathsf{K}\nabla\phi_\mu, \nabla q\right)_0 + \left(\gamma\phi_\mu, q\right)_{0,\Gamma_R} = 0.$$

Thus the above Definition 5.1.5 applies and $\mu = T_\Gamma^R\phi_\mu$, i.e., $T_\Gamma^R \circ K_R$ is the identity on $\left(H^{1/2}_{\Gamma_D}(\Gamma)\right)^*$. At the same time the above equation means that the same ϕ_μ also is $K_D(T_\Gamma\phi_\mu)$. One way to understand this reasoning is that on the image of K_R the operator $K_D \circ T_\Gamma$ is the identity. Combining these ideas leads to

$$T_\Gamma^R \circ K_D \circ T_\Gamma \circ K_R = \mathrm{id} \qquad \text{in } \left(H^{1/2}_{\Gamma_D}(\Gamma)\right)^*,$$

$$T_\Gamma \circ K_R \circ T_\Gamma^R \circ K_D = \mathrm{id} \qquad \text{in } H^{1/2}_{\Gamma_D}(\Gamma)$$

where the second equation can be derived in a similar fashion.

5.2 Saddle Point Problems

The special structure of Stokes (and other) equations, leads to an abstract formulation which requires a few tailored tools. The theory is commonly known as that of saddle point problems and well established. Classical literature includes [BBF13, BF91, GR86], see also [Joh16].

5.2.1 *Notation and Formulation of the Abstract Problem*

Let V and Q be Hilbert spaces with inner products $(\cdot,\cdot)_V$ and $(\cdot,\cdot)_Q$, norms $\|\cdot\|_V$ and $\|\cdot\|_Q$, and duals V^* and Q^* respectively. As before, the duality pairings between elements $f \in V^*$ or $g \in Q^*$ and $v \in V$ or $q \in Q$ are denoted by $(f, v)_{V^*\times V}$ or $(g, q)_{Q^*\times Q}$ respectively. Furthermore, let $a(\cdot,\cdot) : V \times V \to \mathbb{R}$ and $b(\cdot,\cdot) : V \times Q \to \mathbb{R}$ be two bilinear forms which are assumed to be continuous, i.e., there are positive constants c_a and c_b such that for all $v, w \in V$ and $q \in Q$ it holds

$$a(v, w) \le c_a \|v\|_V \|w\|_V, \tag{5.11}$$

$$b(v, q) \le c_b \|v\|_V \|q\|_Q. \tag{5.12}$$

Associated with these two bilinear forms define two linear operators $A : V \to V^*$ and $B : V \to Q^*$ by

$$(Av, w)_{V^*\times V} := a(v, w) \qquad \text{for all } v, w \in V, \tag{5.13}$$

$$(Bv, q)_{Q^*\times Q} = \left(B^*q, v\right)_{V^*\times V} := b(v, q) \qquad \text{for all } v \in V \text{ and } q \in Q. \tag{5.14}$$

Here $B^* : Q \to V^*$ is the dual operator of B, where the identification $Q^{**} = Q$ is used.

The abstract problem to be solved is the following: Let $f \in V^*$ and $g \in Q^*$ be given. Find $u \in V$ and $p \in Q$ such that for all $v \in V$ and all $q \in Q$ it holds

$$\begin{cases} a(u, v) + b(v, p) = (f, v)_{V^*\times V}, \\ \qquad\qquad b(u, q) = (g, q)_{Q^*\times Q}. \end{cases} \tag{5.15}$$

This can be rewritten as a system of equations in V^* and Q^* using the operators A and B: Find $u \in V$ and $p \in Q$ such that

$$\begin{cases} Au + B^*p = f, \\ \qquad\quad Bu = g. \end{cases} \tag{5.16}$$

Clearly, existence of a solution to the above set of equations can only be expected for $g \in \operatorname{Im} B$. Further necessary assumptions to guarantee existence and uniqueness are discussed in the next part.

Remark 5.2.1 If the bilinear form $a(\cdot,\cdot)$ is symmetric, Eq. (5.15) are optimality conditions of the saddle point problem

$$\inf_{v\in V}\sup_{q\in Q}\frac{1}{2}a(v,v)+b(v,q)-(f,v)_{V^*\times V}-(g,q)_{Q^*\times Q}.$$

Even if $a(\cdot,\cdot)$ is not symmetric, problems of the form (5.15) or (5.16) are usually called saddle point problems in the literature, [BF91].

5.2.2 Existence and Uniqueness

There are two essential conditions to prove existence and uniqueness of a solution of problem (5.16), one on each of the two operators A and B. In the literature one can find different assumptions which are equivalent or stronger than the ones needed. To clarify this, two propositions showing the relations between some popular formulations are given after the main theorem of this section. The proofs are mainly taken from [BF91] and [GR86].

Theorem 5.2.2 (Existence and Uniqueness) *For any $f \in V^*$ and $g \in \operatorname{Im} B \subset Q^*$ problem* (5.16) *admits a solution $(u,p) \in V \times Q$, where u is unique and p is unique up to an element of* $\ker B^*$, *if*

(i) The restricted operator $A_0 : \ker B \to (\ker B)^*$, defined for all $v_0, w_0 \in \ker B \subset V$ by $(A_0v_0, w_0)_{(\ker B)^*\times\ker B} = (Av_0, w_0)_{V^*\times V}$, is an isomorphism,
(ii) The image of B is closed.

Proof Assume (i) and (ii) are true and $f \in V^*$ and $g \in \operatorname{Im} B \subset Q^*$ are given. Since $g \in \operatorname{Im} B$, there is a $u_g \in V$ such that $Bu_g = g$. Consider $f - Au_g$ as an element in $(\ker B)^*$. Then through (i) there exists $u_0 \in \ker B$ such that $A_0u_0 = f - Au_g$. Denoting $u = u_0 + u_g$ it holds

$$(f - Au, v_0)_{(\ker B)^*\times\ker B} = 0 \quad \text{for all } v_0 \in \ker B.$$

That means $f - Au \in (\ker B)^\circ$. Since $\operatorname{Im} B$ is closed, the closed range Theorem 2.1.9 states $(\ker B)^\circ = \operatorname{Im} B^*$. Therefore, there exists $p \in Q$, which is unique up to an element of $\ker B^*$, such that $B^*p = f - Au$ in V^*. The pair (u,p) solves problem (5.16). In order to prove uniqueness, i.e., that the pair (u,p) does not depend on the choice u_g, it is shown that a solution (u,p) of the homogeneous problem

$$\begin{cases} Au + B^*p = 0, \\ Bu = 0 \end{cases}$$

must be trivial. Testing the first equation of this homogeneous system with $v_0 \in \ker B$, it follows (note $u \in \ker B$ and $B^* p \in \operatorname{Im} B^* = (\ker B)^\circ$)

$$\begin{aligned} 0 &= (Au, v_0)_{V^* \times V} + \left(B^* p, v_0\right)_{V^* \times V} \\ &= (Au, v_0)_{(\ker B)^* \times \ker B} \\ &= (A_0 u, v_0)_{(\ker B)^* \times \ker B}. \end{aligned}$$

The operator A_0 is assumed to be invertible, and especially has trivial kernel so that u must vanish. Now it holds $B^* p = 0$ in V^*, i.e., p is zero up to an element of $\ker B^*$. Therefore, Eq. (5.16) has a solution (u, p) which is unique up to an element of $\ker B^*$ for p. □

Remark 5.2.3 From the last part of the proof one can see that if the operator B is assumed to be surjective, according to Proposition 2.1.8, its transposed B^* is injective, implying $p = 0$. In conclusion the solution of (5.16) is unique if, additionally, B is surjective.

Remark 5.2.4 The conditions in the previous Theorem 5.2.2 are not just sufficient but also necessary, see for example the Remarks 4.1 in [GR86], Chapter 1.

The first statement in the following proposition is exactly the first condition in the previous Theorem 5.2.2 showing existence and uniqueness.

Proposition 5.2.5 *Consider the following four statements.*

a) *The operator $A_0 : \ker B \to (\ker B)^*$, defined by $(A_0 v_0, w_0)_{\ker B^* \times \ker B} = a(v_0, w_0)$ for all $v_0, w_0 \in \ker B \subset V$, is an isomorphism.*
b) *There exists $\alpha > 0$ such that*

$$\inf_{u_0 \in \ker B} \sup_{v_0 \in \ker B} \frac{a(u_0, v_0)}{\|u_0\|_V \|v_0\|_V} \geq \alpha, \tag{5.17a}$$

$$\inf_{v_0 \in \ker B} \sup_{u_0 \in \ker B} \frac{a(u_0, v_0)}{\|u_0\|_V \|v_0\|_V} \geq \alpha. \tag{5.17b}$$

c) *The operator A is coercive on* $\ker B$, *i.e., there exists $\alpha > 0$ such that $a(v_0, v_0) \geq \alpha \|v_0\|_V^2$ for all $v_0 \in \ker B$.*
d) *The operator A is coercive on V, i.e., there exists $\alpha > 0$ such that $a(v, v) \geq \alpha \|v\|_V^2$ for all $v \in V$.*

Then the following implications hold

$$a) \iff b) \impliedby c) \impliedby d)$$

Proof "d) $\Longrightarrow$ c)": This follows immediately because $\ker B \subset V$.

"c) $\Longrightarrow$ b)": For all $v_0 \in \ker B$ it is

$$\sup_{u_0 \in \ker B} \frac{a(u_0, v_0)}{\|u_0\|_V \|v_0\|_V} \geq \frac{a(v_0, v_0)}{\|v_0\|_V^2} \geq \alpha.$$

This implies the inequality (5.17b). Analogously one can show (5.17a).

"b) $\Longrightarrow$ a)": Using the operator A_0 inequality (5.17a) reads

$$\inf_{u_0 \in \ker B} \sup_{v_0 \in \ker B} \frac{(A_0 u_0, v_0)}{\|u_0\|_V \|v_0\|_V} \geq \alpha,$$

which means A_0 must have a trivial kernel, $\ker A_0 = \{0\}$, i.e., A_0 is injective. Furthermore, it means

$$\inf_{u_0 \in \ker B} \frac{\|A_0 u_0\|_{(\ker B)^*}}{\|u_0\|_V} \geq \alpha. \tag{5.18}$$

Note that the norm of $A_0 u_0$ is the standard norm of the dual space $(\ker B)^*$, i.e., $\|A_0 u_0\|_{(\ker B)^*} = \sup_{v_0 \in \ker B} (A_0 u_0, v_0)/\|v_0\|_{\ker B}$. According to Corollary 2.1.12 of the open mapping theorem, Theorem 2.1.10, the image of A_0 is closed. Similarly, inequality (5.17b) shows that the transposed operator $A_0^* : \ker B \to (\ker B)^*$ of A_0 is as well injective, and therefore its image is dense, see Proposition 2.1.8. This proves the bijectivity and, together with (5.18), the continuous invertibility of A_0.

"a) $\Longrightarrow$ b)": Continuity of the operator $A_0^{-1} : (\ker B)^* \to \ker B \subset V$ means there exists a constant $1/\alpha \geq 0$ such that for all $v_0^* \in (\ker B)^*$ it holds $\left\|A_0^{-1} v_0^*\right\|_V \leq \frac{1}{\alpha} \left\|v_0^*\right\|_{(\ker B)^*}$. Now let $u_0 \in \ker B \subset V$ be given and denote $v_0^* = A_0 u_0$, then it holds

$$\|u_0\|_V \leq \frac{1}{\alpha} \|A_0 u_0\|_{(\ker B)^*} = \frac{1}{\alpha} \sup_{v_0 \in \ker B} \frac{a(u_0, v_0)}{\|v_0\|_V}$$

$$\Longrightarrow \sup_{v_0 \in \ker B} \frac{a(u_0, v_0)}{\|v_0\|_V \|u_0\|_V} \geq \alpha \qquad \text{for all } u_0 \in \ker B.$$

That is equivalent to inequality (5.17a). Analogously one can show (5.17b). □

Proposition 5.2.6 *Consider the following nine statements.*

a) There exists a constant $\beta > 0$ *such that*

$$\inf_{q \in Q} \sup_{v \in V} \frac{b(v, q)}{\|v\|_V \|q\|_Q} \geq \beta. \tag{5.19}$$

b) The operator $B : (\ker B)^{\perp} \to Q^*$ *is an isomorphism and there exists a constant* $\beta > 0$ *such that for all* $v \in (\ker B)^{\perp}$ *it holds* $\|Bv\|_{Q^*} \geq \beta \|v\|_V$.

c) The operator $B^* : Q \to (\ker B)^{\circ}$ *is an isomorphism and there exists a constant* $\beta > 0$ *such that for all* $q \in Q$ *it holds* $\|B^*q\|_{V^*} \geq \beta \|q\|_Q$.

d) There exists a constant $\beta > 0$ *such that for all* $q \in Q$ *it holds*

$$\sup_{v \in V} \frac{b(v, q)}{\|v\|_V} \geq \beta \|q\|_{Q/\ker B^*}. \tag{5.20}$$

e) There exists a constant $\beta > 0$ *such that for all* $v \in V$ *it holds*

$$\sup_{q \in Q} \frac{b(v, q)}{\|q\|_Q} \geq \beta \|v\|_{V/\ker B}. \tag{5.21}$$

f) The image of B, $\operatorname{Im} B$, *is closed in* Q^*.

g) The image of B^*, $\operatorname{Im} B^*$, *is closed in* V^*.

h) $(\ker B)^{\circ} = \operatorname{Im} B^*$.

i) $(\ker B^*)_{\circ} = \operatorname{Im} B$.

The following implications hold

$$a) \iff b) \iff c) \quad \implies \quad d) \iff e) \iff f) \iff g) \iff h) \iff i).$$

Proof "a) $\implies$ c)": Using the operator norm in V^* Eq. (5.19) reads

$$\|B^*q\|_{V^*} \geq \beta \|q\|_Q \tag{5.22}$$

for all $q \in Q$, i.e., B^* is injective. Furthermore, it implies the continuity of the inverse of B^* (defined on $\operatorname{Im} B^*$). Corollary 2.1.12 states that $\operatorname{Im} B^*$ is closed and Theorem 2.1.7 shows $\operatorname{Im} B^* = \overline{\operatorname{Im} B^*} = (\ker B)^{\circ}$, i.e., $B^* : Q \to (\ker B)^{\circ}$ is an isomorphism.

"c) $\implies$ a)": Rewriting the equation $\|B^*q\|_{V^*} \geq \beta \|q\|_Q$ gives

$$\sup_{v \in V} \frac{b(v, q)}{\|v\|_V \|q\|_Q} \geq \beta$$

for all $q \in Q$. This implies (5.19).

"b) $\implies$ c)": Due to Proposition 2.1.8 and B being surjective, B^* is injective. Furthermore, $\operatorname{Im} B$ is closed because B is bounded away from zero, see Corollary 2.1.12. Therefore, the closed range theorem, Theorem 2.1.9, implies $\operatorname{Im} B^* = \overline{\operatorname{Im} B^*} = (\ker B)^{\circ}$, i.e., B^* is surjective. As a consequence of the open mapping theorem, Corollary 2.1.11, the inverse $(B^*)^{-1}$ is continuous, i.e., for all $q \in Q$ it holds $\|B^*q\|_{V^*} \geq \beta \|q\|_Q$. Since $B \mapsto B^*$ is isometric, the constant β is the same for b) and c).

"c) $\implies$ b)": Interchanging B and B^* in "b) $\implies$ c)" proves the claim.

"f) $\iff$ g) $\iff$ h) $\iff$ i)": This is the closed range theorem, Theorem 2.1.9.

"d) $\implies$ g)": Equation (5.20) implies $\|B^*q\|_{V^*} \geq \beta\|q\|_{Q/\ker B^*}$ for all $q \in Q$. Then Corollary 2.1.12 proves the claim, noting that the quotient space $Q/\ker B^*$ is a Banach space.

"g) $\implies$ d)": Corollary 2.1.12 shows $\|B^*q\|_{V^*} \geq \beta\|q\|_Q$. Additionally, note that for all $q \in Q$ it holds $\|q\|_Q \geq \|q\|_{Q/\ker B^*}$.

"e) $\implies$ f)": Equation (5.21) implies $\|Bv\|_{Q^*} \geq \beta\|v\|_{V/\ker B}$. Then copy the second part of the proof of Corollary 2.1.12 and note that the quotient space $V/\ker B$ is a Banach space.

"f) $\implies$ e)": Corollary 2.1.12 shows $\|Bv\|_{Q^*} \geq \beta\|v\|_V$. Additionally note that for all $v \in V$ it holds $\|v\|_V \geq \|v\|_{V/\ker B}$.

"a) $\implies$ d)": This follows directly, just note that $\|q\|_Q \geq \|q\|_{Q/\ker B^*}$ for all $q \in Q$. □

Remark 5.2.7 If the operator $B : V \to Q^*$ is surjective, then all of the statements in Proposition 5.2.6 are equivalent. One often chooses $Q/\ker B^*$ instead of Q so that B is naturally surjective. Together with Remark 5.2.3 this means that any one of the statements a)–i) of the previous Proposition 5.2.6 implies unique solvability of the system (5.16) for u and p.

Remark 5.2.8 The first statement in Proposition 5.2.6 is the celebrated inf–sup condition which one finds most often in the literature. It is also called Babuška–Brezzi, Ladyzhenskaya–Babuška–Brezzi or just LBB condition.

In conclusion, existence and uniqueness is guaranteed if in each Propositions 5.2.5 and 5.2.6 any one of the conditions is satisfied. Using the constants $\alpha > 0$ and $\beta > 0$ from these two propositions an estimate of the norm of the solution can be obtained: From the equation $A_0u_0 = f - Au_g$ in the proof of the main Theorem 5.2.2 it follows

$$\|u_0\|_V \leq \frac{1}{\alpha}\big(\|f\|_{V^*} + c_a\|u_g\|_V\big)$$

and therefore

$$\|u\|_V \leq \|u_g\|_V + \frac{1}{\alpha}\big(\|f\|_{V^*} + c_a\|u_g\|_V\big).$$

From Proposition 5.2.6, e), one can choose u_g such that $\|u_g\|_V \leq \frac{1}{\beta}\|g\|_{Q^*}$ to yield

$$\|u\|_V \leq \frac{1}{\alpha}\|f\|_{V^*} + \Big(\frac{c_a}{\alpha} + 1\Big)\frac{1}{\beta}\|g\|_{Q^*}, \tag{5.23}$$

$$\|p\|_{Q/\ker B^*} \leq \frac{1}{\beta}\Big(\frac{c_a}{\alpha} + 1\Big)\|f\|_{V^*} + \frac{c_a}{\beta^2}\Big(\frac{c_a}{\alpha} + 1\Big)\|g\|_{Q^*}. \tag{5.24}$$

The estimate on the pressure uses Proposition 5.2.6, c), together with $B^*p = f - Au$, i.e., $\|p\|_{Q/\ker B^*} \leq \frac{1}{\beta}\|f - Au\|_{V^*} \leq \frac{1}{\beta}\|f\|_{V^*} + \frac{c_a}{\beta}\|u\|_V$.

5.3 Application to the Stokes Problem

The Stokes equations are a linear system describing the flow of a fluid. They can be deduced from the nonlinear Navier–Stokes equations when a slow and incompressible fluid is assumed. Then the convective term is small and can be neglected.

The Stokes problem is given as follows: Find the velocity $\mathbf{u} : \Omega \to \mathbb{R}^d$ and the pressure $p : \Omega \to \mathbb{R}$ such that

$$\begin{cases} -\nabla \cdot \mathsf{T}(\mathbf{u}, p) = \mathbf{f} & \text{in } \Omega, \\ \nabla \cdot \mathbf{u} = 0 & \text{in } \Omega, \end{cases} \tag{5.25}$$

where $\mathsf{T}(\mathbf{u}, p) = 2\nu \mathsf{D}(\mathbf{u}) - p\,\mathrm{id}$ is the Cauchy stress tensor, id the identity tensor, $\nu > 0$ the kinematic viscosity, and $\mathsf{D}(\mathbf{u}) = \frac{1}{2}\left(\nabla \mathbf{u} + \nabla \mathbf{u}^T\right)$ the symmetric part of $\nabla \mathbf{u}$ also called the deformation tensor. The right-hand side vector field $\mathbf{f} : \Omega \to \mathbb{R}^d$ is called source term.

Additionally, a solution $(\mathbf{u}, p)$ has to satisfy boundary conditions. The focus here is on Dirichlet and Robin type boundary conditions similar to the case of the Laplace equation in Sect. 5.1. For this reason let $\partial\Omega$ be divided into two disjoint (relatively open) subsets Γ_D and Γ_R such that $\overline{\Gamma_\mathrm{D}} \cup \overline{\Gamma_\mathrm{R}} = \partial\Omega$. The boundary conditions then read

$$\begin{cases} \mathbf{u} = \mathbf{u}_\mathrm{D} & \text{on } \Gamma_\mathrm{D}, \\ \gamma \mathbf{u} + \mathsf{T}(\mathbf{u}, p) \cdot \boldsymbol{n} = \boldsymbol{g}_\mathrm{R} & \text{on } \Gamma_\mathrm{R}. \end{cases}$$

Here $\mathbf{u}_\mathrm{D} : \Gamma_\mathrm{D} \to \mathbb{R}^d$ and $\boldsymbol{g}_\mathrm{R} : \Gamma_\mathrm{R} \to \mathbb{R}^d$ are given functions on the respective boundary parts. The function $\gamma : \Gamma_\mathrm{R} \to \mathbb{R}$ is assumed to be not negative. In case $\gamma = 0$, Neumann boundary conditions are recovered. If furthermore also $\boldsymbol{g}_\mathrm{R}$ vanishes, the term outflow or do-nothing boundary condition is widely used.

Remark 5.3.1 In the literature the Cauchy stress tensor T is often defined as $\mathsf{T} = \nu \nabla \mathbf{u} - p\,\mathrm{id}$. The resulting formulation (5.25) is equivalent for a smooth velocity solution $\mathbf{u}$, which can be shown using the divergence constraint $\nabla \cdot \mathbf{u} = 0$. After discretization using, for example, the finite element method, this is no longer true. It is assumed that using the deformation tensor is the appropriate approach, see [LI06].

Remark 5.3.2 It is possible to vary boundary conditions not just on different parts of the boundary but also by components of the solution. For example one could want a no penetration boundary condition $\mathbf{u} \cdot \boldsymbol{n} = 0$ and on the same boundary part a Neumann type condition in the tangential direction, $\boldsymbol{\tau} \cdot \mathsf{T}(\mathbf{u}, p) \cdot \boldsymbol{n} = 0$, with $\boldsymbol{\tau}$ being a tangential vector. For smooth boundaries this does not considerably complicate the analysis, but requires the introduction of much more notation.

5.3.1 Weak Form

In order to construct weak formulations of the Stokes problem assume $(\mathbf{u}, p)$ is a smooth solution to (5.25), e.g., $\mathbf{u} \in (\mathcal{D}(\overline{\Omega}))^d$ and $p \in \mathcal{D}(\overline{\Omega})$. Then multiply the first equation of (5.25) with a test function $\boldsymbol{v} \in (C^\infty(\Omega))^d$ which vanishes in a neighborhood of the Dirichlet boundary part Γ_D,

$$\boldsymbol{v} \in \left(\mathcal{D}(\Omega \cup \partial\Omega \setminus \Gamma_\mathrm{D})\right)^d.$$

The second equation of (5.25) is multiplied by a test function $q \in \mathcal{D}(\overline{\Omega})$. Both equations are then integrated over the domain Ω and the first is integrated by parts yielding

$$\begin{cases} (\mathsf{T}(\mathbf{u}, p), \mathsf{D}(\boldsymbol{v}))_0 - (\mathsf{T}(\mathbf{u}, p) \cdot \boldsymbol{n}, \boldsymbol{v})_{0,\Gamma_\mathrm{R}} = (\mathbf{f}, \boldsymbol{v})_0, \\ \qquad\qquad\qquad\qquad\qquad\qquad (\nabla \cdot \mathbf{u}, q) = 0. \end{cases}$$

Inserting the boundary conditions on Γ_R this can be rewritten as

$$\begin{cases} a(\mathbf{u}, \boldsymbol{v}) + b(\boldsymbol{v}, p) = \ell(\boldsymbol{v}), \\ \qquad\qquad\quad b(\mathbf{u}, q) = 0, \end{cases} \tag{5.26}$$

where the bilinear forms $a(\cdot, \cdot)$ and $b(\cdot, \cdot)$ and the right-hand side ℓ are defined as

$$\begin{aligned} a(\mathbf{u}, \boldsymbol{v}) &= (2\nu\mathsf{D}(\mathbf{u}), \mathsf{D}(\boldsymbol{v}))_0 + (\gamma\mathbf{u}, \boldsymbol{v})_{0,\Gamma_\mathrm{R}}, \\ b(\boldsymbol{v}, p) &= -(\nabla \cdot \boldsymbol{v}, p)_0, \\ \ell(\boldsymbol{v}) &= (\mathbf{f}, \boldsymbol{v})_0 + \left(g_\mathrm{R}, \boldsymbol{v}\right)_{0,\Gamma_\mathrm{R}}. \end{aligned}$$

As in the derivation of a weak form to the Laplace equation in Sect. 5.1, through approximation one can choose

$$\begin{aligned} &\boldsymbol{V} := \mathbf{H}^1_{\Gamma_\mathrm{D}}(\Omega) \qquad \text{and} \\ &Q := L^2(\Omega) \end{aligned}$$

as test spaces and Eq. (5.26) stay meaningful, see Theorem 3.4.5 and Eq. (4.3) for each velocity component and Theorem 3.4.5 also for the pressure. The data have to satisfy the following smoothness properties:

$$\begin{aligned} \mathbf{f} &\in \boldsymbol{V}^*, \\ \mathbf{u}_\mathrm{D} &\in \mathbf{H}^{1/2}(\Gamma_\mathrm{D}), \\ \gamma &\in L^\infty(\Gamma_\mathrm{R}), \\ g_\mathrm{R} &\in \left(\mathbf{H}^{1/2}_{00}(\Gamma_\mathrm{R})\right)^*. \end{aligned}$$

As in the case of the Laplace problem the definition of ℓ involves duality pairings

$$\ell(\boldsymbol{v}) = (\mathbf{f}, \boldsymbol{v})_{\boldsymbol{V}^* \times \boldsymbol{V}} + \left(g_{\mathrm{R}}, \boldsymbol{v}\right)_{\left(\mathbf{H}_{00}^{1/2}(\Gamma_{\mathrm{R}})\right)^* \times \mathbf{H}_{00}^{1/2}(\Gamma_{\mathrm{R}})}$$

The weak formulation of the Stokes problem (5.25) is: Find $(\mathbf{u}, p) \in \mathbf{H}^1(\Omega) \times Q$ such that $\mathbf{u} - \mathbf{u}_{\mathrm{D}} \in \boldsymbol{V}$ and for all $\boldsymbol{v} \in \boldsymbol{V}$ and all $q \in Q$ it holds

$$\begin{cases} a(\mathbf{u}, \boldsymbol{v}) + b(\boldsymbol{v}, p) = \ell(\boldsymbol{v}), \\ \qquad\qquad b(\mathbf{u}, q) = 0. \end{cases} \tag{5.27}$$

Remark 5.3.3 In the expression $\mathbf{u} - \mathbf{u}_{\mathrm{D}} \in \boldsymbol{V}$ the term $\mathbf{u}_{\mathrm{D}}$ has to be understood as a vector of extensions according to Theorem 4.2.4 for each of its components. See also Remark 5.1.1 in the case of the Laplace problem.

5.3.2 Existence and Uniqueness

In order to put this formulation into the abstract setting for saddle point problems, define $\mathbf{U} = \mathbf{u} - \mathbf{u}_{\mathrm{D}}$ and consider

$$\begin{cases} a(\mathbf{U}, \boldsymbol{v}) + b(\boldsymbol{v}, p) = \ell(\boldsymbol{v}) - a(\mathbf{u}_{\mathrm{D}}, \boldsymbol{v}) =: L(\boldsymbol{v}), \\ \qquad\qquad b(\mathbf{U}, q) = -b(\mathbf{u}_{\mathrm{D}}, q). \end{cases}$$

This way the bilinear form $a(\cdot, \cdot)$ acts on $\boldsymbol{V} \times \boldsymbol{V}$ instead of $\mathbf{H}^1(\Omega) \times \boldsymbol{V}$. Furthermore, these equations are equivalent to the original weak formulation (5.27), the solution $\mathbf{u}$ can be recovered as $\mathbf{U} + \mathbf{u}_{\mathrm{D}}$. To prove existence and uniqueness of such solutions the coercivity of $a(\cdot, \cdot)$ and the inf–sup condition for $b(\cdot, \cdot)$ is shown and then Theorem 5.2.2 can be applied. The bilinear forms and the right-hand side are continuous with constants c_a and c_b respectively, which can be shown in the same way as in the Laplace case, see Sect. 5.1.2.

Remark 5.3.4 The operator B of the abstract saddle point problem is the divergence:

$$B : \boldsymbol{V} \to Q^*, \qquad (B\mathbf{u}, q)_{Q^* \times Q} = -(\nabla \cdot \mathbf{u}, q)_0.$$

The dual operator $B^* : Q \to \boldsymbol{V}^*$ of B is related to the gradient. In case $\Gamma_{\mathrm{D}} = \partial\Omega$, i.e., $\Gamma_{\mathrm{R}} = \emptyset$ and $V = \mathbf{H}_0^1(\Omega)$, it is an extension of

$$\nabla : Q \supset H^1(\Omega) \to \boldsymbol{V}^*, \qquad (\nabla q, \boldsymbol{v})_{\boldsymbol{V}^* \times \boldsymbol{V}} = (\nabla q, \boldsymbol{v})_0,$$

and integration by parts, see Corollary 4.4.1, results in

$$\left(B^* q, \boldsymbol{v}\right)_{\boldsymbol{V}^* \times \boldsymbol{V}} = (\nabla q, \boldsymbol{v})_0 = -(\nabla \cdot \boldsymbol{v}, q)_0 = (B\boldsymbol{v}, q)_{Q^* \times Q},$$

because $\boldsymbol{v}$ vanishes on the boundary. If $\Gamma_{\mathrm{R}} \neq \emptyset$, boundary terms may enter and B^* can not be interpreted as (an extension of) a gradient.

5.3.2.1 Coercivity of *a*

Let $\boldsymbol{v} \in V$ be given, then using Eq. (4.13) on page 82, which is based on Korn's inequality, Lemma 3.5.3, it is

$$\begin{aligned} a(\boldsymbol{v}, \boldsymbol{v}) &= 2\nu(\mathsf{D}(\boldsymbol{v}), \mathsf{D}(\boldsymbol{v}))_0 + (\gamma \boldsymbol{v}, \boldsymbol{v})_{0,\Gamma_{\mathrm{R}}} \\ &\geq 2\nu c \|\boldsymbol{v}\|_1 + \inf_{\boldsymbol{x} \in \Gamma_{\mathrm{R}}} \gamma(\boldsymbol{x}) \; \|\boldsymbol{v}\|_{0,\Gamma_{\mathrm{R}}}^2 \\ &\geq \alpha \|\boldsymbol{v}\|_1. \end{aligned}$$

Therefore, $a(\cdot, \cdot)$ is coercive on $\boldsymbol{V}$ with a constant α, depending on the domain and ν, and according to Proposition 5.2.5 the first requirement of Theorem 5.2.2 is fulfilled. If γ is not zero and $\boldsymbol{v}$ has a nonzero trace on the Robin boundary Γ_{R} the constant α could be chosen larger with an additional dependency on the boundary coefficient γ.

5.3.2.2 Inf–sup Condition for *b*

The following proof of the inf–sup condition is from [QV99, Proposition 5.3.2]. Let $q \in Q$ be given. Split q into a constant part $\overline{q} := \frac{1}{|\Omega|} \int_\Omega q(\boldsymbol{x}) \, \mathrm{d}\boldsymbol{x}$ and the rest $\widetilde{q} := q - \overline{q} \in L_0^2(\Omega)$ which has zero mean value over Ω. According to Theorem 4.5.2 choose $\boldsymbol{v}_0 \in \mathbf{H}_0^1(\Omega)$ such that $\nabla \cdot \boldsymbol{v}_0 = \widetilde{q}$ and $\|\boldsymbol{v}_0\|_V \leq C \|\widetilde{q}\|_Q$ to yield

$$\sup_{\boldsymbol{v} \in V} \frac{b(\boldsymbol{v}, q)}{\|\boldsymbol{v}\|_V} \geq \frac{(\nabla \cdot \boldsymbol{v}_0, q)_0}{\|\boldsymbol{v}_0\|_V} = \frac{\|\widetilde{q}\|_Q^2}{\|\boldsymbol{v}_0\|_V} + \frac{(\widetilde{q}, \overline{q})_0}{\|\boldsymbol{v}_0\|_V} = \frac{\|\widetilde{q}\|_Q^2}{\|\boldsymbol{v}_0\|_V} \geq \frac{1}{C} \|\widetilde{q}\|_Q,$$

where it is used that $(\widetilde{q}, \overline{q})_0 = \overline{q} \int_\Omega \widetilde{q}(\boldsymbol{x}) \, \mathrm{d}\boldsymbol{x} = 0$. In the case of Dirichlet conditions on the entire boundary, $\Gamma_{\mathrm{D}} = \partial\Omega$, one typically chooses $Q = L_0^2(\Omega)$ and the inf–sup condition (5.19) is established by the above inequality with $\beta = 1/C$, where C is the continuity constant of the divergence operator. In case $Q = L^2(\Omega)$ it is $\|\widetilde{q}\|_Q = \|q\|_{Q/\ker B^*}$, i.e., Proposition 5.2.6 (d) is fulfilled again with $\beta = 1/C$.

If $\Gamma_{\mathrm{D}} \neq \partial\Omega$ the idea is to consider a larger Lipschitz domain $\Omega' \supsetneq \Omega$ such that $\partial\Omega \cap \partial\Omega' = \Gamma_{\mathrm{D}}$, i.e., the boundaries of the two domains share only the Dirichlet part of $\partial\Omega$, while $\Gamma_{\mathrm{R}} \cap \partial\Omega' = \emptyset$, see Fig. 5.2 for an example. Let q' be the extension

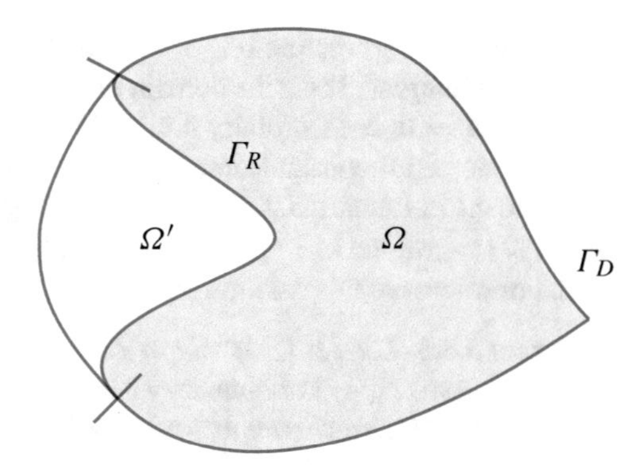

Fig. 5.2 Sketch of Ω and Ω' inspired by Figure 5.3.2 in [QV99]

of q by zero to Ω', then the above reasoning applied to q', the inclusion $\Omega \subset \Omega'$, and the triangle inequality yield

$$\frac{(\nabla \cdot \boldsymbol{v}_0, q')_{0,\Omega'}}{\|\boldsymbol{v}_0\|_{1,\Omega'}} \geq \frac{1}{C'}\left\|q' - \overline{q'}\right\|_{0,\Omega'} \geq \frac{1}{C'}\left\|q - \overline{q'}\right\|_{0,\Omega} \geq \frac{1}{C'}\left|\|q\|_{0,\Omega} - \left\|\overline{q'}\right\|_{0,\Omega}\right|,$$

with $\boldsymbol{v}_0 \in \mathbf{H}_0^1(\Omega')$ and a possibly different constant C'. The norm of $\overline{q'}$ can be bounded using the Hölder inequality, Theorem 2.3.1, as follows

$$\left\|\overline{q'}\right\|_{0,\Omega}^2 = \frac{|\Omega|}{|\Omega'|^2}\left(\int_{\Omega'} q(\boldsymbol{x})\,\mathrm{d}\boldsymbol{x}\right)^2 = \frac{|\Omega|}{|\Omega'|^2}\left(\int_{\Omega} q(\boldsymbol{x})\,\mathrm{d}\boldsymbol{x}\right)^2 \leq \frac{|\Omega|^2}{|\Omega'|^2}\|q\|_{0,\Omega}^2.$$

Combining these results and noting that $\|\boldsymbol{v}_0\|_{1,\Omega'} \geq \|\boldsymbol{v}_0\|_{1,\Omega}$ gives

$$\frac{(\nabla \cdot \boldsymbol{v}_0, q)_{0,\Omega}}{\|\boldsymbol{v}_0\|_{1,\Omega}} \geq \frac{(\nabla \cdot \boldsymbol{v}_0, q')_{0,\Omega'}}{\|\boldsymbol{v}_0\|_{1,\Omega'}} \geq \underbrace{\frac{1}{C'}\left(1 - \frac{|\Omega|}{|\Omega'|}\right)}_{:=\beta}\|q\|_{0,\Omega}$$

and therefore

$$\inf_{q \in Q} \sup_{\boldsymbol{v} \in V} \frac{b(\boldsymbol{v}, q)}{\|\boldsymbol{v}\|_V \|q\|_Q} \geq \beta,$$

i.e., the inf–sup condition (5.19).

5.3.2.3 Application of the Saddle Point Theory to the Stokes Problem

Thanks to Propositions 5.2.5 and 5.2.6 together with Theorem 5.2.2 the weak formulation (5.27) of the Stokes Problem has a unique solution, at least up to an element of $\ker B^*$ in the pressure. In the case of Dirichlet conditions on the entire boundary, a function $\boldsymbol{v} \in V = \mathbf{H}_0^1(\Omega)$ has vanishing traces and therefore its divergence integrated over the domain Ω is also zero, $(B^*1, \boldsymbol{v})_{V^* \times V} = -\int_\Omega \nabla \cdot \boldsymbol{v} = -\int_{\partial\Omega} \boldsymbol{v} \cdot \boldsymbol{n} = 0$, see Corollary 4.4.1 with a constant $\varphi = 1$. Hence, the kernel of B^* consists of all constant functions and the pressure is only unique up to some constant. As in Remark 5.2.3 one can choose $Q = L_0^2(\Omega)$ (instead of $Q = L^2(\Omega)$) which is isomorphic to $L^2(\Omega)/\ker B^*$. Within this space the pressure is unique as well. For reference this result is repeated here in detail:

Theorem 5.3.5 *Let $\Omega \subset \mathbb{R}^d$ be a Lipschitz domain together with two disjoint subsets Γ_D and Γ_R of the boundary $\partial\Omega$ which are themselves Lipschitz, such that $\overline{\Gamma_\mathrm{D}} \cup \overline{\Gamma_\mathrm{R}} = \partial\Omega$. Furthermore, let $\boldsymbol{V} := \mathbf{H}_{\Gamma_\mathrm{D}}^1(\Omega)$ and $Q = L^2(\Omega)$ (if $\Gamma_\mathrm{R} = \emptyset$, set $Q = L_0^2(\Omega)$) and*

$$\mathbf{f} \in \boldsymbol{V}^*, \qquad \mathbf{u}_\mathrm{D} \in \mathbf{H}^{1/2}(\Gamma_\mathrm{D}), \qquad \gamma \in L^\infty(\Gamma_\mathrm{R}), \qquad \mathbf{g}_\mathrm{R} \in \left(H_{00}^{1/2}(\Gamma_\mathrm{R})\right)^*,$$

where γ is nonnegative. Then there exist unique $\mathbf{u} \in \mathbf{H}^1(\Omega)$ and $q \in Q$ such that $T_{\Gamma_\mathrm{D}}\mathbf{u} = \mathbf{u}_\mathrm{D}$, i.e., $\mathbf{u} = \mathbf{u}_\mathrm{D}$ on Γ, and for all $\boldsymbol{v} \in \boldsymbol{V}$ and $q \in Q$ it is

$$\begin{cases} a(\mathbf{u}, \boldsymbol{v}) + b(\boldsymbol{v}, p) = \ell(\boldsymbol{v}), \\ \qquad\qquad b(\mathbf{u}, q) = 0, \end{cases} \tag{5.27}$$

with

$$\begin{aligned} a(\mathbf{u}, \boldsymbol{v}) &= (2\nu \mathsf{D}(\mathbf{u}), \mathsf{D}(\boldsymbol{v}))_0 + (\gamma \mathbf{u}, \boldsymbol{v})_{0,\Gamma_\mathrm{R}}, \\ b(\boldsymbol{v}, p) &= -(\nabla \cdot \boldsymbol{v}, p)_0, \\ \ell(\boldsymbol{v}) &= (\mathbf{f}, \boldsymbol{v})_{\boldsymbol{V}^* \times \boldsymbol{V}} + (\mathbf{g}_\mathrm{R}, \boldsymbol{v})_{\left(\mathbf{H}_{00}^{1/2}(\Gamma_\mathrm{R})\right)^* \times \mathbf{H}_{00}^{1/2}(\Gamma_\mathrm{R})}. \end{aligned}$$

An a priori estimate on the solutions $\mathbf{U} = \mathbf{u} - \mathbf{u}_\mathrm{D}$ and p of the Stokes equations is then (see also Eq. (5.23) and (5.24) at the end of Sect. 5.2)

$$\begin{aligned} \|\mathbf{U}\|_V &\le \frac{1}{\alpha}\|L\|_{V^*} + \left(\frac{c_a}{\alpha} + 1\right)\frac{1}{\beta}\|g\|_Q, \\ \|p\|_{Q/\ker B^*} &\le \frac{1}{\beta}\left(\frac{c_a}{\alpha} + 1\right)\|L\|_{V^*} + \frac{c_a}{\beta^2}\left(\frac{c_a}{\alpha} + 1\right)\|g\|_Q, \end{aligned}$$

with

$$\begin{aligned} \|L\|_{V^*} &= \|\mathbf{f}\|_{V^*} + c_{T_{\Gamma_\mathrm{R}}}\|\mathbf{g}_\mathrm{R}\|_{\left(H_{00}^{1/2}(\Gamma_\mathrm{R})\right)^*} + c_a c_{E_{\Gamma_\mathrm{D}}}\|\mathbf{u}_\mathrm{D}\|_{H^{1/2}(\Gamma_\mathrm{D})}, \\ \|g\|_Q &= c_{E_{\Gamma_\mathrm{D}}}\|\mathbf{u}_\mathrm{D}\|_{H^{1/2}(\Gamma_\mathrm{D})}. \end{aligned}$$

The constants $c_{T_{\Gamma_\mathrm{R}}}$ and $c_{E_{\Gamma_\mathrm{D}}}$ are the continuity constants of the operators T_{Γ_R} and E_{Γ_D}, respectively, see also Theorems 4.2.4 and 4.3.4. An estimate for $\mathbf{u}$ then is

$$\begin{aligned}\|\mathbf{u}\|_1 &\le \frac{1}{\alpha}\|L\|_{V^*} + \left(\frac{c_a}{\alpha}+1\right)\frac{1}{\beta}\|g\|_Q + c_{E_{\Gamma_\mathrm{D}}}\|\mathbf{u}_\mathrm{D}\|_{H^{1/2}(\Gamma_\mathrm{D})}\\ &= \frac{\tilde{c}_L}{\alpha} + \left(1+\frac{c_a}{\alpha}\right)\left(1+\frac{1}{\beta}\right)c_{E_{\Gamma_\mathrm{D}}}\|\mathbf{u}_\mathrm{D}\|_{H^{1/2}(\Gamma_\mathrm{D})}\end{aligned} \tag{5.28}$$

with

$$\tilde{c}_L = \|\mathbf{f}\|_{V^*} + c_{T_{\Gamma_\mathrm{R}}}\|\mathbf{g}_\mathrm{R}\|_{\left(H^{1/2}_{00}(\Gamma_\mathrm{R})\right)^*}.$$

In terms of $\tilde{c}_L$ the bound on the pressure can be expressed as

$$\|p\|_{Q/\ker B^*} \le \frac{1}{\beta}\left(\frac{c_a}{\alpha}+1\right)\tilde{c}_L + \frac{c_a}{\beta}\left(\frac{c_a}{\alpha}+1\right)\left(1+\frac{1}{\beta}\right)c_{E_{\Gamma_\mathrm{D}}}\|\mathbf{u}_\mathrm{D}\|_{H^{1/2}(\Gamma_\mathrm{D})}. \tag{5.29}$$

Note that the right-hand sides in both Eqs. (5.28) and (5.29) above include inverses of the viscosity ν, i.e., it is large for small ν.

Remark 5.3.6 It is assumed that the divergence of $\mathbf{u}$ vanishes, i.e., the second equation in the Stokes problem (5.25) reads $\nabla\cdot\mathbf{u} = 0$. If a nonzero divergence is prescribed via $\nabla\cdot\mathbf{u} = g_\mathrm{div} \in L^2(\Omega)$, the norm of g in the above theorem changes to $\|g\|_Q = c_{E_{\Gamma_\mathrm{D}}}\|\mathbf{u}_\mathrm{D}\|_{H^{1/2}(\Gamma_\mathrm{D})} + \|g_\mathrm{div}\|_Q$. While in fluid dynamics the divergence is typically zero, this can be helpful to construct particular functions in $\mathbf{H}^1(\Omega)$ with a prescribed divergence g_div and a continuous dependence on g_div where the boundary conditions and all data (including ν, γ, $\mathbf{f}$) are free to choose. In Theorem 4.5.2 such a construction is available only in $\mathbf{H}^1_0(\Omega)$ and $g_\mathrm{div} \in L^2_0(\Omega)$.

5.3.3 View as an Operator on a Part of the Boundary

Very similar to the Laplace case one can view the above theorem as the definition of an operator which maps the data $\mathbf{f}$, $\mathbf{u}_\mathrm{D}$, and $\mathbf{g}_\mathrm{R}$ to a solution $(\mathbf{u}, p)$. And again setting all but one of these to zero results in a linear operator. Of special interest in the coupling of Stokes and Darcy equations is the case where data is only given at a particular part of the boundary. For that purpose let $\partial\Omega$ be split into three disjoint parts $\partial\Omega = \overline{\Gamma_\mathrm{D}} \cup \overline{\Gamma_\mathrm{R}} \cup \overline{\Gamma}$. For simplicity assume that $\Gamma_\mathrm{R} \ne \emptyset$, otherwise the adjustments mentioned above apply. Then the operators introduced below are well defined according to Theorem 5.3.5.

The presentation is not as detailed as in the case of the Laplace equation in Sect. 5.1.3.

5.3.3.1 Dirichlet Operator on Γ

Denote the space $T_\Gamma\left(\mathbf{H}^1_{\Gamma_\mathrm{D}}(\Omega)\right)$ as $\mathbf{H}^{1/2}_{\Gamma_\mathrm{D}}(\Gamma)$.

Proposition 5.3.7 *Define the operator* $K_\mathrm{D} : \mathbf{H}^{1/2}_{\Gamma_\mathrm{D}}(\Gamma) \to \mathbf{H}^1_{\Gamma_\mathrm{D}}(\Omega) \times L^2(\Omega)$*, where for all* $\boldsymbol{v} \in \mathbf{H}^1_{\Gamma_\mathrm{D}\cup\Gamma}(\Omega)$ *and all* $q \in L^2(\Omega)$ *the pair* $(\mathbf{u}, p) = K_\mathrm{D}(\lambda)$ *solves*

$$\begin{cases} (2\nu \mathsf{D}(\mathbf{u}), \mathsf{D}(\boldsymbol{v}))_0 + (\gamma \mathbf{u}, \boldsymbol{v})_{0,\Gamma_\mathrm{R}} - (\nabla \cdot \boldsymbol{v}, p)_0 = 0, \\ (\nabla \cdot \mathbf{u}, q)_0 = 0 \end{cases} \tag{5.30}$$

and $T_\Gamma \mathbf{u} = \lambda$*. This operator is linear, continuous and positive, i.e., there exist constants* $c_{K_\mathrm{D}} > 0$ *and* $\alpha_{K_\mathrm{D}} > 0$ *such that*

$$\|K_\mathrm{D}(\lambda)\|_{\mathbf{H}^1(\Omega)\times L^2(\Omega)} \le c_{K_\mathrm{D}} \|\lambda\|_{\mathbf{H}^{1/2}_{\Gamma_\mathrm{D}}(\Gamma)},$$

$$\|K_\mathrm{D}(\lambda)\|_{\mathbf{H}^1(\Omega)\times L^2(\Omega)} \ge \alpha_{K_\mathrm{D}} \|\lambda\|_{\mathbf{H}^{1/2}_{\Gamma_\mathrm{D}}(\Gamma)}.$$

Proof The linearity of K_D follows directly from the linearity of Eq. (5.30) and of the extension operator E_Γ. Continuity is shown in Sect. 5.3.2 with a constant depending on the inverse of the coercivity constant α_a, the continuity constant of c_a of the bilinear form a, and the inverse of the inf–sup constant β. In particular, Eqs. (5.28) and (5.29) apply with $\tilde{c}_L = 0$. The continuity of the trace operator T_Γ yields

$$\|\lambda\|_{\mathbf{H}^{1/2}_{\Gamma_\mathrm{D}}(\Gamma)} = \|T_\Gamma(K_\mathrm{D}(\lambda))\|_{\mathbf{H}^{1/2}_{\Gamma_\mathrm{D}}(\Gamma)} \le c_{T_\Gamma} \|K_\mathrm{D}(\lambda)\|_{\mathbf{H}^1(\Omega)\times L2(\Omega)}$$

and therefore the positivity of K_D with $\alpha = c_{T_\Gamma}^{-1}$. □

The operator K_D restricted to its first component (the velocity) is another linear continuous right inverse of T_Γ on $\mathbf{H}^{1/2}_{\Gamma_\mathrm{D}}(\Gamma)$, see also Theorem 4.3.4 and the subsequent discussion.

5.3.3.2 Robin Operator on Γ

Consider now the case where only Robin data is prescribed on Γ. Then define the Robin operator for the Stokes equations as follows:

Proposition 5.3.8 *Define the operators* $K_\mathrm{R} : \left(\mathbf{H}^{1/2}_{\Gamma_\mathrm{D}}(\Gamma)\right)^* \to \mathbf{H}^1_{\Gamma_\mathrm{D}}(\Omega) \times L^2(\Omega)$*, where for all* $\boldsymbol{v} \in \mathbf{H}^1_{\Gamma_\mathrm{D}}(\Omega)$ *and all* $q \in L^2(\Omega)$ *the pair* $(\mathbf{u}, p) = K_\mathrm{R}(\mu)$ *solves*

$$\begin{cases} (2\nu \mathsf{D}(\mathbf{u}), \mathsf{D}(\boldsymbol{v}))_0 + (\gamma \mathbf{u}, \boldsymbol{v})_{0,\Gamma_\mathrm{R}\cup\Gamma} - (\nabla \cdot \boldsymbol{v}, p)_0 = (\mu, \boldsymbol{v})_{\left(\mathbf{H}^{1/2}_{\Gamma_\mathrm{D}}(\Gamma)\right)^* \times \mathbf{H}^{1/2}_{\Gamma_\mathrm{D}}(\Gamma)}, \\ (\nabla \cdot \mathbf{u}, q)_0 = 0. \end{cases} \tag{5.31}$$

This operator is linear, continuous and positive, i.e., there exist constants $c_{K_\mathrm{R}} > 0$ *and* $\alpha_{K_\mathrm{R}} > 0$ *such that*

$$\|K_\mathrm{R}(\mu)\|_{\mathbf{H}^1(\Omega)\times L^2(\Omega)} \leq c_{K_\mathrm{R}} \|\mu\|_{\left(\mathbf{H}^{1/2}_{\Gamma_\mathrm{D}}(\Gamma)\right)^*},$$

$$\|K_\mathrm{R}(\mu)\|_{\mathbf{H}^1(\Omega)\times L^2(\Omega)} \geq \alpha_{K_\mathrm{R}} \|\mu\|_{\left(\mathbf{H}^{1/2}_{\Gamma_\mathrm{D}}(\Gamma)\right)^*}.$$

Proof Since Eq. (5.31) are linear, so is K_R. As before, continuity is shown in Sect. 5.3.2 with a constant depending on the inverse of the coercivity constant α_a, the continuity constant of c_a of the bilinear form a, and the inverse of the inf–sup constant β. More precisely, here Eqs. (5.28) and (5.29) reduce to

$$\|\mathbf{u}\|_1 \leq \frac{c_{T_\Gamma}}{\alpha} \|\mu\|_{\left(\mathbf{H}^{1/2}_{\Gamma_\mathrm{D}}(\Gamma)\right)^*} \qquad \text{and} \qquad \|p\|_0 \leq \frac{1}{\beta}\left(\frac{c_a}{\alpha}+1\right) c_{T_\Gamma} \|\mu\|_{\left(\mathbf{H}^{1/2}_{\Gamma_\mathrm{D}}(\Gamma)\right)^*}.$$

To show positivity, for all $\lambda \in \mathbf{H}^{1/2}_{\Gamma_\mathrm{D}}(\Gamma)$ denote $(\mathbf{u}_\lambda, p_\lambda) = K_\mathrm{D}(\lambda)$ and $(\mathbf{u}_\mu, p_\mu) = K_\mathrm{R}(\mu)$ for all $\mu \in \left(\mathbf{H}^{1/2}_{\Gamma_\mathrm{D}}(\Gamma)\right)^*$, then it is

$$\begin{aligned}
(\mu, \lambda)_{\left(\mathbf{H}^{1/2}_{\Gamma_\mathrm{D}}(\Gamma)\right)^* \times \mathbf{H}^{1/2}_{\Gamma_\mathrm{D}}(\Gamma)} &= (\mu, T_\Gamma \mathbf{u}_\lambda)_{\left(\mathbf{H}^{1/2}_{\Gamma_\mathrm{D}}(\Gamma)\right)^* \times \mathbf{H}^{1/2}_{\Gamma_\mathrm{D}}(\Gamma)} \\
&= a(\mathbf{u}_\mu, \mathbf{u}_\lambda) \\
&\leq c_a c_{K_\mathrm{D}} \left\|\mathbf{u}_\mu\right\|_1 \|\lambda\|_{\mathbf{H}^{1/2}_{\Gamma_\mathrm{D}}(\Gamma)} \\
&\leq c_a c_{K_\mathrm{D}} \|K_\mathrm{R}(\mu)\|_{\mathbf{H}^1(\Omega)\times L^2(\Omega)} \|\lambda\|_{\mathbf{H}^{1/2}_{\Gamma_\mathrm{D}}(\Gamma)}
\end{aligned}$$

and hence

$$\begin{aligned}
\|\mu\|_{\left(\mathbf{H}^{1/2}_{\Gamma_\mathrm{D}}(\Gamma)\right)^*} &= \sup_{\lambda \in \mathbf{H}^{1/2}_{\Gamma_\mathrm{D}}(\Gamma)} \frac{(\mu, \lambda)_{\left(\mathbf{H}^{1/2}_{\Gamma_\mathrm{D}}(\Gamma)\right)^* \times \mathbf{H}^{1/2}_{\Gamma_\mathrm{D}}(\Gamma)}}{\|\lambda\|_{\mathbf{H}^{1/2}_{\Gamma_\mathrm{D}}(\Gamma)}} \\
&\leq c_a c_{K_\mathrm{D}} \|K_\mathrm{R}(\mu)\|_{\mathbf{H}^1(\Omega)\times L^2(\Omega)},
\end{aligned}$$

i.e., $\alpha_{K_\mathrm{R}} = 1/(c_a c_{K_\mathrm{D}})$. □

Just as the corresponding operator in the Laplace case, Sect. 5.1.3, the first component of the Robin operator K_R is an extension from $\left(\mathbf{H}^{1/2}_{\Gamma_\mathrm{D}}(\Gamma)\right)^*$ to $\mathbf{H}^1_{\Gamma_\mathrm{D}}(\Omega)$ and a left inverse T^R_Γ can be defined:

Definition 5.3.9 Let $(\mathbf{u}, p) \in \mathbf{H}^1_{\Gamma_\mathrm{D}} \times L^2(\Omega)$ be given such that for all $(\boldsymbol{v}, q) \in \mathbf{H}^1_{\Gamma_\mathrm{D}\cup\Gamma} \times L^2(\Omega)$ it is

$$\begin{cases}
(2\nu \mathsf{D}(\mathbf{u}), \mathsf{D}(\boldsymbol{v}))_0 + (\gamma \mathbf{u}, \boldsymbol{v})_{0,\Gamma_\mathrm{R}} - (\nabla \cdot \boldsymbol{v}, p)_0 = 0, \\
\qquad\qquad\qquad\qquad\qquad\qquad (\nabla \cdot \mathbf{u}, q)_0 = 0.
\end{cases}$$

Then its Robin data $T_\Gamma^{\mathrm{R}}(\mathbf{u}, p) \in \left(\mathbf{H}^{1/2}_{\Gamma_\mathrm{D}}(\Gamma)\right)^*$ on Γ is defined to be

$$\left(T_\Gamma^{\mathrm{R}}(\mathbf{u}, p), \xi\right)_{\left(\mathbf{H}^{1/2}_{\Gamma_\mathrm{D}}(\Gamma)\right)^* \times \mathbf{H}^{1/2}_{\Gamma_\mathrm{D}}(\Gamma)}$$
$$:= (2\nu \mathsf{D}(\mathbf{u}), \mathsf{D}(E_\Gamma \xi))_0 + (\gamma \mathbf{u}, E_\Gamma \xi)_{0,\Gamma_\mathrm{R} \cup \Gamma} - (\nabla \cdot (E_\Gamma \xi), p)_0,$$

where $E_\Gamma \xi$ is an extension of (each component of) $\xi \in \mathbf{H}^{1/2}_{\Gamma_\mathrm{D}}(\Gamma)$, see Theorem 4.3.4 and in its subsequent discussion.

The above definition is well posed in the sense that it does not depend on the particular extension E_Γ used. Additionally, it is a generalization of the operator $(\mathbf{u}, p) \mapsto \mathsf{T}(\mathbf{u}, p) \cdot \boldsymbol{n} + \gamma \mathbf{u}$ on the Robin boundary Γ_R.

5.3.3.3 A Connection of the Dirichlet and Robin Operator on Γ

With the same reasoning as in the Laplace case the following identities hold:

$$T_\Gamma^{\mathrm{R}} \circ K_\mathrm{D} \circ T_\Gamma \circ K_\mathrm{R} = \mathrm{id} \qquad \text{in } \left(\mathbf{H}^{1/2}_{\Gamma_\mathrm{D}}(\Gamma)\right)^*,$$
$$T_\Gamma \circ K_\mathrm{R} \circ T_\Gamma^{\mathrm{R}} \circ K_\mathrm{D} = \mathrm{id} \qquad \text{in } \mathbf{H}^{1/2}_{\Gamma_\mathrm{D}}(\Gamma).$$

In fact, the notation in this section is chosen such that these equations are exactly the same as in the Laplace case.

Chapter 6
Stokes–Darcy Equations

6.1 The Setting

Let $\Omega \subset \mathbb{R}^d$ be a Lipschitz domain split into two disjoint nonempty subdomains Ω_p and Ω_f which are Lipschitz, too. The index $_\mathrm{p}$ refers to the Darcy subdomain where a porous medium is modeled, while the index $_\mathrm{f}$ refers to the free flow domain with a Stokes model.

The intersection $\Gamma_\mathrm{I} := \overline{\Omega_\mathrm{p}} \cap \overline{\Omega_\mathrm{f}}$ is called interface in the following. The boundary of the domain Ω is split into four relatively open parts $\Gamma_{\mathrm{f,N}} \subset \partial\Omega_\mathrm{f}$, $\Gamma_{\mathrm{f,D}} \subset \partial\Omega_\mathrm{f}$, $\Gamma_{\mathrm{p,N}} \subset \partial\Omega_\mathrm{p}$, and $\Gamma_{\mathrm{p,D}} \subset \partial\Omega_\mathrm{p}$ where Dirichlet (essential) and Neumann (natural) boundary conditions, respectively, are prescribed. Further, let $\boldsymbol{n}$ be the unit normal vector pointing outward of Ω on $\partial\Omega$ and pointing outward of Ω_f on the interface Γ_I. Figure 6.2 shows such a setting as an example.

Consider a Darcy problem in Ω_p and a Stokes problem in Ω_f,

$$-\nabla \cdot \mathsf{T}(\boldsymbol{u}_\mathrm{f}, p_\mathrm{f}) = \mathbf{f}_\mathrm{f} \quad \text{in } \Omega_\mathrm{f}, \tag{6.1a}$$

$$\nabla \cdot \boldsymbol{u}_\mathrm{f} = 0 \quad \text{in } \Omega_\mathrm{f}, \tag{6.1b}$$

$$\boldsymbol{u}_\mathrm{p} + \mathsf{K}\nabla\varphi_\mathrm{p} = 0 \quad \text{in } \Omega_\mathrm{p}, \tag{6.1c}$$

$$\nabla \cdot \boldsymbol{u}_\mathrm{p} = f_\mathrm{p} \quad \text{in } \Omega_\mathrm{p}, \tag{6.1d}$$

together with boundary conditions on the outer boundary $\partial\Omega$

$$\boldsymbol{u}_\mathrm{f} = \mathbf{u}_\mathrm{b} \text{ on } \Gamma_{\mathrm{f,D}}, \tag{6.2a}$$

$$\mathsf{T}(\boldsymbol{u}_\mathrm{f}, p_\mathrm{f}) \cdot \boldsymbol{n} = \mathbf{g}_\mathrm{f} \text{ on } \Gamma_{\mathrm{f,N}}, \tag{6.2b}$$

$$\boldsymbol{u}_\mathrm{p} \cdot \boldsymbol{n} = g_\mathrm{p} \text{ on } \Gamma_{\mathrm{p,N}}, \tag{6.2c}$$

$$\varphi_\mathrm{p} = \varphi_\mathrm{b} \text{ on } \Gamma_{\mathrm{p,D}}. \tag{6.2d}$$

U. Wilbrandt, *Stokes–Darcy Equations*, Advances in Mathematical Fluid Mechanics,
https://doi.org/10.1007/978-3-030-02904-3_6

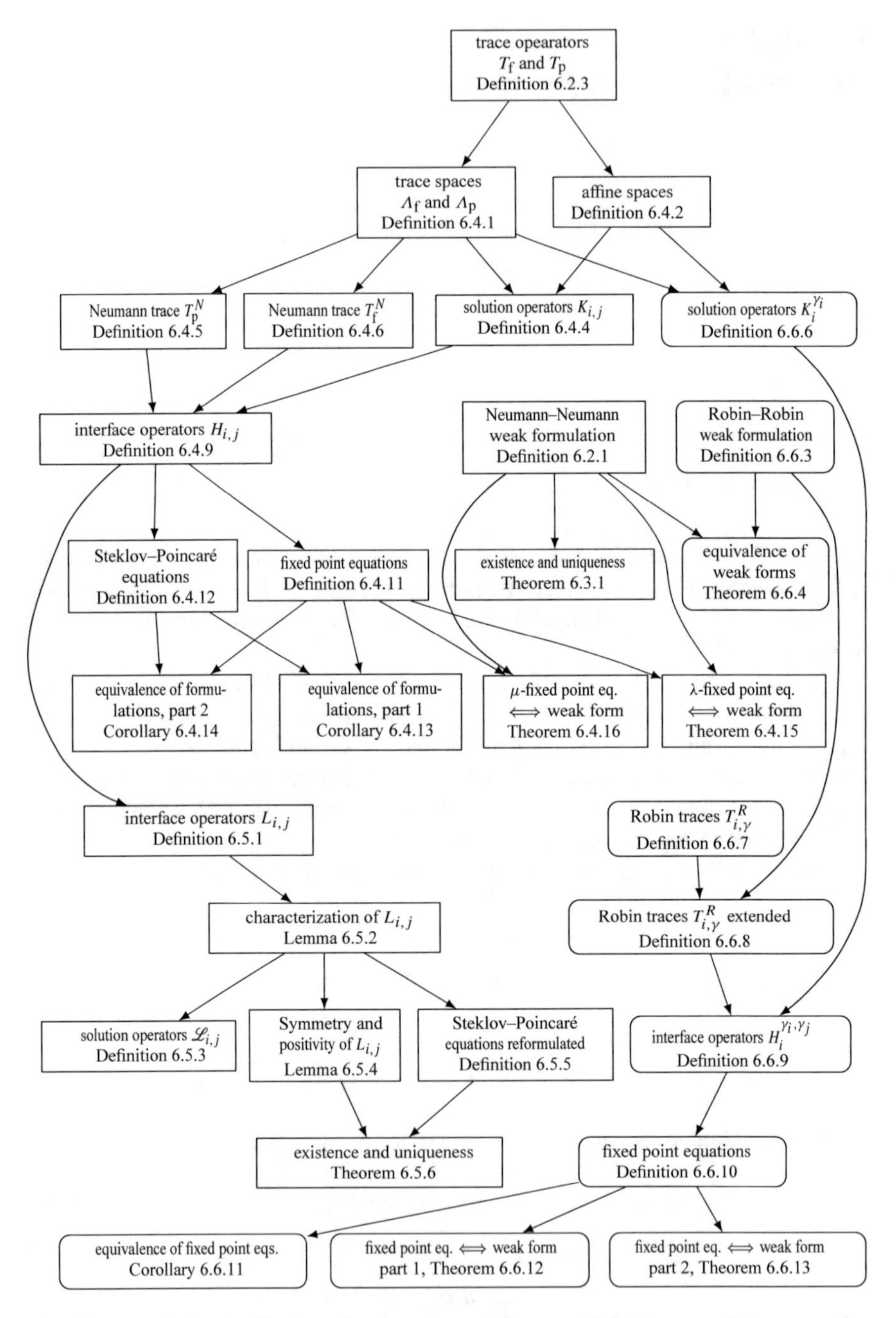

Fig. 6.1 A graph showing the dependencies among all theorems, definitions, corollaries, as well as propositions in this chapter. Rounded corners indicate items related to the Robin–Robin approach from Sect. 6.6. Note that implicit dependencies are not always shown

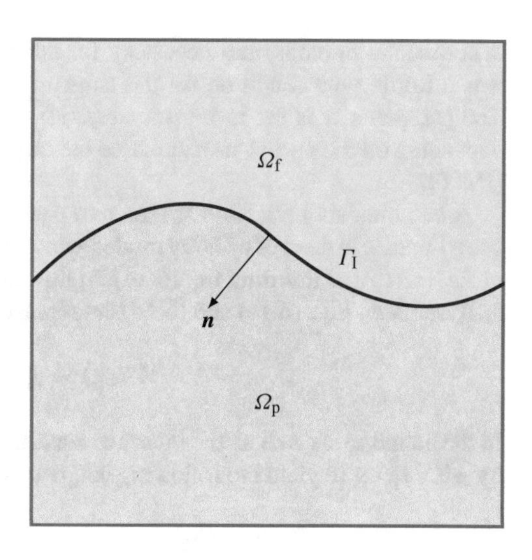

Fig. 6.2 Sketch of the general setting

Here $\boldsymbol{u}_\mathrm{f} : \Omega_\mathrm{f} \to \mathbb{R}^d$ and $\boldsymbol{u}_\mathrm{p} : \Omega_\mathrm{p} \to \mathbb{R}^d$ are vector-valued functions, called the Stokes velocity and Darcy velocity (also hydraulic discharge), respectively. The other two unknowns $p_\mathrm{f} : \Omega_\mathrm{f} \to \mathbb{R}$ and $\varphi_\mathrm{p} : \Omega_\mathrm{p} \to \mathbb{R}$ are scalar functions which are called Stokes and Darcy pressure, respectively. Often φ_p is also referred to as hydraulic head. The vector-valued function $\mathbf{f}_\mathrm{f} : \Omega_\mathrm{f} \to \mathbb{R}^d$ and the scalar $f_\mathrm{p} : \Omega_\mathrm{p} \to \mathbb{R}$ are the right-hand sides and K is the hydraulic conductivity tensor, which is assumed to be symmetric and positive definite. Furthermore, as in the previous chapter, $\mathsf{T}(\mathbf{u}, p) = 2\nu\mathsf{D}(\mathbf{u}) - p\,\mathrm{id}$ is the Cauchy stress tensor, $\nu > 0$ the kinematic viscosity, and $\mathsf{D}(\mathbf{u}) = \frac{1}{2}\left(\nabla\mathbf{u} + \nabla\mathbf{u}^T\right)$ the symmetric part of $\nabla\mathbf{u}$ or the deformation tensor. The functions prescribing the Dirichlet and Neumann boundary conditions are given. Other types such as Robin or periodic boundary conditions are possible with the usual modifications.

To complete the description of the general setting these equations are equipped with conditions on the interface Γ_I:

$$\boldsymbol{u}_\mathrm{f} \cdot \boldsymbol{n} = \boldsymbol{u}_\mathrm{p} \cdot \boldsymbol{n} \qquad \text{on } \Gamma_\mathrm{I}, \tag{6.3a}$$

$$-\boldsymbol{n} \cdot \mathsf{T}(\boldsymbol{u}_\mathrm{f}, p_\mathrm{f}) \cdot \boldsymbol{n} = g\varphi_\mathrm{p} \qquad \text{on } \Gamma_\mathrm{I}, \tag{6.3b}$$

$$\boldsymbol{u}_\mathrm{f} \cdot \boldsymbol{\tau}_i + \alpha\boldsymbol{\tau}_i \cdot \mathsf{T}(\boldsymbol{u}_\mathrm{f}, p_\mathrm{f}) \cdot \boldsymbol{n} = 0 \qquad \text{on } \Gamma_\mathrm{I}, i = 1, \dots, d-1. \tag{6.3c}$$

Here $(\boldsymbol{\tau}_i)_{i=1,\dots d-1}$ is an orthonormal basis of the tangential space on Γ_I and g the gravitational acceleration which is omitted, i.e., assumed to be one in the following.

The first interface condition (6.3a) describes the mass conservation and the second equation (6.3b) represents the balance of momentum. The third interface condition (6.3c) is called Beavers–Joseph–Saffman[1] condition, based on

[1] Sometimes also called Beavers–Joseph–Saffman–Jones condition.

experimental findings, and necessary for the problem to be well posed. Note that it is a Robin type condition for the tangential components of the Stokes velocity. The parameter α is set to be $\alpha = \alpha_0\sqrt{(\tau_i \cdot \mathsf{K} \cdot \tau_i)/(\nu g)}$, where α_0 is a constant depending on the porous medium. See the original works [Jon73, Saf71] as well as [JM00].

A common simplification for the two equations in the porous medium Ω_p is to use a Laplace instead of a Darcy model, which is derived from taking the divergence of Eq. (6.1c) and inserting Eq. (6.1d). In this monograph, only this simplification is analyzed, i.e., Eqs. (6.1c) and (6.1d) are replaced by

$$-\nabla \cdot \big(\mathsf{K}\nabla\varphi_\mathrm{p}\big) = f_\mathrm{p} \qquad \text{in } \Omega_\mathrm{p}. \tag{6.4}$$

In the boundary as well as the interface conditions the normal flux $\boldsymbol{u}_\mathrm{p} \cdot \boldsymbol{n}$ is replaced by $-\mathsf{K}\nabla\varphi_\mathrm{p} \cdot \boldsymbol{n}$ to yield the following coupled system:

$$\left\{\begin{aligned}
-\nabla \cdot \mathsf{T}(\boldsymbol{u}_\mathrm{f}, p_\mathrm{f}) &= \mathbf{f}_\mathrm{f} && \text{in } \Omega_\mathrm{f},\\
\nabla \cdot \boldsymbol{u}_\mathrm{f} &= 0 && \text{in } \Omega_\mathrm{f},\\
-\nabla \cdot \big(\mathsf{K}\nabla\varphi_\mathrm{p}\big) &= f_\mathrm{p} && \text{in } \Omega_\mathrm{p},\\
\boldsymbol{u}_\mathrm{f} &= \mathbf{u}_\mathrm{b} && \text{on } \Gamma_{\mathrm{f,D}},\\
\mathsf{T}(\boldsymbol{u}_\mathrm{f}, p_\mathrm{f}) \cdot \boldsymbol{n} &= \mathbf{g}_\mathrm{f} && \text{on } \Gamma_{\mathrm{f,N}},\\
-\mathsf{K}\nabla\varphi_\mathrm{p} \cdot \boldsymbol{n} &= g_\mathrm{p} && \text{on } \Gamma_{\mathrm{p,N}},\\
\varphi_\mathrm{p} &= \varphi_\mathrm{b} && \text{on } \Gamma_{\mathrm{p,D}},\\
\boldsymbol{u}_\mathrm{f} \cdot \boldsymbol{n} &= -\mathsf{K}\nabla\varphi_\mathrm{p} \cdot \boldsymbol{n} && \text{on } \Gamma_\mathrm{I},\\
-\boldsymbol{n} \cdot \mathsf{T}(\boldsymbol{u}_\mathrm{f}, p_\mathrm{f}) \cdot \boldsymbol{n} &= \varphi_\mathrm{p} && \text{on } \Gamma_\mathrm{I},\\
\boldsymbol{u}_\mathrm{f} \cdot \boldsymbol{\tau}_i + \alpha\boldsymbol{\tau}_i \cdot \mathsf{T}(\boldsymbol{u}_\mathrm{f}, p_\mathrm{f}) \cdot \boldsymbol{n} &= 0 && \text{on } \Gamma_\mathrm{I}, i = 1, \ldots, d-1.
\end{aligned}\right. \tag{6.5}$$

This model is analyzed in detail in this chapter. Before weak formulations together with existence and uniqueness results are shown, some possible extensions to this model are mentioned.

6.1.1 Possible Extensions of the Model

The coupled system (6.5) consists of rather simple models but already proves to be numerically challenging. Various extensions and modifications are studied in the literature.

In the free flow domain, the Stokes equations describe a slow incompressible flow. This model is often replaced by the nonlinear Navier–Stokes equations, see for example Section 6 in [DQ09] or the shallow water equations, [DMQ02]. The mixed Darcy equations (6.1c) and (6.1d) can be used instead of the simpler Laplace formulation (6.4) used in this work, see [LSY02, MMS15, GS07]. The porous medium is modeled as fully saturated in the equations above, a two-phase approach is considered for example in [RGM15]. The Brinkman model combines both Stokes and Darcy and can therefore be used in either subdomain (e.g., [Ang11]) or as a unified approach, where coefficients in the partial differential equations determine the subdomains and no particular interface conditions are needed, see for example [BC09]. The system (6.5) presented here is stationary, for the time dependent case, see [DZ11, CGHW14, RM14]. The interface conditions are in fact a simplification of the more difficult Beavers–Joseph condition, see for example [JM00, CGHW10]. Furthermore, the Stokes–Darcy system can be extended by a transport equation, [VY09].

6.2 Weak Formulation

The weak formulation is derived as in Sects. 5.1.1 and 5.3.1: multiplication with a test function (denoted by $\boldsymbol{v}$, q, and ψ) and integration over the respective subdomain, followed by an integration by parts and insertion of boundary and interface conditions. The involved spaces are

$$\begin{aligned} \boldsymbol{V}_\mathrm{f} &= \mathbf{H}^1_{\Gamma_\mathrm{f,D}}(\Omega_\mathrm{f}), \\ Q_\mathrm{f} &= L^2(\Omega_\mathrm{f}), \\ Q_\mathrm{p} &= H^1_{\Gamma_\mathrm{p,D}}(\Omega_\mathrm{p}), \end{aligned} \tag{6.6}$$

together with the standard norms:

$$\|\boldsymbol{v}\|_{\boldsymbol{V}_\mathrm{f}} := \|\boldsymbol{v}\|_{1,\Omega_\mathrm{f}}, \qquad \|\mathrm{p}\|_{Q_\mathrm{f}} := \|p\|_{0,\Omega_\mathrm{f}}, \qquad \|\psi\|_{Q_\mathrm{p}} := \|\psi\|_{1,\Omega_\mathrm{p}}.$$

The term $(\mathsf{T}(\boldsymbol{u}_\mathrm{f}, p_\mathrm{f}) \cdot \boldsymbol{n}, \boldsymbol{v})_{0,\Gamma_\mathrm{I}}$ which enters the equations after integration by parts in the Stokes subdomain, is decomposed in normal and tangential parts as follows

$$\begin{aligned} (\mathsf{T}(\boldsymbol{u}_\mathrm{f}, p_\mathrm{f}) \cdot \boldsymbol{n}, \boldsymbol{v})_{0,\Gamma_\mathrm{I}} &= (\boldsymbol{n} \cdot \mathsf{T}(\boldsymbol{u}_\mathrm{f}, p_\mathrm{f}) \cdot \boldsymbol{n}, \boldsymbol{v} \cdot \boldsymbol{n})_{0,\Gamma_\mathrm{I}} \\ &\quad + \sum_{i=1}^{d-1} (\boldsymbol{\tau}_i \cdot \mathsf{T}(\boldsymbol{u}_\mathrm{f}, p_\mathrm{f}) \cdot \boldsymbol{n}, \boldsymbol{v} \cdot \boldsymbol{\tau}_i)_{0,\Gamma_\mathrm{I}}, \end{aligned}$$

where the identity $\boldsymbol{v} = (\boldsymbol{v} \cdot \boldsymbol{n})\boldsymbol{n} + \sum_{i=1}^{d-1} (\boldsymbol{v} \cdot \boldsymbol{\tau}_i)\boldsymbol{\tau}_i$ is employed. In the two terms above the conservation of momentum (6.3b) and the Beavers–Joseph–Saffman condition (6.3c) are inserted. The mass conservation (6.3a) replaces $\mathsf{K}\nabla\varphi_\mathrm{p} \cdot \boldsymbol{n}$ in the term $\left(\mathsf{K}\nabla\varphi_\mathrm{p} \cdot \boldsymbol{n}, \psi\right)_{0,\Gamma_\mathrm{I}}$ which enters due to integration by parts in Ω_p. This leads to the equations

$$
\begin{aligned}
(2\nu \mathsf{D}(\mathbf{u}), \mathsf{D}(\boldsymbol{v}))_{0,\Omega_\mathrm{f}} &+ \frac{1}{\alpha}\sum_{i=1}^{d-1}(\mathbf{u}\cdot\boldsymbol{\tau}_i, \boldsymbol{v}\cdot\boldsymbol{\tau}_i)_{0,\Gamma_\mathrm{I}} \\
-(\nabla\cdot\boldsymbol{v}, p)_{0,\Omega_\mathrm{f}} + \left(\varphi_\mathrm{p}, \boldsymbol{v}\cdot\boldsymbol{n}\right)_{0,\Gamma_\mathrm{I}} &= (\mathbf{f}_\mathrm{f}, \boldsymbol{v})_0 + (\mathbf{g}_\mathrm{f}, \boldsymbol{v})_{0,\Gamma_{\mathrm{f,N}}}, \\
(\nabla\cdot\boldsymbol{u}_\mathrm{f}, q)_{0,\Omega_\mathrm{f}} &= 0, \\
(\mathsf{K}\nabla\varphi, \nabla\psi)_{0,\Omega_\mathrm{p}} - (\boldsymbol{u}_\mathrm{f}\cdot\boldsymbol{n}, \psi)_{0,\Gamma_\mathrm{I}} &= \left(f_\mathrm{p}, \psi\right)_0 - \left(g_\mathrm{p}, \psi\right)_{0,\Gamma_{\mathrm{p,N}}}.
\end{aligned}
$$

The terms on the right-hand sides can be extended to respective duals as before in Sects. 5.1.1 and 5.3.1. The following definition additionally groups the terms above and serves as a summary and reference. Since in each subdomain a Neumann-type problem is posed, this approach is referred to as the Neumann–Neumann weak formulation. Another suggestion with Robin type problems is introduced in Sect. 6.6.

Definition 6.2.1 (Neumann–Neumann Weak Formulation) With the spaces from Eq. (6.6), let $\nu > 0$, $\alpha > 0$, $\mathsf{K} \in \left(L^\infty(\Omega_\mathrm{p})\right)^{d\times d}$ positive definite, as well as

$$
\mathbf{f}_\mathrm{f} \in \boldsymbol{V}_\mathrm{f}^*, \qquad f_\mathrm{p} \in Q_\mathrm{p}^*, \qquad \mathbf{g}_\mathrm{f} \in \left(T_{\Gamma_{\mathrm{f,N}}}(\boldsymbol{V}_\mathrm{f})\right)^*, \qquad g_\mathrm{p} \in \left(T_{\Gamma_{\mathrm{p,N}}}(Q_\mathrm{p})\right)^*
$$

and

$$
\mathbf{u}_\mathrm{b} \in \mathbf{H}^{1/2}(\Gamma_{\mathrm{f,D}}), \qquad \varphi_\mathrm{b} \in H^{1/2}(\Gamma_{\mathrm{p,D}}),
$$

be given. Define the bilinear and linear forms $a_\mathrm{f} : \mathbf{H}^1(\Omega_\mathrm{f}) \times \mathbf{H}^1(\Omega_\mathrm{f}) \to \mathbb{R}$, $b_\mathrm{f} : \mathbf{H}^1(\Omega_\mathrm{f}) \times L^2(\Omega_\mathrm{f}) \to \mathbb{R}$, $a_\mathrm{p} : H^1(\Omega_\mathrm{p}) \times H^1(\Omega_\mathrm{p}) \to \mathbb{R}$, $\ell_\mathrm{f} : \boldsymbol{V}_\mathrm{f} \to \mathbb{R}$, and $\ell_\mathrm{p} : Q_\mathrm{p} \to \mathbb{R}$, as

$$
\begin{aligned}
a_\mathrm{f}(\mathbf{u}, \boldsymbol{v}) &= (2\nu \mathsf{D}(\mathbf{u}), \mathsf{D}(\boldsymbol{v}))_{0,\Omega_\mathrm{f}} + \frac{1}{\alpha}\sum_{i=1}^{d-1}(\mathbf{u}\cdot\boldsymbol{\tau}_i, \boldsymbol{v}\cdot\boldsymbol{\tau}_i)_{0,\Gamma_\mathrm{I}}, \\
b_\mathrm{f}(\boldsymbol{v}, p) &= -(\nabla\cdot\boldsymbol{v}, p)_{0,\Omega_\mathrm{f}}, \\
a_\mathrm{p}(\varphi, \psi) &= (\mathsf{K}\nabla\varphi, \nabla\psi)_{0,\Omega_\mathrm{p}}, \\
\ell_\mathrm{f}(\boldsymbol{v}) &= (\mathbf{f}_\mathrm{f}, \boldsymbol{v})_{\boldsymbol{V}_\mathrm{f}^*\times\boldsymbol{V}_\mathrm{f}} + (\mathbf{g}_\mathrm{f}, \boldsymbol{v})_{\left(T_{\Gamma_{\mathrm{f,N}}}(\boldsymbol{V}_\mathrm{f})\right)^*\times\left(T_{\Gamma_{\mathrm{f,N}}}(\boldsymbol{V}_\mathrm{f})\right)}, \\
\ell_\mathrm{p}(\psi) &= \left(f_\mathrm{p}, \psi\right)_{Q_\mathrm{p}^*\times Q_\mathrm{p}} - \left(g_\mathrm{p}, \psi\right)_{\left(T_{\Gamma_{\mathrm{p,N}}}(Q_\mathrm{p})\right)^*\times\left(T_{\Gamma_{\mathrm{p,N}}}(Q_\mathrm{p})\right)}.
\end{aligned}
$$

Then the weak formulation of the Stokes–Darcy coupled problem is: Find $(\boldsymbol{u}_\mathrm{f}, p_\mathrm{f}, \varphi_\mathrm{p}) \in \mathbf{H}^1(\Omega_\mathrm{f}) \times L^2(\Omega_\mathrm{f}) \times H^1(\Omega_\mathrm{p})$, such that for all $\boldsymbol{v} \in \boldsymbol{V}_\mathrm{f}$, $q \in Q_\mathrm{f}$, and all $\psi \in Q_\mathrm{p}$ it is

$$\begin{cases} a_\mathrm{f}(\boldsymbol{u}_\mathrm{f}, \boldsymbol{v}) + b_\mathrm{f}(\boldsymbol{v}, p_\mathrm{f}) + \big(\varphi_\mathrm{p}, \boldsymbol{v} \cdot \boldsymbol{n}\big)_{0,\Gamma_\mathrm{I}} = \ell_\mathrm{f}(\boldsymbol{v}) \\ b_\mathrm{f}(\boldsymbol{u}_\mathrm{f}, q) = 0 \\ a_\mathrm{p}\big(\varphi_\mathrm{p}, \psi\big) - (\boldsymbol{u}_\mathrm{f} \cdot \boldsymbol{n}, \psi)_{0,\Gamma_\mathrm{I}} = \ell_\mathrm{p}(\psi), \end{cases} \tag{6.7}$$

together with $\boldsymbol{u}_\mathrm{f} = \mathbf{u}_\mathrm{b}$ on $\Gamma_{\mathrm{f,D}}$ and $\varphi_\mathrm{p} = \varphi_\mathrm{b}$ on $\Gamma_{\mathrm{p,D}}$.

The bilinear and linear forms are essentially the ones introduced in the sections on the respective subproblems, see Sects. 5.1 and 5.3. The main difference is that the boundary integral related to the Robin condition on the interface Γ_I in the definition of a_f only includes tangential components.

Remark 6.2.2 Let Γ be a Lipschitz continuous part of the boundary of a domain Ω. A vector-valued function $\boldsymbol{v} \in \mathbf{H}^{1/2}(\Gamma)$ is defined such that each component is in $H^{1/2}(\Gamma)$. However, this does not imply that also $\boldsymbol{v} \cdot \boldsymbol{n} \in H^{1/2}(\Gamma)$, since $\boldsymbol{n}$ may have jumps. In fact, considering a piecewise smooth part Γ, then the normal component is in $H^{1/2}$ only piecewise, and the same is valid for the tangential ones. However all components are in $L^2(\Gamma)$, so that all integrals over Γ_I above are well defined.

As a part of a Lipschitz boundary with a localization as in Definition 2.2.1, the interface Γ_I is represented as a set of Lipschitz continuous functions h_i which, due to Rademacher's theorem ([Eva98], Theorem 6 in Section 5.8.3), are differentiable almost everywhere. To avoid technical difficulties, it is assumed that the h_i are not differentiable at only finitely many points (and lines for $d = 3$). In practice, especially in geometries which are resolved by a finite element grid, the interface is often piecewise smooth so that this assumption is no restriction.

The (normal) traces of functions in $\mathbf{H}^1(\Omega_\mathrm{f})$ and $H^1(\Omega_\mathrm{p})$ onto the interface Γ play a central role in the following analysis and are therefore introduced additionally in a separate definition. For notational consistency, the normal trace of a vector field in the Stokes subdomain is defined on the entire solution space, including the pressure space.

Definition 6.2.3 (Trace Operators) Let $T_\mathrm{f} : (\mathbf{H}^1(\Omega_\mathrm{f}) \times L^2(\Omega_\mathrm{f})) \to L^2(\Gamma_\mathrm{I})$, $(\boldsymbol{v}, q) \mapsto \boldsymbol{v} \cdot \boldsymbol{n}$, and $T_\mathrm{p} : H^1(\Omega_\mathrm{p}) \to H^{1/2}(\Gamma_\mathrm{I})$ be the trace operators onto the interface. According to Theorems 4.1.1 and 4.2.4 they are continuous. The image space of T_f is in fact piecewise $H^{1/2}$ on the interface, see the previous Remark 6.2.2. The continuity constants of the two trace operators are denoted by c_{T_f} and c_{T_p}, respectively.

Note that the continuity of T_f holds even on the piecewise $H^{1/2}$ space on the interface Γ_I, because it is a composition of local (piecewise) traces. As before, the explicit statement of the trace operators is sometimes omitted for better readability, for example $\boldsymbol{v} \cdot \boldsymbol{n}$ on Γ_I rather than $T_f(\boldsymbol{v}, q)$ in the weak formulation (6.7). The trace spaces are studied in more detail in Sect. 6.4.1.

6.3 Existence and Uniqueness

The theory of the Neumann–Neumann formulation has been studied in the literature, see for example [DQ09] and the references therein.

To put the weak formulation (6.7) of the coupled Stokes–Darcy equations (6.5) into the setting of abstract saddle point theory, see Sect. 5.2, for all $\underline{w} = (\boldsymbol{w}, \omega) \in \mathbf{H}^1(\Omega_\mathrm{f}) \times H^1(\Omega_\mathrm{p})$, $\underline{v} = (\boldsymbol{v}, \psi) \in \boldsymbol{V}_\mathrm{f} \times Q_\mathrm{p}$ and all $q \in Q_\mathrm{f}$ define

$$\begin{aligned}\mathcal{A}(\underline{w}, \underline{v}) &= a_\mathrm{f}(\boldsymbol{w}, \boldsymbol{v}) + a_\mathrm{p}(\omega, \psi) - (\boldsymbol{w} \cdot \boldsymbol{n}, \psi)_{0,\Gamma_\mathrm{I}} + (\omega, \boldsymbol{v} \cdot \boldsymbol{n})_{0,\Gamma_\mathrm{I}}, \\ \mathcal{B}(\underline{w}, q) &= b_\mathrm{f}(\boldsymbol{w}, q), \\ \mathcal{F}(\underline{v}) &= \ell_\mathrm{f}(\boldsymbol{v}) + \ell_\mathrm{p}(\psi) - \mathcal{A}\big((\mathbf{E}_{\Gamma_{\mathrm{f,D}}}\mathbf{u}_\mathrm{b}, E_{\Gamma_{\mathrm{p,D}}}\varphi_\mathrm{b}), \underline{v}\big), \\ \mathcal{G}(q) &= -b_\mathrm{f}(\mathbf{E}_{\Gamma_{\mathrm{f,D}}}\mathbf{u}_\mathrm{b}, q).\end{aligned}$$

The operators $\mathbf{E}_{\Gamma_{\mathrm{f,D}}} : \mathbf{H}^{1/2}(\Gamma_{\mathrm{f,D}}) \to \mathbf{H}^1(\Omega_\mathrm{f})$ and $E_{\Gamma_{\mathrm{p,D}}} : H^{1/2}(\Gamma_{\mathrm{p,D}}) \to H^1(\Omega_\mathrm{p})$ extend Dirichlet boundary data into the respective subdomain, see Theorem 4.2.4. Then the weak formulation (6.7) can be rewritten to: Find $\underline{u} = (\boldsymbol{u}_\mathrm{f}^0, \varphi_\mathrm{p}^0) \in \boldsymbol{V}_\mathrm{f} \times Q_\mathrm{p}$ and $p_\mathrm{f} \in Q_\mathrm{f}$ such that for all $\underline{v} = (\boldsymbol{v}, \psi) \in \boldsymbol{V}_\mathrm{f} \times Q_\mathrm{p}$ and $q \in Q_\mathrm{f}$ it holds

$$\begin{cases} \mathcal{A}(\underline{u}, \underline{v}) + \mathcal{B}(\underline{v}, p_\mathrm{f}) = \mathcal{F}(\underline{v}) \\ \qquad\qquad\quad \mathcal{B}(\underline{u}, q) = \mathcal{G}(q). \end{cases} \tag{6.8}$$

The solution can be recovered via $\boldsymbol{u}_\mathrm{f} = \boldsymbol{u}_\mathrm{f}^0 + \mathbf{E}_{\Gamma_{\mathrm{f,D}}}\mathbf{u}_\mathrm{b}$ and $\varphi_\mathrm{p} = \varphi_\mathrm{p}^0 + E_{\Gamma_{\mathrm{p,D}}}\varphi_\mathrm{b}$. In order to apply Theorem 5.2.2 together with Propositions 5.2.5 and 5.2.6 to these bilinear forms, the following properties are proved in the next three subsections.

1. continuity of $\mathcal{A}$, $\mathcal{B}$, $\mathcal{F}$, and $\mathcal{G}$ with constants $c_\mathcal{A}$, $c_\mathcal{B}$, $c_\mathcal{F}$, and $c_\mathcal{G}$, respectively
2. coercivity of $\mathcal{A}$ on $\{\underline{v} \in \boldsymbol{V}_\mathrm{f} \times Q_\mathrm{p} \mid \mathcal{B}(\underline{v}, q) = 0 \text{ for all } q \in Q_\mathrm{f}\}$ with constant $\alpha_\mathcal{A}$
3. inf–sup condition for $\mathcal{B}$ with constant β.

Once these properties are proved, the existence and uniqueness can be established:

Theorem 6.3.1 (Existence and Uniqueness) *The Neumann–Neumann weak formulation* (6.8) *of the Stokes–Darcy problem admits a unique solution* $(\underline{u}, p_\mathrm{f})$, $\underline{u} = (\boldsymbol{u}_\mathrm{f}^0, \varphi_\mathrm{p}^0) \in \boldsymbol{V}_\mathrm{f} \times Q_\mathrm{p}$, $p_\mathrm{f} \in Q_\mathrm{f}$. *Moreover, the following a priori estimates hold:*

$$\begin{aligned}\left\|\left(\boldsymbol{u}_\mathrm{f}^0, \varphi_\mathrm{p}^0\right)\right\|_W &\le \frac{1}{\alpha_\mathcal{A}}\left(\sqrt{2}c_\mathcal{F} + \frac{\alpha + 2c_\mathcal{A}}{\beta}c_\mathcal{G}\right), \\ \|p_\mathrm{f}\|_{Q_\mathrm{f}} &\le \frac{1}{\beta}\left(\left(1 + \frac{2c_\mathcal{A}}{\alpha}\right)\sqrt{2}c_\mathcal{F} + \frac{2c_\mathcal{A}(\alpha + 2c_\mathcal{A})}{\alpha\beta}c_\mathcal{G}\right).\end{aligned}$$

6.3.1 Continuity of $\mathcal{A}$, $\mathcal{B}$, $\mathcal{F}$, and $\mathcal{G}$

The bilinear forms are continuous with constants $c_{a_\mathrm{f}} = 2\nu + \frac{1}{\alpha}(d-1)c_{T_\mathrm{f}}^2$, $c_{b_\mathrm{f}} = 1$, and $c_{a_\mathrm{p}} = \|\mathsf{K}\|_{(L^\infty(\Omega))^{d\times d}}$, respectively, see Sects. 5.3 and 5.1. Let $\underline{w} = (\boldsymbol{w}, \omega)$, $\underline{v} = (\boldsymbol{v}, \psi) \in W = \boldsymbol{V}_\mathrm{f} \times Q_\mathrm{p}$ and $q \in Q_\mathrm{f}$ be given. Then

$$
\begin{aligned}
\mathcal{A}(\underline{w}, \underline{v}) &= a_\mathrm{f}(\boldsymbol{w}, \boldsymbol{v}) + a_\mathrm{p}(\omega, \psi) - (\boldsymbol{w}\cdot\boldsymbol{n}, \psi)_{\Gamma_\mathrm{I}} + (\omega, \boldsymbol{v}\cdot\boldsymbol{n})_{\Gamma_\mathrm{I}} \\
&\leq c_{a_\mathrm{f}}\|\boldsymbol{w}\|_{\boldsymbol{V}_\mathrm{f}}\|\boldsymbol{v}\|_{\boldsymbol{V}_\mathrm{f}} + c_{a_\mathrm{p}}\|\omega\|_{Q_\mathrm{p}}\|\psi\|_{Q_\mathrm{p}} \\
&\quad + c_{T_\mathrm{f}}c_{T_\mathrm{p}}\|\boldsymbol{w}\|_{\boldsymbol{V}_\mathrm{f}}\|\psi\|_{Q_\mathrm{p}} + c_{T_\mathrm{f}}c_{T_\mathrm{p}}\|\boldsymbol{v}\|_{\boldsymbol{V}_\mathrm{f}}\|\omega\|_{Q_\mathrm{p}} \\
&\leq \frac{c_\mathcal{A}}{2}\Big(\|\boldsymbol{w}\|_{\boldsymbol{V}_\mathrm{f}} + \|\omega\|_{Q_\mathrm{p}}\Big)\Big(\|\boldsymbol{v}\|_{\boldsymbol{V}_\mathrm{f}} + \|\psi\|_{Q_\mathrm{p}}\Big) \\
&\leq c_\mathcal{A}\Big(\|\boldsymbol{w}\|_{\boldsymbol{V}_\mathrm{f}}^2 + \|\omega\|_{Q_\mathrm{p}}^2\Big)^{1/2}\Big(\|\boldsymbol{v}\|_{\boldsymbol{V}_\mathrm{f}}^2 + \|\psi\|_{Q_\mathrm{p}}^2\Big)^{1/2} \\
&= c_\mathcal{A}\big\|\underline{w}\big\|_W\big\|\underline{v}\big\|_W,
\end{aligned}
$$

where

$$
c_\mathcal{A} = 2\max\left\{c_{a_\mathrm{f}},\ c_{a_\mathrm{p}},\ c_{T_\mathrm{f}}c_{T_\mathrm{p}}\right\}
$$

and the continuity constants of the respective traces onto the interface are c_{T_f} and c_{T_p}, see Definition 6.2.3. This proves the continuity of $\mathcal{A}$. For $\mathcal{B}$ it holds

$$
\mathcal{B}(\underline{w}, q) = b_\mathrm{f}(\boldsymbol{w}, q) \leq c_{b_\mathrm{f}}\|\boldsymbol{w}\|_{\boldsymbol{V}_\mathrm{f}}\|q\|_{Q_\mathrm{f}} \leq c_{b_\mathrm{f}}\big\|\underline{w}\big\|_W\|q\|_{Q_\mathrm{f}},
$$

which means $c_\mathcal{B} = c_{b_\mathrm{f}} = 1$. The linear forms $\mathcal{F}$ and $\mathcal{G}$ are continuous due to

$$
\begin{aligned}
&\mathcal{F}(\underline{v}) \leq c_{\ell_\mathrm{f}}\|\boldsymbol{v}\|_{\boldsymbol{V}_\mathrm{f}} + c_{\ell_\mathrm{p}}\|\psi\|_{Q_\mathrm{p}} + c_\mathcal{A}\big\|(\mathbf{E}_{\Gamma_{\mathrm{f,D}}}\mathbf{u}_\mathrm{b}, E_{\Gamma_{\mathrm{p,D}}}\varphi_\mathrm{b})\big\|_W\big\|\underline{v}\big\|_W = c_\mathcal{F}\big\|\underline{v}\big\|_W, \\
&\mathcal{G}(q) \leq c_{\mathbf{E}_{\Gamma_{\mathrm{f,D}}}}\|\mathbf{u}_\mathrm{b}\|_{H^{1/2}(\Gamma_{\mathrm{f,D}})}\|q\|_{Q_\mathrm{f}} = c_\mathcal{G}\|q\|_{Q_\mathrm{f}},
\end{aligned}
$$

with

$$
\begin{aligned}
&c_\mathcal{F} = \max\left\{c_{\ell_\mathrm{f}},\ c_{\ell_\mathrm{p}},\ c_\mathcal{A}\Big(c_{\mathbf{E}_{\Gamma_{\mathrm{f,D}}}}^2\|\mathbf{u}_\mathrm{b}\|_{H^{1/2}(\Gamma_{\mathrm{f,D}})}^2 + c_{E_{\Gamma_{\mathrm{p,D}}}}^2\|\varphi_\mathrm{b}\|_{H^{1/2}(\Gamma_{\mathrm{p,D}})}^2\Big)^{1/2}\right\}, \\
&c_\mathcal{G} = c_{\mathbf{E}_{\Gamma_{\mathrm{f,D}}}}\|\mathbf{u}_\mathrm{b}\|_{H^{1/2}(\Gamma_{\mathrm{f,D}})},
\end{aligned}
$$

the continuity constants of the right-hand sides

$$c_{\ell_\mathrm{f}} = \|\mathbf{f}_\mathrm{f}\|_{V_\mathrm{f}^*} + \|\mathbf{g}_\mathrm{f}\|_{\left(T_{\Gamma_{\mathrm{f,N}}}(V_\mathrm{f})\right)^*},$$

$$c_{\ell_\mathrm{p}} = \|f_\mathrm{p}\|_{Q_\mathrm{p}^*} + \|g_\mathrm{p}\|_{\left(T_{\Gamma_{\mathrm{p,N}}}(Q_\mathrm{p})\right)^*},$$

and of the extension operators $\mathbf{E}_{\Gamma_{\mathrm{f,D}}}$ and $E_{\Gamma_{\mathrm{p,D}}}$.

6.3.2 Coercivity of $\mathcal{A}$

Let $\underline{v} = (\boldsymbol{v}, \psi) \in W = \boldsymbol{V}_\mathrm{f} \times Q_\mathrm{p}$ be given. The two interface terms in $\mathcal{A}(\underline{v}, \underline{v})$ cancel and it is

$$\mathcal{A}(\underline{v}, \underline{v}) = a_\mathrm{f}(\boldsymbol{v}, \boldsymbol{v}) + a_\mathrm{p}(\psi, \psi) \geq \alpha_{a_\mathrm{f}} \|\boldsymbol{v}\|_{\boldsymbol{V}_\mathrm{f}}^2 + \alpha_{a_\mathrm{p}} \|\psi\|_{Q_\mathrm{p}}^2 = \alpha_\mathcal{A} \left\|\underline{v}\right\|_W^2,$$

with $\alpha_\mathcal{A} = \min\left\{\alpha_{a_\mathrm{f}}, \alpha_{a_\mathrm{p}}\right\}$ and the coercivity constants α_{a_f} and α_{a_p} of the two subproblems, see also Sects. 5.1.2 and 5.3.2.

6.3.3 Inf–sup Condition

The bilinear form b_f is inf–sup stable, i.e., there is a constant $\beta > 0$ such that for all $q \in Q_\mathrm{f}$ there exists a $\boldsymbol{w} \in \boldsymbol{V}_\mathrm{f}$ such that

$$\frac{b_\mathrm{f}(\boldsymbol{w}, q)}{\|\boldsymbol{w}\|_{\boldsymbol{V}_\mathrm{f}}} \geq \beta \|q\|_{Q_\mathrm{f}/M},$$

with the set $M := \{q \in Q_\mathrm{f} \mid b_\mathrm{f}(\boldsymbol{v}, q) = 0 \text{ for all } \boldsymbol{v} \in \boldsymbol{V}_\mathrm{f}\}$, see Sect. 5.3.2.2. Choosing $\underline{w} = (\boldsymbol{w}, 0)$ for a given q yields

$$\frac{\mathcal{B}(\underline{w}, q)}{\left\|\underline{w}\right\|_W} \geq \beta \|q\|_{Q_\mathrm{f}/M}.$$

Note that the set M equals $\left\{q \in Q_\mathrm{f} \;\middle|\; \mathcal{B}(\underline{v}, q) = 0 \text{ for all } \underline{v} \in W\right\}$ which implies that the same β can be used as the inf–sup constant for $\mathcal{B}$.

6.4 Weak Formulation Rewritten

While in Sect. 6.3 the weak formulation (6.7) is rewritten as a saddle point problem where the bilinear form $\mathcal{A}$ included both a_{f} and a_{p}, i.e., parts from both subdomains, instead, in this section a different but equivalent formulation is introduced which acts on suitable spaces on the interface Γ_{I}.

6.4.1 *Spaces*

On the interface the trace spaces play a central role in the analysis.

Definition 6.4.1 (Trace Spaces) The restrictions of the test spaces onto the interface via the trace operators from Definition 6.2.3 are denoted by

$$\Lambda_{\mathrm{p}} := T_{\mathrm{p}}(Q_{\mathrm{p}}),$$
$$\Lambda_{\mathrm{f}} := T_{\mathrm{f}}(\boldsymbol{V}_{\mathrm{f}}, Q_{\mathrm{f}}).$$

While the first space (Λ_{p}) is a subspace of $H^{1/2}(\Gamma_{\mathrm{I}})$ containing $H^{1/2}_{00}(\Gamma_{\mathrm{I}})$, the second ($\Lambda_{\mathrm{f}}$) is a subspace of $L^2(\Gamma_{\mathrm{I}})$ and only included in $H^{1/2}(\Gamma_{\mathrm{I}})$ if Γ is smooth, see Remark 6.2.2. The norms can therefore be the natural ones or, equivalently,

$$\|\mu\|_{\Lambda_{\mathrm{p}}} := \inf_{\substack{\psi \in Q_{\mathrm{p}} \\ T_{\mathrm{p}}\psi = \mu}} \|\psi\|_{Q_{\mathrm{p}}},$$

$$\|\mu\|_{\Lambda_{\mathrm{f}}} := \inf_{\substack{(\boldsymbol{v},q) \in \boldsymbol{V}_{\mathrm{f}} \times Q_{\mathrm{f}} \\ T_{\mathrm{f}}(\boldsymbol{v},q) = \mu}} \|\boldsymbol{v}\|_{\boldsymbol{V}_{\mathrm{f}}}.$$

In case Γ_{I} is smooth, both of these two spaces therefore contain $H^{1/2}_{00}(\Gamma)$ and can but need not be equal, depending on where the Dirichlet boundary conditions are prescribed. If $\overline{\Gamma_{\mathrm{p,D}}} \cap \overline{\Gamma_{\mathrm{I}}} = \overline{\Gamma_{\mathrm{f,D}}} \cap \overline{\Gamma_{\mathrm{I}}}$ then $\Lambda_{\mathrm{p}} = \Lambda_{\mathrm{f}}$ holds true. A longer discussion on this issue follows in Sect. 6.6.2.

The solution to the coupled problem, Eq. (6.5), in general has nonzero Dirichlet boundary data, it is sought in affine spaces. Consequently, the corresponding restrictions to the interface (through T_{f} and T_{p}) do not necessarily belong to Λ_{p} or Λ_{f}.

Definition 6.4.2 (Affine Spaces) The solution spaces for the Stokes velocity and the Darcy pressure are affine spaces

$$\widetilde{Q}_{\mathrm{p}} = \left\{ \varphi \in H^1(\Omega_{\mathrm{p}}) \;\middle|\; T_{\Gamma_{\mathrm{p,D}}}\varphi = \varphi_{\mathrm{b}} \right\},$$

$$\widetilde{\boldsymbol{V}}_{\mathrm{f}} = \left\{ \boldsymbol{v} \in \mathbf{H}^1(\Omega_{\mathrm{f}}) \;\middle|\; T_{\Gamma_{\mathrm{f,D}}}\boldsymbol{v} = \mathbf{u}_{\mathrm{b}} \right\}.$$

Additionally their restrictions onto the interface Γ_{I} via the trace operators from Definition 6.2.3 are also affine and denoted by

$$\widetilde{\Lambda}_{\mathrm{p}} := T_{\mathrm{p}}(\widetilde{Q}_{\mathrm{p}}),$$
$$\widetilde{\Lambda}_{\mathrm{f}} := T_{\mathrm{f}}\Big(\widetilde{\mathbf{V}}_{\mathrm{f}} \times L^2(\Omega_{\mathrm{f}})\Big).$$

In fact, these trace spaces could be defined with the help of two extensions $E_{\Gamma_{\mathrm{p,D}}}\varphi_{\mathrm{b}} \in H^1(\Omega_{\mathrm{p}})$ and $E_{\Gamma_{\mathrm{f,D}}}\mathbf{u}_{\mathrm{b}} \in \mathbf{H}^1(\Omega_{\mathrm{f}})$ of the Dirichlet boundary data in the respective subdomains. Then it is $\widetilde{\Lambda}_{\mathrm{p}} = T_{\mathrm{p}}(E_{\Gamma_{\mathrm{p,D}}}\varphi_{\mathrm{b}}) + \Lambda_{\mathrm{p}}$ and $\widetilde{\Lambda}_{\mathrm{f}} = T_{\mathrm{f}}(E_{\Gamma_{\mathrm{f,D}}}\mathbf{u}_{\mathrm{b}}, 0) + \Lambda_{\mathrm{f}}$. The affine space $\widetilde{\Lambda}_{\mathrm{p}}$ is different from the linear space Λ_{p} if, and only if, the boundary data cannot be extended by 0 to a function in $H^{1/2}(\Gamma_{\mathrm{I}} \cup \Gamma_{\mathrm{p,D}})$. This occurs if the Dirichlet boundary $\Gamma_{\mathrm{p,D}}$ touches the interface Γ_{I} and nonzero data is prescribed in the vicinity of $\overline{\Gamma_{\mathrm{p,D}}} \cap \overline{\Gamma_{\mathrm{I}}} \neq \emptyset$. Similar ideas apply to $\widetilde{\Lambda}_{\mathrm{f}}$ and Λ_{f} where also the normal has to be taken into account.

Remark 6.4.3 In general, the restriction spaces on the interface are not equal, however at least for a smooth interface Γ_{I} the following inclusions hold, $i, j \in \{\mathrm{p}, \mathrm{f}\}$,

$$H_{00}^{1/2}(\Gamma_{\mathrm{I}}) \subset \Lambda_i \subset H^{1/2}(\Gamma_{\mathrm{I}}) \subset L^2(\Gamma_{\mathrm{I}}) \subset \left(H^{1/2}(\Gamma_{\mathrm{I}})\right)^* \subset \Lambda_j^* \subset \left(H_{00}^{1/2}(\Gamma_{\mathrm{I}})\right)^*,$$

where the dual spaces are denoted by the superscript *. Furthermore, even if the interface is only Lipschitz continuous the affine spaces are subspaces of $L^2(\Gamma_{\mathrm{I}})$ implying

$$\widetilde{\Lambda}_{\mathrm{p}} \subset \Lambda_{\mathrm{f}}^* \qquad \text{and} \qquad \widetilde{\Lambda}_{\mathrm{f}} \subset \Lambda_{\mathrm{p}}^*.$$

The case $\Lambda_{\mathrm{p}} \neq \Lambda_{\mathrm{f}}$ is discussed in more detail in Sect. 6.6.2.

6.4.2 Revisiting Operators on the Boundary

In Chap. 5 the Laplace as well as the Stokes problem is introduced and analyzed. At the end of Sects. 5.1.3 and 5.3.3 these subproblems are reinterpreted as operators which take data on a particular boundary part and return the solution in the respective subdomain. In this section these ideas are applied to the Stokes–Darcy case. First, only Dirichlet (index $_{\mathrm{D}}$) and Neumann (index $_{\mathrm{N}}$) conditions on the interface Γ_{I} are of interest.

Definition 6.4.4 (Neumann and Dirichlet Operators) With the spaces from Definitions 6.4.1 and 6.4.2 let $K_{\mathrm{p,N}} : \Lambda_{\mathrm{p}}^* \to \widetilde{Q}_{\mathrm{p}}$, $K_{\mathrm{p,D}} : \widetilde{\Lambda}_{\mathrm{p}} \to \widetilde{Q}_{\mathrm{p}}$, $K_{\mathrm{f,N}} : \Lambda_{\mathrm{f}}^* \to \widetilde{\mathbf{V}}_{\mathrm{f}} \times Q_{\mathrm{f}}$, and $K_{\mathrm{f,D}} : \widetilde{\Lambda}_{\mathrm{f}} \to \widetilde{\mathbf{V}}_{\mathrm{f}} \times Q_{\mathrm{f}}$ be operators which map given Neumann (N) or Dirichlet (D) data on the interface Γ_{I} to solutions of the subproblems. In strong form these are described by:

$K_{\mathrm{p,N}}(\mu) = \varphi$ solves

$$\begin{cases} -\nabla \cdot (\mathsf{K}\nabla\varphi) = f_{\mathrm{p}} & \text{in } \Omega_{\mathrm{p}}, \\ -\mathsf{K}\nabla\varphi \cdot \boldsymbol{n} = g_{\mathrm{p}} & \text{on } \Gamma_{\mathrm{p,N}}, \\ \varphi = \varphi_{\mathrm{b}} & \text{on } \Gamma_{\mathrm{p,D}}, \\ -\mathsf{K}\nabla\varphi \cdot \boldsymbol{n} = \mu & \text{on } \Gamma_{\mathrm{I}}. \end{cases}$$

$K_{\mathrm{f,N}}(\mu) = (\mathbf{u}, p)$ solves

$$\begin{cases} -\nabla \cdot \mathsf{T}(\mathbf{u}, p) = \mathbf{f}_{\mathrm{f}} & \text{in } \Omega_{\mathrm{f}}, \\ \nabla \cdot \mathbf{u} = 0 & \text{in } \Omega_{\mathrm{f}}, \\ \mathsf{T}(\mathbf{u}, p) \cdot \boldsymbol{n} = \mathbf{g}_{\mathrm{f}} & \text{on } \Gamma_{\mathrm{f,N}}, \\ \mathbf{u} = \mathbf{u}_{\mathrm{b}} & \text{on } \Gamma_{\mathrm{f,D}}, \\ \mathbf{u} \cdot \boldsymbol{\tau}_i + \alpha \boldsymbol{\tau}_i \cdot \mathsf{T}(\mathbf{u}, p) \cdot \boldsymbol{n} = 0 & \text{on } \Gamma_{\mathrm{I}}, \\ -\boldsymbol{n} \cdot \mathsf{T}(\mathbf{u}, p) \cdot \boldsymbol{n} = \mu & \text{on } \Gamma_{\mathrm{I}}. \end{cases}$$

$K_{\mathrm{p,D}}(\lambda) = \varphi$ solves

$$\begin{cases} -\nabla \cdot (\mathsf{K}\nabla\varphi) = f_{\mathrm{p}} & \text{in } \Omega_{\mathrm{p}}, \\ -\mathsf{K}\nabla\varphi \cdot \boldsymbol{n} = g_{\mathrm{p}} & \text{on } \Gamma_{\mathrm{p,N}}, \\ \varphi = \varphi_{\mathrm{b}} & \text{on } \Gamma_{\mathrm{p,D}}, \\ \varphi = \lambda & \text{on } \Gamma_{\mathrm{I}}. \end{cases}$$

$K_{\mathrm{f,D}}(\lambda) = (\mathbf{u}, p)$ solves

$$\begin{cases} -\nabla \cdot \mathsf{T}(\mathbf{u}, p) = \mathbf{f}_{\mathrm{f}} & \text{in } \Omega_{\mathrm{f}}, \\ \nabla \cdot \mathbf{u} = 0 & \text{in } \Omega_{\mathrm{f}}, \\ \mathsf{T}(\mathbf{u}, p) \cdot \boldsymbol{n} = \mathbf{g}_{\mathrm{f}} & \text{on } \Gamma_{\mathrm{f,N}}, \\ \mathbf{u} = \mathbf{u}_{\mathrm{b}} & \text{on } \Gamma_{\mathrm{f,D}}, \\ \mathbf{u} \cdot \boldsymbol{\tau}_i + \alpha \boldsymbol{\tau}_i \cdot \mathsf{T}(\mathbf{u}, p) \cdot \boldsymbol{n} = 0 & \text{on } \Gamma_{\mathrm{I}}, \\ \mathbf{u} \cdot \boldsymbol{n} = \lambda & \text{on } \Gamma_{\mathrm{I}}. \end{cases}$$

Transformed to the weak formulations, the above operators are characterized by

- $K_{\mathrm{p,N}}(\mu) = \varphi$ such that $T_{\Gamma_{\mathrm{p,D}}}\varphi = \varphi_{\mathrm{b}}$ and for all $\psi \in Q_p$ it is

$$a_{\mathrm{p}}(\varphi, \psi) = \ell_{\mathrm{p}}(\psi) + \left(\mu, T_{\mathrm{p}}\psi\right)_{\Lambda_{\mathrm{p}}^* \times \Lambda_{\mathrm{p}}},$$

- $K_{\mathrm{p,D}}(\lambda) = \varphi$ such that $T_{\Gamma_{\mathrm{p,D}}}\varphi = \varphi_{\mathrm{b}}$, $T_{\mathrm{p}}\varphi = \lambda$, and for all $\psi \in Q_p$ with $T_{\mathrm{p}}\psi = 0$ (i.e., $\psi \in H^1_{\Gamma_{\mathrm{p,D}} \cup \Gamma_{\mathrm{I}}}(\Omega_{\mathrm{p}})$) it is

$$a_{\mathrm{p}}(\varphi, \psi) = \ell_{\mathrm{p}}(\psi),$$

- $K_{\mathrm{f,N}}(\mu) = (\mathbf{u}, p)$ such that $T_{\Gamma_{\mathrm{f,D}}}\mathbf{u} = \mathbf{u}_{\mathrm{b}}$ and for all $(\boldsymbol{v}, q) \in \boldsymbol{V}_{\mathrm{f}} \times Q_{\mathrm{f}}$ it is

$$\begin{cases} a_{\mathrm{f}}(\mathbf{u}, \boldsymbol{v}) + b_{\mathrm{f}}(\boldsymbol{v}, p) = \ell_{\mathrm{f}}(\boldsymbol{v}) - (\mu, T_{\mathrm{f}}\boldsymbol{v})_{\Lambda_{\mathrm{f}}^* \times \Lambda_{\mathrm{f}}} \\ b_{\mathrm{f}}(\mathbf{u}, q) = 0, \end{cases}$$

- $K_{\mathrm{f,D}}(\lambda) = (\mathbf{u}, p)$ such that $T_{\Gamma_{\mathrm{f,D}}}\mathbf{u} = \mathbf{u}_{\mathrm{b}}$, $T_{\mathrm{f}}(\mathbf{u}, p) = \lambda$, and for all $(\boldsymbol{v}, q) \in \boldsymbol{V}_{\mathrm{f}} \times Q_{\mathrm{f}}$ with $T_{\mathrm{f}}(\boldsymbol{v}, q) = 0$ it is

$$\begin{cases} a_{\mathrm{f}}(\mathbf{u}, \boldsymbol{v}) + b_{\mathrm{f}}(\boldsymbol{v}, p) = \ell_{\mathrm{f}}(\boldsymbol{v}) \\ \qquad\qquad b_{\mathrm{f}}(\mathbf{u}, q) = 0. \end{cases}$$

Note that the ranges of the respective Neumann and Dirichlet operators are the same, i.e., $\operatorname{Im} K_{i,\mathrm{N}} = \operatorname{Im} K_{i,\mathrm{D}}$, $i \in \{\mathrm{p}, \mathrm{f}\}$. For example the trace (using T_{p} or T_{f}) of a solution to a Neumann problem can be used as the argument to $K_{i,\mathrm{D}}$. The uniqueness of the solutions of these equations assures that the result is that exact same solution from before. Instead of recovering the Dirichlet data on the interface Γ_{I} via the two trace operators T_{p} and T_{f}, it is possible to recover the Neumann data as well. A similar construction as in Definitions 5.1.5 and 5.3.9 is made individually for each subproblem.

Definition 6.4.5 Let $\phi \in H^1(\Omega_{\mathrm{p}})$ be given such that for all $\psi \in Q_{\mathrm{p}} \cap \ker T_{\mathrm{p}} = H^1_{\Gamma_{\mathrm{p,D}} \cup \Gamma_{\mathrm{I}}}(\Omega)$ it is

$$a_{\mathrm{p}}(\phi, \psi) = \ell_{\mathrm{p}}(\psi).$$

Then its Neumann data $T_{\mathrm{p}}^{\mathrm{N}}\phi \in \Lambda_{\mathrm{p}}^*$ on Γ_{I} is defined as

$$\left(T_{\mathrm{p}}^{\mathrm{N}}\phi, \xi\right)_{\Lambda_{\mathrm{p}}^* \times \Lambda_{\mathrm{p}}} := a_{\mathrm{p}}\big(\phi, \psi_\xi\big) - \ell_{\mathrm{p}}(\psi_\xi)$$

with an extension $\psi_\xi \in Q_{\mathrm{p}}$ defined in Theorem 4.3.4 or in its subsequent discussion, respectively.

Next, Neumann data for the Stokes subdomain is introduced.

Definition 6.4.6 Let $(\mathbf{u}, p) \in \mathbf{H}^1(\Omega_{\mathrm{f}}) \times L^2(\Omega_{\mathrm{f}})$ be given such that for all $(\boldsymbol{v}, q) \in (\boldsymbol{V}_{\mathrm{f}} \cap \ker T_{\mathrm{f}}) \times Q_{\mathrm{p}}$ it is

$$\begin{cases} a_{\mathrm{f}}(\mathbf{u}, \boldsymbol{v}) + b_{\mathrm{f}}(\boldsymbol{v}, p) = \ell_{\mathrm{f}}(\boldsymbol{v}), \\ \qquad\qquad b_{\mathrm{f}}(\mathbf{u}, q) = 0. \end{cases}$$

Then its Neumann data $T_{\mathrm{f}}^{\mathrm{N}}(\mathbf{u}, p) \in \Lambda_{\mathrm{f}}^*$ on Γ_{I} is defined as

$$\left(T_{\mathrm{f}}^{\mathrm{N}}(\mathbf{u}, p), \xi\right)_{\Lambda_{\mathrm{f}}^* \times \Lambda_{\mathrm{f}}} := \ell_{\mathrm{f}}(\boldsymbol{v}_\xi) - a_{\mathrm{f}}\big(\mathbf{u}, \boldsymbol{v}_\xi\big) - b(\boldsymbol{v}_\xi, p),$$

with an extension $\boldsymbol{v}_\xi \in \boldsymbol{V}_{\mathrm{f}}$ defined in Theorem 4.3.4 or in its subsequent discussion, respectively.

Both of the above definitions do not depend on the particular choice of the extension, because any two extensions differ by an element in $Q_{\mathrm{p}} \cap \ker T_{\mathrm{p}}$ or

$(V_{\mathrm{f}} \times Q_{\mathrm{f}}) \cap \ker T_{\mathrm{f}}$, respectively, for which the defining expression is zero by assumption. Furthermore, this means that the Neumann data can be defined in a meaningful way only if the domain of definition is suitably restricted. In fact, the assumptions on ϕ and $(\mathbf{u}, p)$ above are satisfied for any Darcy or Stokes solution including functions from the image spaces $\mathrm{Im}(K_{\mathrm{p,N}}) = \mathrm{Im}(K_{\mathrm{p,D}}) \subset \widetilde{Q}_{\mathrm{p}}$ and $\mathrm{Im}(K_{\mathrm{f,N}}) = \mathrm{Im}(K_{\mathrm{f,D}}) \subset \widetilde{V}_{\mathrm{f}} \times Q_{\mathrm{f}}$, respectively. These Neumann data operators T_i^{N} are left inverses to the respective Neumann operators $K_{i,\mathrm{N}}$ in the subdomains, very similar to the trace and the respective Dirichlet operator. Additionally, on the image spaces $\mathrm{Im}(K_{i,j})$ they are also right inverses, i.e., for $i \in \{\mathrm{p}, \mathrm{f}\}$ it is

$$T_i \circ K_{i,\mathrm{D}} = \mathrm{id} \quad \text{in } \widetilde{\Lambda}_i, \tag{6.9a}$$

$$K_{i,\mathrm{D}} \circ T_i = \mathrm{id} \quad \text{in } \mathrm{Im}(K_{i,\mathrm{N}}) = \mathrm{Im}(K_{i,\mathrm{D}}), \tag{6.9b}$$

$$T_i^{\mathrm{N}} \circ K_{i,\mathrm{N}} = \mathrm{id} \quad \text{in } \Lambda_i^*, \tag{6.9c}$$

$$K_{i,\mathrm{N}} \circ T_i^{\mathrm{N}} = \mathrm{id} \quad \text{in } \mathrm{Im}(K_{i,\mathrm{N}}) = \mathrm{Im}(K_{i,\mathrm{D}}). \tag{6.9d}$$

In case the data (ℓ_{f} and ℓ_{p}) is zero, taking the Neumann data is a linear operation.

Remark 6.4.7 The operators T_i^{N} are indeed an extension of the Neumann data: For smooth functions ϕ and $(\mathbf{u}, p)$ integration by parts shows $T_{\mathrm{p}}^{\mathrm{N}}\phi = -\mathsf{K}\nabla\phi \cdot \boldsymbol{n}$ and $T_{\mathrm{f}}^{\mathrm{N}}(\mathbf{u}, p) = -\boldsymbol{n} \cdot \mathsf{T}(\mathbf{u}, p) \cdot \boldsymbol{n}$ on Γ_{I}. While $T_p^N\varphi$ has the usual sign, because $\boldsymbol{n}$ is pointing into the Darcy subdomain Ω_{p}, the Neumann data in the Stokes part has a minus sign. This matches the definition of $K_{\mathrm{f,N}}(\mu)$ where also it is $-\boldsymbol{n} \cdot \mathsf{T}(\mathbf{u}, p) \cdot \boldsymbol{n} = \mu$. The reason for this notational choice is mainly the conservation of momentum (6.3b).

Remark 6.4.8 The above identities (6.9) are an elaborated way of stating that it is possible to recover

- the data on the interface with which a subproblem is solved, and
- a solution by extracting its Dirichlet/Neumann data and solving a suitable Dirichlet/Neumann problem with it.

6.4.3 Operators Acting Only on the Interface

The weak form (6.8) of the Stokes–Darcy coupled problem is rewritten in terms of operators acting solely on spaces on the interface Γ_{I}.

Definition 6.4.9 (Operators on the Interface) With the solution operators $K_{i,j}$ from Definition 6.4.4 and the traces from Definitions 6.2.3, 6.4.5 and 6.4.6 define

$$H_{\mathrm{p,N}} : \Lambda_{\mathrm{p}}^* \to \widetilde{\Lambda}_{\mathrm{p}}, \qquad H_{\mathrm{p,N}} = T_{\mathrm{p}} \circ K_{\mathrm{p,N}}, \tag{6.10a}$$

$$H_{\mathrm{p,D}} : \widetilde{\Lambda}_{\mathrm{p}} \to \Lambda_{\mathrm{p}}^* \qquad H_{\mathrm{p,D}} = T_{\mathrm{p}}^{\mathrm{N}} \circ K_{\mathrm{p,D}}, \tag{6.10b}$$

$$H_{\mathrm{f,N}} : \Lambda_{\mathrm{f}}^* \to \widetilde{\Lambda}_{\mathrm{f}} \qquad H_{\mathrm{f,N}} = T_{\mathrm{f}} \circ K_{\mathrm{f,N}}, \tag{6.10c}$$

$$H_{\mathrm{f,D}} : \widetilde{\Lambda}_{\mathrm{f}} \to \Lambda_{\mathrm{f}}^* \qquad H_{\mathrm{f,D}} = T_{\mathrm{f}}^{\mathrm{N}} \circ K_{\mathrm{f,D}}, \tag{6.10d}$$

essentially mapping Dirichlet to Neumann data ($H_{i,\mathrm{D}}$) or vice versa ($H_{i,\mathrm{N}}$).

As discussed in Sects. 5.1.3.3 and 5.3.3.3, and also from Eqs. (6.9), these operators are related to each other through the identities

$$H_{\mathrm{p,N}} \circ H_{\mathrm{p,D}} = \mathrm{id} \quad \text{in } \widetilde{\Lambda}_{\mathrm{p}}, \tag{6.11a}$$

$$H_{\mathrm{p,D}} \circ H_{\mathrm{p,N}} = \mathrm{id} \quad \text{in } \Lambda_{\mathrm{p}}^*, \tag{6.11b}$$

$$H_{\mathrm{f,N}} \circ H_{\mathrm{f,D}} = \mathrm{id} \quad \text{in } \widetilde{\Lambda}_{\mathrm{f}}, \tag{6.11c}$$

$$H_{\mathrm{f,D}} \circ H_{\mathrm{f,N}} = \mathrm{id} \quad \text{in } \Lambda_{\mathrm{f}}^*. \tag{6.11d}$$

Remark 6.4.10 These operators are therefore bijective maps from $\widetilde{\Lambda}_i$ to Λ_i^* ($H_{i,\mathrm{D}}$) and Λ_i^* to $\widetilde{\Lambda}_i$ ($H_{i,\mathrm{N}}$), respectively.

6.4.4 Equations on the Interface

The solutions of the following equations on the interface are related to the solution of the Stokes–Darcy coupled system which is made precise in Sect. 6.4.5. Here only the equations and a few reformulations are presented.

Definition 6.4.11 (Fixed Point Equations) Using the operators on the interface from Definition 6.4.9 define the following two fixed point equations

$$H_{\mathrm{f,N}}\big(H_{\mathrm{p,N}}(\lambda)\big) = \lambda, \tag{6.12}$$

$$H_{\mathrm{p,N}}\big(H_{\mathrm{f,N}}(\mu)\big) = \mu. \tag{6.13}$$

Using the invertibility properties (6.11a)–(6.11d), the solution to the fixed point equation (6.12) also solves

$$H_{\mathrm{p,D}}\big(H_{\mathrm{f,D}}(\lambda)\big) = \lambda \tag{6.14}$$

and similarly the solution to (6.13) also solves

$$H_{\mathrm{f,D}}\big(H_{\mathrm{p,D}}(\mu)\big) = \mu. \tag{6.15}$$

Note that the nested operators on the left-hand sides of (6.14) and (6.15) are not defined for all $\lambda \in \widetilde{\Lambda}_{\mathrm{f}}$ and $\mu \in \widetilde{\Lambda}_{\mathrm{p}}$, since the range of the inner operator is larger than the domain of definition of the outer one. These two equations are therefore inappropriate in an iterative scheme trying to find the fixed point, i.e., a Dirichlet–Dirichlet method is not studied here. Therefore, from now on, the focus is on the other two formulations (6.12) and (6.13). Note that in (6.12) $\lambda \in \widetilde{\Lambda}_{\mathrm{f}}$ plays the role of the normal velocity on the interface while $\mu \in \widetilde{\Lambda}_{\mathrm{p}}$ in (6.13) is the pressure (or normal stress).

Another possible formulation is introduced in the following definition:

Definition 6.4.12 (Steklov–Poincaré Equations) Using the operators on the interface from Definition 6.4.9 define the following Steklov–Poincarè equations

$$H_{\mathrm{p,N}}(\lambda) - H_{\mathrm{f,D}}(\lambda) = 0, \tag{6.16}$$

$$H_{\mathrm{f,N}}(\mu) - H_{\mathrm{p,D}}(\mu) = 0. \tag{6.17}$$

The first equation (6.16) is derived from Eq. (6.12) through multiplying with $H_{\mathrm{f},D}$ and Eq. (6.11d). Similarly, Eqs. (6.13) and (6.11b) yield (6.17). Already from here one can see that the operators solving Stokes and Darcy systems are not a composition in the Steklov–Poincarè equation, while they are in the fixed point formulation. This means evaluating the left-hand side for a given λ or μ can be done in parallel only for the former. Note that in other publications, e.g. [DQ09], some of the above operators are defined with an additional minus sign so that the Steklov–Poincaré equations are sums rather than differences.

The operators involved in the Steklov–Poincaré equations (6.16) and (6.17) have different ranges and domains of definition. An iterative scheme therefore needs to apply another operator, e.g., a preconditioner, which corrects this drawback. Solving a Neumann problem in one of the subdomains (depending on which of the two Steklov–Poincaré equations is considered) has the desired property.

The mentioned formulations are summarized in the following:

Corollary 6.4.13 (Equivalence of Formulations, Part 1) *The fixed point and Steklov–Poincaré equations are equivalent:*

$$H_{\mathrm{f,N}}\big(H_{\mathrm{p,N}}(\lambda)\big) = \lambda \in \widetilde{\Lambda}_{\mathrm{f}} \quad \Longleftrightarrow \quad H_{\mathrm{p,N}}(\lambda) - H_{\mathrm{f,D}}(\lambda) = 0 \in \widetilde{\Lambda}_{\mathrm{p}}, \tag{6.18}$$

$$H_{\mathrm{p,N}}\big(H_{\mathrm{f,N}}(\mu)\big) = \mu \in \widetilde{\Lambda}_{\mathrm{p}} \quad \Longleftrightarrow \quad H_{\mathrm{f,N}}(\mu) - H_{\mathrm{p,D}}(\mu) = 0 \in \widetilde{\Lambda}_{\mathrm{f}}. \tag{6.19}$$

Additionally, with a solution of either one of the fixed point equations from Definition 6.4.11, one can find a solution to the respective other using the identities (6.11):

Corollary 6.4.14 (Equivalence of Formulations, Part 2) *If λ solves* (6.12), $\mu := H_{\mathrm{p,N}}(\lambda)$ *solves* (6.13). *Conversely, if μ solves* (6.13), $\lambda := H_{\mathrm{f,N}}(\mu)$ *solves* (6.12).

Combining this last result with Corollary 6.4.13, means that when λ is the solution of the fixed point equation (6.12), then $\mu := H_{\mathrm{p,N}}(\lambda)$ equals $H_{\mathrm{f,D}}(\lambda)$. Similarly, it holds $\lambda := H_{\mathrm{f,N}}(\mu) = H_{\mathrm{p,D}}(\mu)$ for the solution μ of the second fixed point equation (6.13).

6.4.5 Equivalence to the Weak Formulation

In this subsection the connection between the weak solution of the coupled system, Definition 6.2.1, and the interface equations, Definitions 6.4.11 and 6.4.12, is established.

Theorem 6.4.15 (Equivalence of Fixed Point Equation and Weak Formulation, Part 1) *If $\lambda \in \widetilde{\Lambda}_{\mathrm{f}}$ solves the fixed point equation* (6.12)*, then*

$$\varphi_{\mathrm{p}} := K_{\mathrm{p,N}}(\lambda) \qquad \textit{and} \qquad (\boldsymbol{u}_{\mathrm{f}}, p_{\mathrm{f}}) := K_{\mathrm{f,N}}(H_{\mathrm{p,N}}(\lambda))$$

solve (6.7)*. Conversely if $(\varphi_{\mathrm{p}}, \boldsymbol{u}_{\mathrm{f}}, p_{\mathrm{f}}) \in \widetilde{Q}_{\mathrm{p}} \times \widetilde{\boldsymbol{V}}_{\mathrm{f}} \times Q_{\mathrm{f}}$ solves* (6.7)*, then $\lambda = T_{\mathrm{f}}(\boldsymbol{u}_{\mathrm{f}}, p_{\mathrm{f}}) \in \widetilde{\Lambda}_{\mathrm{f}}$ solves* (6.12)*.*

Proof First let $\lambda \in \widetilde{\Lambda}_{\mathrm{f}}$ solve (6.12). By definition $(\varphi_{\mathrm{p}}, \boldsymbol{u}_{\mathrm{f}}, p_{\mathrm{f}})$ satisfy the correct boundary data away from the interface and, using Eq. (6.12), $\boldsymbol{u}_{\mathrm{f}} \cdot \boldsymbol{n} = T_{\mathrm{f}}(\boldsymbol{u}_{\mathrm{f}}, p_{\mathrm{f}}) = H_{\mathrm{f,N}}(H_{\mathrm{p,N}}(\lambda)) = \lambda$ on Γ_{I}. Let $\boldsymbol{v} \in \boldsymbol{V}_{\mathrm{f}}$, $q \in Q_{\mathrm{f}}$, and $\psi \in Q_{\mathrm{p}}$ be given. Then, using $H_{\mathrm{p,N}}(\lambda) = T_{\mathrm{p}}(\varphi)$, the definitions of $K_{\mathrm{f,N}}$ and $K_{\mathrm{p,N}}$ read

$$\begin{aligned}
a_{\mathrm{f}}(\boldsymbol{u}_{\mathrm{f}}, \boldsymbol{v}) + b_{\mathrm{f}}(\boldsymbol{v}, p_{\mathrm{f}}) &= \ell_{\mathrm{f}}(\boldsymbol{v}) - \big(\varphi_{\mathrm{p}}, \boldsymbol{v} \cdot \boldsymbol{n}\big)_{0,\Gamma_{\mathrm{I}}}, \\
b_{\mathrm{f}}(\boldsymbol{u}_{\mathrm{f}}, q) &= 0, \\
a_{\mathrm{p}}\big(\varphi_{\mathrm{p}}, \psi\big) &= \ell_{\mathrm{p}}(\psi) + (\lambda, \psi)_{\Lambda_{\mathrm{p}}^* \times \Lambda_{\mathrm{p}}} = \ell_{\mathrm{p}}(\psi) + (\boldsymbol{u}_{\mathrm{f}} \cdot \boldsymbol{n}, \psi)_{0,\Gamma_{\mathrm{I}}},
\end{aligned}$$

which is precisely Eq. (6.7).

Conversely, let $(\varphi_{\mathrm{p}}, \boldsymbol{u}_{\mathrm{f}}, p_{\mathrm{f}})$ solve (6.7) and $\lambda := T_{\mathrm{f}}(\boldsymbol{u}_{\mathrm{f}}, p_{\mathrm{f}}) = \boldsymbol{u}_{\mathrm{f}} \cdot \boldsymbol{n}$. Then by definition of $K_{\mathrm{p,N}}$ and $K_{\mathrm{f,N}}$ it is $\varphi_{\mathrm{p}} = K_{\mathrm{p,N}}(\lambda)$ and $(\boldsymbol{u}_{\mathrm{f}}, p_{\mathrm{f}}) = K_{\mathrm{f,N}}(T_{\mathrm{p}}(\varphi_{\mathrm{p}}))$. Hence,

$$\begin{aligned}
\lambda = T_{\mathrm{f}}(K_{\mathrm{f,N}}(T_{\mathrm{p}}(\varphi_{\mathrm{p}}))) &= H_{\mathrm{f,N}}(T_{\mathrm{p}}(\varphi_{\mathrm{p}})) \\
&= H_{\mathrm{f,N}}(T_{\mathrm{p}}(K_{\mathrm{p,N}}(\lambda))) = H_{\mathrm{f,N}}(H_{\mathrm{p,N}}(\lambda)),
\end{aligned}$$

i.e., Eq. (6.12) holds. □

Theorem 6.4.16 (Equivalence of Fixed Point Equation and Weak Formulation, Part 2) *If $\mu \in \widetilde{\Lambda}_{\mathrm{p}}$ solves the interface equation* (6.13)*, then*

$$(\boldsymbol{u}_{\mathrm{f}}, p_{\mathrm{f}}) := K_{\mathrm{f,N}}(\mu) \qquad \textit{and} \qquad \varphi_{\mathrm{p}} := K_{\mathrm{p,N}}(H_{\mathrm{f,N}}(\mu))$$

solve (6.7)*. Conversely if $(\varphi_{\mathrm{p}}, \boldsymbol{u}_{\mathrm{f}}, p_{\mathrm{f}}) \in \widetilde{Q}_{\mathrm{p}} \times \widetilde{\boldsymbol{V}}_{\mathrm{f}} \times Q_{\mathrm{f}}$ solves* (6.7)*, then $\mu = T_{\mathrm{p}}(\varphi_{\mathrm{p}}) \in \widetilde{\Lambda}_{\mathrm{p}}$ solves* (6.13)*.*

Proof Combining Corollary 6.4.14 and Theorem 6.4.15 shows the claim. Alternatively, a very similar proof as in Theorem 6.4.15 can be used directly. □

Remark 6.4.17 Using the identities (6.9) and (6.11), in Theorem 6.4.15 it also is $(\boldsymbol{u}_\mathrm{f}, p_\mathrm{f}) = K_{\mathrm{f,D}}(\lambda)$ and $\varphi_\mathrm{p} = K_{\mathrm{p,D}}(H_{\mathrm{f,D}}(\lambda))$. Similarly in Theorem 6.4.16 it is $\varphi_\mathrm{p} = K_{\mathrm{p,D}}(\mu)$ and $(\boldsymbol{u}_\mathrm{f}, p_\mathrm{f}) = K_{\mathrm{f,D}}(H_{\mathrm{p,D}}(\mu))$.

Once either one of the fixed point equations (6.12) and (6.13) or one of the Steklov–Poincaré equations (6.16) and (6.17) is solved, Corollaries 6.4.13 and 6.4.14 together with the previous theorems provide a solution to the coupled weak formulation (6.7). Thus, it is possible to approach the coupled Stokes–Darcy problem through these interface equations and develop algorithms for them.

6.4.6 *Decoupling the Equations Further*

In this subsection another view on the presented equations is developed. While in the weak formulation (6.7) the solution variables are u, p, and φ, in the interface equations these are λ and μ. Combining these to one formulation where all the above variables are the solutions at once formally decouples the problems in the subdomains further. For all $\boldsymbol{v} \in \boldsymbol{V}_\mathrm{f}$, $q \in Q_\mathrm{f}$, $\psi \in Q_\mathrm{p}$, $\lambda_\mathrm{f} \in \Lambda_\mathrm{f}$, and $\lambda_\mathrm{p} \in \Lambda_\mathrm{p}$ it is

$$\begin{aligned} a_\mathrm{f}(\boldsymbol{u}_\mathrm{f}, \boldsymbol{v}) + b_\mathrm{f}(\boldsymbol{v}, p_\mathrm{f}) + (\mu, \boldsymbol{v}\cdot\boldsymbol{n})_{0,\Gamma_\mathrm{I}} &= \ell_f(\boldsymbol{v}), \\ b_\mathrm{f}(\boldsymbol{u}_\mathrm{f}, q) &= 0, \\ \left(\lambda, \lambda_\mathrm{p}\right)_{0,\Gamma_\mathrm{I}} - \left(\boldsymbol{u}_\mathrm{f}\cdot\boldsymbol{n}, \lambda_\mathrm{p}\right)_{0,\Gamma_\mathrm{I}} &= 0, \\ a_\mathrm{p}(\varphi_\mathrm{p}, \psi) - (\lambda, \psi)_{0,\Gamma_\mathrm{I}} &= \ell_\mathrm{p}(\psi), \\ (\mu, \lambda_\mathrm{f})_{0,\Gamma_\mathrm{I}} - \left(\varphi_\mathrm{p}, \lambda_\mathrm{f}\right)_{0,\Gamma_\mathrm{I}} &= 0. \end{aligned} \tag{6.20}$$

These equations are further decoupled in the sense that the equations in the subdomains do not directly include the solutions from the other subdomain.

6.5 Linear Operators on the Interface

The operators $H_{i,j}$, $i \in \{\mathrm{f},\mathrm{p}\}$, $j \in \{\mathrm{N},\mathrm{D}\}$, have some interesting properties which are inherited by the respective subproblems. However the affine spaces $\widetilde{Q}_\mathrm{p}$, $\widetilde{\boldsymbol{V}}_\mathrm{f}$, $\widetilde{\Lambda}_\mathrm{p}$, and $\widetilde{\Lambda}_f$ are not suitable to define linear operators solving certain subproblems. For $i \in \{\mathrm{f},\mathrm{p}\}$, let $\lambda_{0,i} = H_{i,\mathrm{N}}(0) \in \widetilde{\Lambda}_i$ be in the domain of definition of $H_{i,\mathrm{D}}$. In fact, any $\lambda_{0,i} \in \widetilde{\Lambda}_i$ can be chosen here. The following definition introduces the linear parts of the interface operators $H_{i,j}$ from Definition 6.4.9. While the Neumann operators $H_{i,\mathrm{N}}$ are affine linear this is not possible for the Dirichlet operators $H_{i,\mathrm{D}}$, because they are defined on an affine space. However also in that case similar linear operators exist:

Definition 6.5.1 (Linear Operators on the Interface) Using the operators $H_{i,j}$ from Definition 6.4.9 and $\lambda_{0,i} = H_{i,\mathrm{N}}(0) \in \widetilde{\Lambda}_i$, define

$$
\begin{aligned}
L_{\mathrm{p,N}} &: \Lambda_p^* \to \Lambda_{\mathrm{p}}, & L_{\mathrm{p,N}}(\mu) &= H_{\mathrm{p,N}}(\mu) - \lambda_{0,\mathrm{p}},\\
L_{\mathrm{f,N}} &: \Lambda_f^* \to \Lambda_{\mathrm{f}}, & L_{\mathrm{f,N}}(\mu) &= H_{\mathrm{f,N}}(\mu) - \lambda_{0,\mathrm{f}}.\\
L_{\mathrm{p,D}} &: \Lambda_p \to \Lambda_{\mathrm{p}}^*, & L_{\mathrm{p,D}}(\lambda) &= H_{\mathrm{p,D}}(\lambda + \lambda_{0,\mathrm{p}}) - H_{\mathrm{p,D}}(\lambda_{0,\mathrm{p}}),\\
L_{\mathrm{f,D}} &: \Lambda_{\mathrm{f}} \to \Lambda_{\mathrm{f}}^*, & L_{\mathrm{f,D}}(\lambda) &= H_{\mathrm{f,D}}(\lambda + \lambda_{0,\mathrm{f}}) - H_{\mathrm{f,D}}(\lambda_{0,\mathrm{f}}).
\end{aligned}
$$

The claim that these operators are indeed linear is shown together with other properties next.

6.5.1 Linearity, Invertibility, Symmetry, and Positivity

The operators $L_{i,j}$ from Definition 6.5.1 in fact represent Neumann/Dirichlet data of a solution in the respective subdomain which has zero data everywhere except on the interface Γ_{I}:

Lemma 6.5.2 (Characterization of $L_{i,j}$) *The operators $L_{i,j}$ can be characterized as follows:*

- *for all $\mu \in \Lambda_{\mathrm{p}}^*$ there exists a unique function $\varphi_\mu \in Q_{\mathrm{p}}$ such that $L_{\mathrm{p,N}}(\mu) = T_{\mathrm{p}}\varphi_\mu$ and it holds for all $\psi \in Q_{\mathrm{p}}$*

$$a_{\mathrm{p}}(\varphi_\mu, \psi) = (\mu, \psi)_{\Lambda_{\mathrm{p}}{}^* \times \Lambda_{\mathrm{p}}},$$

- *for all $\lambda \in \Lambda_{\mathrm{p}}$ there exists a unique function $\varphi_\lambda \in Q_{\mathrm{p}}$ such that $L_{\mathrm{p,D}}(\lambda) = T_{\mathrm{p}}^{\mathrm{N}}\varphi_\lambda$, $T_{\mathrm{p}}\varphi_\lambda = \lambda$, and for all $\psi \in Q_{\mathrm{p}}$ with $T_{\mathrm{p}}\psi = 0$ it is*

$$a_{\mathrm{p}}(\varphi_\lambda, \psi) = 0,$$

- *for all $\mu \in \Lambda_{\mathrm{f}}^*$ there exists a unique pair $(\mathbf{u}_\mu, p_\mu) \in \boldsymbol{V}_{\mathrm{f}} \times Q_{\mathrm{f}}$ such that $L_{\mathrm{f,N}}(\mu) = T_{\mathrm{f}}(\mathbf{u}_\mu, p_\mu)$ and for all $(\boldsymbol{v}, q) \in \boldsymbol{V}_{\mathrm{f}} \times Q_{\mathrm{f}}$ it is*

$$
\begin{cases}
a_{\mathrm{f}}\big(\mathbf{u}_\mu, \boldsymbol{v}\big) + b_{\mathrm{f}}(\boldsymbol{v}, p_\mu) = -(\mu, \boldsymbol{v} \cdot \boldsymbol{n})_{\Lambda_{\mathrm{f}}{}^* \times \Lambda_{\mathrm{f}}},\\
b_{\mathrm{f}}(\mathbf{u}_\mu, q) = 0,
\end{cases}
$$

- *for all $\lambda \in \Lambda_{\mathrm{f}}$ there exists a unique pair $(\mathbf{u}_\lambda, p_\lambda) \in \boldsymbol{V}_{\mathrm{f}} \times Q_{\mathrm{f}}$ such that $L_{\mathrm{f,D}}(\lambda) = T_{\mathrm{f}}^{\mathrm{N}}(\mathbf{u}_\lambda, p_\lambda)$, $T_{\mathrm{f}}(\mathbf{u}_\lambda, p_\lambda) = \lambda$, and for all $(\boldsymbol{v}, q) \in \boldsymbol{V}_{\mathrm{f}} \times Q_{\mathrm{f}}$ with $T_{\mathrm{f}}(\boldsymbol{v}, q) = 0$ it is*

$$
\begin{cases}
a_{\mathrm{f}}(\mathbf{u}_\lambda, \boldsymbol{v}) + b_{\mathrm{f}}(\boldsymbol{v}, p_\lambda) = 0,\\
b_{\mathrm{f}}(\mathbf{u}_\lambda, q) = 0.
\end{cases}
$$

Proof First, let $j = \mathrm{N}$, i.e., consider the case $L_{i,\mathrm{N}}(\cdot) = T_i(K_{i,\mathrm{N}}(\cdot)) - T_i(K_{i,\mathrm{N}}(0))$. Let $\varphi_\mu := K_{\mathrm{p,N}}(\mu) - K_{\mathrm{p,N}}(0)$ and $(\mathbf{u}_\mu, p_\mu) := K_{\mathrm{f,N}}(\mu) - K_{\mathrm{f,N}}(0)$, so that $T_\mathrm{p}\varphi_\mu = L_{\mathrm{p,N}}(\mu)$ and $T_\mathrm{f}(\mathbf{u}_\mu, p_\mu) = L_{\mathrm{f,N}}(\mu)$, because the traces T_p and T_f are linear. Furthermore, for all $\psi \in Q_\mathrm{p}$ and all $(\boldsymbol{v}, q) \in \boldsymbol{V}_\mathrm{f} \times Q_\mathrm{f}$, the definition of $K_{i,\mathrm{N}}$ leads to

$$\begin{aligned}
a_\mathrm{p}\big(\varphi_\mu, \psi\big) &= a_\mathrm{p}\big(K_{\mathrm{p,N}}(\mu), \psi\big) - a_\mathrm{p}\big(K_{\mathrm{p,N}}(0), \psi\big)\\
&= \ell_\mathrm{p}(\psi) + \big(\mu, T_\mathrm{p}\psi\big)_{\Lambda_\mathrm{p}{}^*\times\Lambda_\mathrm{p}} - \ell_\mathrm{p}(\psi) = \big(\mu, T_\mathrm{p}\psi\big)_{\Lambda_\mathrm{p}{}^*\times\Lambda_\mathrm{p}},\\
a_\mathrm{f}\big(\mathbf{u}_\mu, \boldsymbol{v}\big) + b_\mathrm{f}(\boldsymbol{v}, p_\mu) &= \ell_\mathrm{f}(\boldsymbol{v}) - (\mu, \boldsymbol{v}\cdot\boldsymbol{n})_{\Lambda_\mathrm{f}{}^*\times\Lambda_\mathrm{f}} - \ell_\mathrm{f}(\boldsymbol{v}) = -(\mu, \boldsymbol{v}\cdot\boldsymbol{n})_{\Lambda_\mathrm{f}{}^*\times\Lambda_\mathrm{f}},\\
b_\mathrm{f}(\mathbf{u}_\mu, p) &= 0,
\end{aligned}$$

as claimed. For $j = \mathrm{D}$, i.e., the Dirichlet operators, and $\lambda \in \Lambda_i$, let $\varphi_\lambda := K_{\mathrm{p,D}}(\lambda+\lambda_{\mathrm{p},0}) - K_{\mathrm{p,D}}(\lambda_\mathrm{p}, 0)$ and $(\mathbf{u}_\lambda, p_\lambda) := K_{\mathrm{f,D}}(\lambda+\lambda_{\mathrm{f},0}) - K_{\mathrm{f,D}}(\lambda_{\mathrm{f},0})$. Then, the identities (6.9) and the linearity of the traces T_p and T_f yield $T_\mathrm{p}\varphi_\lambda = \lambda+\lambda_{\mathrm{p},0}-\lambda_{\mathrm{p},0} = \lambda$ and $T_\mathrm{f}(\mathbf{u}_\lambda, p_\lambda) = \lambda + \lambda_{\mathrm{f},0} - \lambda_{\mathrm{f},0} = \lambda$. Additionally, for all $\psi \in Q_\mathrm{p}$ with $T_\mathrm{p}\psi = 0$ and all $(\boldsymbol{v}, q) \in \boldsymbol{V}_\mathrm{f} \times Q_\mathrm{f}$ with $T_\mathrm{f}(\boldsymbol{v}, q) = 0$, the definition of $K_{i,\mathrm{D}}$ gives

$$\begin{aligned}
a_\mathrm{p}(\varphi_\lambda, \psi) &= a_\mathrm{p}\big(K_{\mathrm{p,D}}(\lambda + \lambda_{\mathrm{p},0}), \psi\big) - a_\mathrm{p}\big(K_{\mathrm{p,D}}(\lambda_{\mathrm{p},0}), \psi\big)\\
&= \ell_\mathrm{p}(\psi) - \ell_\mathrm{p}(\psi) = 0,\\
a_\mathrm{f}(\mathbf{u}_\lambda, \boldsymbol{v}) + b_\mathrm{f}(\boldsymbol{v}, p_\lambda) &= \ell_\mathrm{f}(\boldsymbol{v}) - \ell_\mathrm{f}(\boldsymbol{v}) = 0,\\
b_\mathrm{f}(\mathbf{u}_\lambda, q) &= 0.
\end{aligned}$$

Finally $T_\mathrm{p}^\mathrm{N}\varphi_\lambda = L_{\mathrm{p,D}}(\lambda)$ and $T_\mathrm{f}^\mathrm{N}(\mathbf{u}_\lambda, p_\lambda) = L_{\mathrm{f,D}}(\lambda)$ because taking the Neumann data is a linear operation in case the data away from the interface is zero. □

Compare the previous characterization of the operators $L_{i,j}$ with the definition of $K_{i,j}$. The only difference is that all data $(f_p, g_\mathrm{p}, \varphi_\mathrm{b}, \mathbf{f}_\mathrm{f}, \mathbf{g}_\mathrm{f}, \mathbf{u}_\mathrm{b})$ vanishes for $L_{i,j}$. This is the same setting as in Propositions 5.1.3, 5.1.4, 5.3.7, and 5.3.8, where instead of Robin here Neumann conditions are considered. Hence, these operators are linear and therefore called the linear parts of $H_{i,j}$ in the following.

Definition 6.5.3 (Solution Operators $\mathscr{L}_{i,j}$) Lemma 6.5.2 assures that the operators

$$\begin{aligned}
\mathscr{L}_{\mathrm{p,N}} : \Lambda_\mathrm{p}^* &\to Q_\mathrm{p}, & \mu &\mapsto \varphi_\mu,\\
\mathscr{L}_{\mathrm{p,D}} : \Lambda_\mathrm{p} &\to Q_\mathrm{p}, & \lambda &\mapsto \varphi_\lambda,\\
\mathscr{L}_{\mathrm{f,N}} : \Lambda_\mathrm{f}^* &\to \boldsymbol{V}_\mathrm{f} \times Q_\mathrm{f}, & \mu &\mapsto (\mathbf{u}_\mu, p_\mu), \text{ and}\\
\mathscr{L}_{\mathrm{f,D}} : \Lambda_\mathrm{f} &\to \boldsymbol{V}_\mathrm{f} \times Q_\mathrm{f}, & \lambda &\mapsto (\mathbf{u}_\lambda, p_\lambda),
\end{aligned}$$

are well defined.

As in Sects. 5.1.3 and 5.3.3 these operators are linear and continuous where the continuity constants $c_{\mathscr{L}_{i,j}}$ reduce to c_{T_i}/α_{a_i} for $j = \mathrm{N}$ and $c_{a_i} c_{E_i}/\alpha_{a_i}$ for $j = \mathrm{D}$. Furthermore, similar identities as in Eqs. (6.9) hold, namely

$$T_i \circ \mathscr{L}_{i,\mathrm{D}} = \mathrm{id} \quad \text{in } \Lambda_i, \tag{6.21a}$$

$$\mathscr{L}_{i,\mathrm{D}} \circ T_i = \mathrm{id} \quad \text{in } \mathrm{Im}(\mathscr{L}_{i,\mathrm{N}}) = \mathrm{Im}(\mathscr{L}_{i,\mathrm{D}}), \tag{6.21b}$$

$$T_i^N \circ \mathscr{L}_{i,\mathrm{N}} = \mathrm{id} \quad \text{in } \Lambda_i^*, \tag{6.21c}$$

$$\mathscr{L}_{i,\mathrm{N}} \circ T_i^{\mathrm{N}} = \mathrm{id} \quad \text{in } \mathrm{Im}(\mathscr{L}_{i,\mathrm{N}}) = \mathrm{Im}(\mathscr{L}_{i,\mathrm{D}}). \tag{6.21d}$$

Here the linear spaces Λ_i replace the affine ones ($\widetilde{\Lambda}_i$) and the image spaces are contained in the respective test spaces, namely $\mathrm{Im}(\mathscr{L}_{\mathrm{p,N}}) = \mathrm{Im}(\mathscr{L}_{\mathrm{p,D}}) \subset Q_{\mathrm{p}}$ and $\mathrm{Im}(\mathscr{L}_{\mathrm{f,N}}) = \mathrm{Im}(\mathscr{L}_{\mathrm{f,D}}) \subset \boldsymbol{V}_{\mathrm{f}} \times Q_{\mathrm{f}}$. Furthermore, the $\mathscr{L}_{i,j}$ are to the operators $L_{i,j}$ what $K_{i,j}$ are to $H_{i,j}$, i.e., the operators $L_{i,j}$ could as well be defined analogously to $H_{i,j}$, see Eqs. (6.10),

$$\begin{aligned} L_{\mathrm{p,N}} &= T_{\mathrm{p}} \circ \mathscr{L}_{\mathrm{p,N}}, \qquad & L_{\mathrm{p,D}} &= T_{\mathrm{p}}^{\mathrm{N}} \circ \mathscr{L}_{\mathrm{p,D}}, \\ L_{\mathrm{f,N}} &= T_{\mathrm{f}} \circ \mathscr{L}_{\mathrm{f,N}}, \qquad & L_{\mathrm{f,D}} &= T_{\mathrm{f}}^{\mathrm{N}} \circ \mathscr{L}_{\mathrm{f,D}}. \end{aligned} \tag{6.22}$$

Following Eqs. (6.21), identities similar to those in (6.11) hold

$$L_{\mathrm{p,N}} \circ L_{\mathrm{p,D}} = \mathrm{id} \quad \text{in } \Lambda_{\mathrm{p}}, \tag{6.23a}$$

$$L_{\mathrm{p,D}} \circ L_{\mathrm{p,N}} = \mathrm{id} \quad \text{in } \Lambda_{\mathrm{p}}^*, \tag{6.23b}$$

$$L_{\mathrm{f,N}} \circ L_{\mathrm{f,D}} = \mathrm{id} \quad \text{in } \Lambda_{\mathrm{f}}, \tag{6.23c}$$

$$L_{\mathrm{f,D}} \circ L_{\mathrm{f,N}} = \mathrm{id} \quad \text{in } \Lambda_{\mathrm{f}}^*. \tag{6.23d}$$

Besides linearity also symmetry and positivity of the two subproblems are inherited by the linear parts $L_{i,j}$ introduced above:

Lemma 6.5.4 (Symmetry and Positivity of $L_{i,j}$) *The linear operators $L_{i,j}$, $i \in \{\mathrm{f}, \mathrm{p}\}$ and $j \in \{\mathrm{N}, \mathrm{D}\}$, are symmetric and positive/negative definite, i.e., it is*

$$\left(\xi, L_{i,\mathrm{N}}(\mu)\right)_{\Lambda_i^* \times \Lambda_i} = \left(\mu, L_{i,\mathrm{N}}(\xi)\right)_{\Lambda_i^* \times \Lambda_i},$$
$$\left(L_{i,\mathrm{D}}(\lambda), \xi\right)_{\Lambda_i^* \times \Lambda_i} = \left(L_{i,\mathrm{D}}(\xi), \lambda\right)_{\Lambda_i^* \times \Lambda_i},$$

and there are positive constants $\alpha_{L_{i,j}} > 0$ such that

$$\begin{aligned} &\left(\xi, L_{\mathrm{p,N}}(\xi)\right)_{\Lambda_{\mathrm{p}}{}^* \times \Lambda_{\mathrm{p}}} \geq \alpha_{L_{\mathrm{p,N}}} \|\xi\|^2_{\Lambda_{\mathrm{p}}^*}, \qquad &\left(\xi, -L_{\mathrm{f,N}}(\xi)\right)_{\Lambda_{\mathrm{f}}{}^* \times \Lambda_{\mathrm{f}}} \geq \alpha_{L_{\mathrm{f,N}}} \|\xi\|^2_{\Lambda_{\mathrm{f}}^*}, \\ &\left(L_{\mathrm{p,D}}(\lambda), \lambda\right)_{\Lambda_{\mathrm{p}}{}^* \times \Lambda_{\mathrm{p}}} \geq \alpha_{L_{\mathrm{p,N}}} \|\lambda\|^2_{\Lambda_{\mathrm{p}}}. \qquad &\left(-L_{\mathrm{f,D}}(\lambda), \lambda\right)_{\Lambda_{\mathrm{f}}{}^* \times \Lambda_{\mathrm{f}}} \geq \alpha_{L_{\mathrm{f,N}}} \|\lambda\|^2_{\Lambda_{\mathrm{f}}}. \end{aligned}$$

Proof Using Eqs. (6.21) and (6.22) and the definitions of T_i^N, it is

$$\begin{aligned}
\big(\xi, L_{\mathrm{p,N}}(\mu)\big)_{\Lambda_{\mathrm{p}}^* \times \Lambda_{\mathrm{p}}} &= \Big(T_{\mathrm{p}}^{\mathrm{N}}(\mathscr{L}_{\mathrm{p,N}}(\xi)), T_{\mathrm{p}}(\mathscr{L}_{\mathrm{p,N}}(\mu))\Big)_{\Lambda_{\mathrm{p}}^* \times \Lambda_{\mathrm{p}}} \\
&= a_{\mathrm{p}}(\mathscr{L}_{\mathrm{p,N}}(\xi), \mathscr{L}_{\mathrm{p,N}}(\mu)), \\
\big(\xi, -L_{\mathrm{f,N}}(\mu)\big)_{\Lambda_{\mathrm{p}}^* \times \Lambda_{\mathrm{p}}} &= -\Big(T_{\mathrm{f}}^{\mathrm{N}}(\mathscr{L}_{\mathrm{f,N}}(\xi)), T_{\mathrm{f}}(\mathscr{L}_{\mathrm{f,N}}(\mu))\Big)_{\Lambda_{\mathrm{p}}^* \times \Lambda_{\mathrm{p}}} \\
&= a_{\mathrm{f}}(\mathbf{u}_\xi, \mathbf{u}_\mu),
\end{aligned}$$

where $\mathscr{L}_{\mathrm{f,N}}(k) = (\mathbf{u}_k, p_k)$, $k \in \{\xi, \mu\}$ (the term $b_{\mathrm{f}}(\mathbf{u}_\mu, p_\xi)$ vanishes). Hence, the symmetry of a_i carries over to $L_{\mathrm{p,N}}$ and $-L_{\mathrm{f,N}}$. Furthermore, with the same reasoning also the positivity is inherited by the respective bilinear form a_i and $\mathscr{L}_{i,\mathrm{N}}$. For the Dirichlet operators it is

$$\begin{aligned}
\big(L_{\mathrm{p,D}}(\lambda), \xi\big)_{\Lambda_{\mathrm{p}}^* \times \Lambda_{\mathrm{p}}} &= \Big(T_{\mathrm{p}}^{\mathrm{N}}(\mathscr{L}_{\mathrm{p,D}}(\lambda)), T_{\mathrm{p}}(\mathscr{L}_{\mathrm{p,D}}(\xi))\Big)_{\Lambda_{\mathrm{p}}^* \times \Lambda_{\mathrm{p}}} \\
&= a_{\mathrm{p}}(\mathscr{L}_{\mathrm{p,D}}(\lambda), \mathscr{L}_{\mathrm{p,D}}(\xi)), \\
\big(-L_{\mathrm{f,D}}(\lambda), \xi\big)_{\Lambda_{\mathrm{p}}^* \times \Lambda_{\mathrm{p}}} &= -\Big(T_{\mathrm{f}}^{\mathrm{N}}(\mathscr{L}_{\mathrm{f,D}}(\lambda)), T_{\mathrm{f}}(\mathscr{L}_{\mathrm{f,D}}(\xi))\Big)_{\Lambda_{\mathrm{p}}^* \times \Lambda_{\mathrm{p}}} \\
&= a_{\mathrm{f}}(\mathbf{u}_\lambda, \mathbf{u}_\xi),
\end{aligned}$$

where this time it is $\mathscr{L}_{\mathrm{f,D}}(k) = (\mathbf{u}_k, p_k)$, $k \in \{\xi, \mu\}$. Hence, also in this case the symmetry and positivity of a_i carries over to $L_{\mathrm{p,D}}$ and $L_{\mathrm{f,N}}$. □

In summary, the four operators $L_{\mathrm{p,D}}$, $-L_{\mathrm{f,D}}$, $L_{\mathrm{p,N}}$, and $-L_{\mathrm{f,N}}$ are linear, symmetric, and positive. This furthermore implies that for all appropriate λ, ξ it is

$$\begin{aligned}
\big(L_{\mathrm{f,D}}(\lambda) - L_{\mathrm{p,N}}(\lambda), \xi\big)_{\Lambda_{\mathrm{f}}^* \times \Lambda_{\mathrm{f}}} &= \big(L_{\mathrm{f,D}}(\xi) - L_{\mathrm{p,N}}(\xi), \lambda\big)_{\Lambda_{\mathrm{f}}^* \times \Lambda_{\mathrm{f}}}, \\
\big(L_{\mathrm{f,N}}(\lambda) - L_{\mathrm{p,D}}(\lambda), \xi\big)_{\Lambda_{\mathrm{p}}^* \times \Lambda_{\mathrm{p}}} &= \big(L_{\mathrm{f,N}}(\xi) - L_{\mathrm{p,D}}(\xi), \lambda\big)_{\Lambda_{\mathrm{p}}^* \times \Lambda_{\mathrm{p}}}, \\
\big(L_{\mathrm{f,N}}\big(L_{\mathrm{p,N}}(\lambda)\big), \xi\big)_{\Lambda_{\mathrm{p}} \cap \Lambda_{\mathrm{f}}} &= \big(\lambda, L_{\mathrm{p,N}}\big(L_{\mathrm{f,N}}(\xi)\big)\big)_{\Lambda_{\mathrm{p}} \cap \Lambda_{\mathrm{f}}}.
\end{aligned}$$

The last equation can be interesting because it states that the operators in the fixed point equations (6.12) and (6.13) below are in fact transposed to each other up to some constant terms. Iterative schemes requiring the transposed operator can therefore be exploited as well.

6.5.2 Equations on the Interface Revisited

The aim of this subsection is to write the interface equations of Sect. 6.4.4 in terms of the linear operators $L_{i,j}$ introduced in Sect. 6.5. The fixed point equation (6.12) reads $H_{\mathrm{f,N}}(H_{\mathrm{p,N}}(\widetilde{\lambda})) = \widetilde{\lambda}$ with $\widetilde{\lambda} = \lambda + \lambda_{0,\mathrm{f}} \in \widetilde{\Lambda}_{\mathrm{f}}$ which together with the definitions of $L_{i,\mathrm{N}}$ leads to

$$L_{\mathrm{f,N}}\big(L_{\mathrm{p,N}}(\widetilde{\lambda}) + \lambda_{0,\mathrm{p}}\big) + \lambda_{0,\mathrm{f}} = \widetilde{\lambda}.$$

Denoting $\chi_{\mathrm{f}} = L_{\mathrm{f,N}}(\lambda_{0,\mathrm{p}}) + L_{\mathrm{f,N}}\big(L_{\mathrm{p,N}}(\lambda_{0,\mathrm{f}})\big) \in \Lambda_{\mathrm{f}}$ this can be reformulated as an equation in Λ_{f}:

$$\lambda - L_{\mathrm{f,N}}\big(L_{\mathrm{p,N}}(\lambda)\big) = \chi_{\mathrm{f}}. \tag{6.24}$$

This is not a fixed point but a linear equation with a nonzero right-hand side. Similarly the other fixed point equation (6.13) can be rewritten as

$$\mu - L_{\mathrm{p,N}}\big(L_{\mathrm{f,N}}(\mu)\big) = \chi_{\mathrm{p}}, \tag{6.25}$$

with right-hand side $\chi_{\mathrm{p}} = L_{\mathrm{p,N}}(\lambda_{0,\mathrm{f}}) + L_{\mathrm{p,N}}\big(L_{\mathrm{f,N}}(\lambda_{0,\mathrm{p}})\big) \in \Lambda_{\mathrm{p}}$. Combining Eqs. (6.24) and (6.25) with the inverse identities (6.23) leads to Steklov–Poincaré type equations:

Definition 6.5.5 (Steklov–Poincaré Equations) With the linear operators $L_{i,j}$ from Definition 6.5.1, define the following two Steklov–Poincaré type equations

$$L_{\mathrm{f,D}}(\lambda) - L_{\mathrm{p,N}}(\lambda) = L_{\mathrm{f,D}}(\chi_{\mathrm{f}}) = \lambda_{0,\mathrm{p}} + L_{\mathrm{p,N}}(\lambda_{0,\mathrm{f}}) \in \widetilde{\Lambda}_{\mathrm{p}} \tag{6.26}$$

$$L_{\mathrm{p,D}}(\mu) - L_{\mathrm{f,N}}(\mu) = L_{\mathrm{p,D}}(\chi_{\mathrm{p}}) = \lambda_{0,\mathrm{f}} + L_{\mathrm{f,N}}(\lambda_{0,\mathrm{p}}) \in \widetilde{\Lambda}_{\mathrm{f}}. \tag{6.27}$$

The linear operators are positive/negative definite and therefore these equations have a unique solution:

Theorem 6.5.6 (Existence and Uniqueness, Steklov–Poincaré Equations) *Both of the two Steklov–Poincaré equations* (6.26) *and* (6.27) *admit unique solutions* $\lambda \in \Lambda_{\mathrm{f}}$ *and* $\mu \in \Lambda_{\mathrm{p}}$.

Proof The operators $L_{\mathrm{p,N}} - L_{\mathrm{f,D}}$ and $L_{\mathrm{p,D}} - L_{\mathrm{f,N}}$ both are positive definite (and symmetric) according to Lemma 6.5.4. Therefore, the bilinear and continuous forms

$$e_{\mathrm{f}} : \Lambda_{\mathrm{f}} \times \Lambda_{\mathrm{f}} \to \mathbb{R}, \qquad e_{\mathrm{f}}(\lambda, \eta) := \big(L_{\mathrm{p,N}}(\lambda) - L_{\mathrm{f,D}}(\lambda), \eta\big)_{\Lambda_{\mathrm{f}}^* \times \Lambda_{\mathrm{f}}},$$

$$\text{and} \quad e_{\mathrm{p}} : \Lambda_{\mathrm{p}} \times \Lambda_{\mathrm{p}} \to \mathbb{R}, \qquad e_{\mathrm{p}}(\mu, \eta) := \big(L_{\mathrm{p,D}}(\mu) - L_{\mathrm{f,N}}(\mu), \eta\big)_{\Lambda_{\mathrm{p}}^* \times \Lambda_{\mathrm{p}}},$$

are both coercive. Furthermore, the right-hand sides $\lambda_{0,\mathrm{p}} + L_{\mathrm{p,N}}(\lambda_{0,\mathrm{f}})$ and $\lambda_{0,\mathrm{f}} + L_{\mathrm{f,N}}(\lambda_{0,\mathrm{p}})$ both define linear continuous functions on the trace spaces

Λ_f and Λ_p respectively, namely $\eta_\mathrm{f} \mapsto \left(\lambda_{0,\mathrm{p}} + L_{\mathrm{p,N}}(\lambda_{0,\mathrm{f}}), \eta_\mathrm{f}\right)_{0,\Gamma_\mathrm{I}}$ and $\eta_\mathrm{p} \mapsto \left(\lambda_{0,\mathrm{f}} + L_{\mathrm{f,N}}(\lambda_{0,\mathrm{p}}), \eta_\mathrm{p}\right)_{0,\Gamma_\mathrm{I}}$. Then, using the theorem of Lax–Milgram, Theorem 2.1.13, there exist unique solutions $\lambda \in \Lambda_\mathrm{f}$ and $\mu \in \Lambda_\mathrm{p}$ such that for all $\eta_\mathrm{f} \in \Lambda_\mathrm{f}$ and $\eta_\mathrm{p} \in \Lambda_\mathrm{p}$ it is

$$e_\mathrm{f}(\lambda, \eta_\mathrm{f}) = \left(\lambda_{0,\mathrm{p}} + L_{\mathrm{p,N}}(\lambda_{0,\mathrm{f}}), \eta_\mathrm{f}\right)_{0,\Gamma_\mathrm{I}},$$

$$e_\mathrm{p}(\mu, \eta_\mathrm{p}) = \left(\lambda_{0,\mathrm{f}} + L_{\mathrm{f,N}}(\lambda_{0,\mathrm{p}}), \eta_\mathrm{p}\right)_{0,\Gamma_\mathrm{I}}.$$

With the positive constants $\alpha_{L_{i,j}}$ from Lemma 6.5.4 together with $c_\mathrm{f} = \left\|\lambda_{0,\mathrm{p}} + L_{\mathrm{p,N}}(\lambda_{0,\mathrm{f}})\right\|_{0,\Gamma_\mathrm{I}}$ and $c_\mathrm{p} = \left\|\lambda_{0,\mathrm{f}} + L_{\mathrm{f,N}}(\lambda_{0,\mathrm{p}})\right\|_{0,\Gamma_\mathrm{I}}$ it is

$$\|\lambda\|_{\Lambda_\mathrm{f}} \leq \frac{c_\mathrm{f}}{\min\left\{\alpha_{L_{\mathrm{p,N}}}, \alpha_{L_{\mathrm{f,D}}}\right\}} \quad \text{and} \quad \|\mu\|_{\Lambda_\mathrm{p}} \leq \frac{c_\mathrm{p}}{\min\left\{\alpha_{L_{\mathrm{p,D}}}, \alpha_{L_{\mathrm{f,N}}}\right\}}.$$

□

Note that if $\lambda \in \Lambda_\mathrm{f}$ and $\mu \in \Lambda_\mathrm{p}$ are the unique solutions to the two Steklov–Poincaré equations (6.26) and (6.27), then they also solve the linear equations (6.24) and (6.25). Furthermore, according to the construction the sums $\widetilde{\lambda} = \lambda + \lambda_{0,\mathrm{f}}$ and $\widetilde{\mu} = \mu + \lambda_{0,\mathrm{p}}$ solve the respective fixed point and Steklov–Poincaré equations (6.18) and (6.19). Therefore, with Theorems 6.4.15 and 6.4.16 the above Theorem 6.5.6 provides an alternative proof for the existence and uniqueness of the Stokes–Darcy coupled problem (6.7). In contrast to the proof given in Sect. 6.3, where a bilinear form $\mathcal{A}$ is defined which includes terms from both subdomains, here only properties of the respective subproblems are used. In Sect. 7.3 the expected convergence behavior of several algorithms are studied. Some minor modifications there also show the existence of solutions to the Stokes–Darcy coupled problem.

6.6 Robin–Robin

In Sect. 6.2 the interface conditions (6.3a) and (6.3b) are inserted directly to get a weak formulation where in each subdomain a Neumann-type problem is solved, see Eq. (6.7). In this section the two conditions are replaced by suitable linear combinations. Let $\gamma_\mathrm{f} \geq 0$ and $\gamma_\mathrm{p} > 0$ be two constants. Together with the Beavers–Joseph–Saffman condition (6.3c) the following interface conditions are imposed:

$$\gamma_\mathrm{f}\boldsymbol{u}_\mathrm{f} \cdot \boldsymbol{n} + \boldsymbol{n} \cdot \mathsf{T}(\boldsymbol{u}_\mathrm{f}, p_\mathrm{f}) \cdot \boldsymbol{n} = \gamma_\mathrm{f}\boldsymbol{u}_\mathrm{p} \cdot \boldsymbol{n} - \varphi_\mathrm{p} \quad \text{on } \Gamma_\mathrm{I}, \tag{6.28a}$$

$$-\gamma_\mathrm{p}\boldsymbol{u}_\mathrm{p} \cdot \boldsymbol{n} - \varphi_\mathrm{p} = -\gamma_\mathrm{p}\boldsymbol{u}_\mathrm{f} \cdot \boldsymbol{n} + \boldsymbol{n} \cdot \mathsf{T}(\boldsymbol{u}_\mathrm{f}, p_\mathrm{f}) \cdot \boldsymbol{n} \quad \text{on } \Gamma_\mathrm{I}. \tag{6.28b}$$

Remark 6.6.1 (Motivation) In Chap. 7 it is shown that iterative methods based on the Neumann–Neumann weak formulation, Definition 6.2.1, are slow or diverging whenever the viscosity ν and hydraulic conductivity K are small. The two modified interface conditions (6.28) above lead to Robin problems in each subdomain and iterative algorithms based on this approach converge reasonably fast even for small ν and K.

The overall problem to be solved is defined by Eqs. (6.1a), (6.1b), and (6.4), together with the boundary and interface conditions (6.2), (6.3c), and (6.28), where $\boldsymbol{u}_\mathrm{p} \cdot \boldsymbol{n}$ is replaced by $-\mathsf{K}\nabla\varphi_\mathrm{p} \cdot \boldsymbol{n}$:

$$\begin{cases} -\nabla \cdot \mathsf{T}(\boldsymbol{u}_\mathrm{f}, p_\mathrm{f}) = \mathbf{f}_\mathrm{f} & \text{in } \Omega_\mathrm{f}, \\ \nabla \cdot \boldsymbol{u}_\mathrm{f} = 0 & \text{in } \Omega_\mathrm{f}, \\ -\nabla \cdot \left(\mathsf{K}\nabla\varphi_\mathrm{p}\right) = f_\mathrm{p} & \text{in } \Omega_\mathrm{p}, \\ \boldsymbol{u}_\mathrm{f} = \mathbf{u}_\mathrm{b} & \text{on } \Gamma_\mathrm{f,D}, \\ \mathsf{T}(\boldsymbol{u}_\mathrm{f}, p_\mathrm{f}) \cdot \boldsymbol{n} = \mathbf{g}_\mathrm{f} & \text{on } \Gamma_\mathrm{f,N}, \\ -\mathsf{K}\nabla\varphi_\mathrm{p} \cdot \boldsymbol{n} = g_\mathrm{p} & \text{on } \Gamma_\mathrm{p,N}, \\ \varphi_\mathrm{p} = \varphi_\mathrm{b} & \text{on } \Gamma_\mathrm{p,D}, \\ \gamma_\mathrm{f}\boldsymbol{u}_\mathrm{f} \cdot \boldsymbol{n} + \boldsymbol{n} \cdot \mathsf{T}(\boldsymbol{u}_\mathrm{f}, p_\mathrm{f}) \cdot \boldsymbol{n} = -\gamma_\mathrm{f}\mathsf{K}\nabla\varphi_\mathrm{p} \cdot \boldsymbol{n} - \varphi_\mathrm{p} & \text{on } \Gamma_\mathrm{I}, \\ \gamma_\mathrm{p}\mathsf{K}\nabla\varphi_\mathrm{p} \cdot \boldsymbol{n} - \varphi_\mathrm{p} = -\gamma_\mathrm{p}\boldsymbol{u}_\mathrm{f} \cdot \boldsymbol{n} + \boldsymbol{n} \cdot \mathsf{T}(\boldsymbol{u}_\mathrm{f}, p_\mathrm{f}) \cdot \boldsymbol{n} & \text{on } \Gamma_\mathrm{I}, \\ \boldsymbol{u}_\mathrm{f} \cdot \boldsymbol{\tau}_i + \alpha\boldsymbol{\tau}_i \cdot \mathsf{T}(\boldsymbol{u}_\mathrm{f}, p_\mathrm{f}) \cdot \boldsymbol{n} = 0 & \text{on } \Gamma_\mathrm{I}, \end{cases} \tag{6.29}$$

with $i = 1, \ldots, d-1$.

6.6.1 Weak Form

The same steps as in the Neumann–Neumann case in Sect. 6.2 are done here, namely multiplication with a suitable test function, integration by parts, followed by insertion of boundary data and interface conditions. The involved spaces do not change, i.e., it remains

$$\begin{aligned} \boldsymbol{V}_\mathrm{f} &= \mathbf{H}^1_{\Gamma_\mathrm{f,D}}(\Omega_\mathrm{f}), \\ Q_\mathrm{f} &= L^2(\Omega_\mathrm{f}), \\ Q_\mathrm{p} &= H^1_{\Gamma_\mathrm{p,D}}(\Omega_\mathrm{p}), \end{aligned}$$

together with the standard norms:

$$\|\boldsymbol{v}\|_{\boldsymbol{V}_\mathrm{f}} := \|\boldsymbol{v}\|_{1,\Omega_\mathrm{f}}, \qquad \|\mathrm{p}\|_{Q_\mathrm{f}} := \|p\|_{0,\Omega_\mathrm{f}}, \qquad \|\psi\|_{Q_\mathrm{p}} := \|\psi\|_{1,\Omega_\mathrm{p}}.$$

The weak formulation is then: Find $(\boldsymbol{u}_\mathrm{f}, p_\mathrm{f}, \varphi_\mathrm{p}) \in \mathbf{H}^1(\Omega_\mathrm{f}) \times L^2(\Omega_\mathrm{f}) \times H^1(\Omega_\mathrm{p})$, such that for all $\boldsymbol{v} \in \boldsymbol{V}_\mathrm{f}$, $q \in Q_\mathrm{f}$ and all $\psi \in Q_\mathrm{p}$ it is

$$\begin{cases} a_\mathrm{f}^R(\boldsymbol{u}_\mathrm{f}, \boldsymbol{v}) + b_\mathrm{f}(\boldsymbol{v}, p_\mathrm{f}) + \big(\varphi_\mathrm{p}, \boldsymbol{v}\cdot\boldsymbol{n}\big)_{0,\Gamma_\mathrm{I}} + \gamma_\mathrm{f}\big(\mathsf{K}\nabla\varphi_\mathrm{p}\cdot\boldsymbol{n}, \boldsymbol{v}\cdot\boldsymbol{n}\big)_{0,\Gamma_\mathrm{I}} = \ell_\mathrm{f}(\boldsymbol{v}) \\ \hfill b_\mathrm{f}(\boldsymbol{u}_\mathrm{f}, q) = 0, \\ a_\mathrm{p}^R(\varphi_\mathrm{p}, \psi) - (\boldsymbol{u}_\mathrm{f}\cdot\boldsymbol{n}, \psi)_{0,\Gamma_\mathrm{I}} + \dfrac{1}{\gamma_\mathrm{p}}(\boldsymbol{n}\cdot\mathsf{T}(\boldsymbol{u}_\mathrm{f}, p_\mathrm{f})\cdot\boldsymbol{n}, \psi)_{0,\Gamma_\mathrm{I}} = \ell_\mathrm{p}(\psi), \end{cases} \tag{6.30}$$

and $\boldsymbol{u}_\mathrm{f} = \mathbf{u}_\mathrm{b}$ on $\Gamma_\mathrm{f,D}$ as well as $\varphi_\mathrm{p} = \varphi_\mathrm{b}$ on $\Gamma_\mathrm{p,D}$. The bilinear forms $a_\mathrm{f}^R : \mathbf{H}^1(\Omega_\mathrm{f}) \times \mathbf{H}^1(\Omega_\mathrm{f}) \to \mathbb{R}$, $b_\mathrm{f} : \mathbf{H}^1(\Omega_\mathrm{f}) \times L^2(\Omega_\mathrm{f}) \to \mathbb{R}$, and $a_\mathrm{p}^R : H^1(\Omega_\mathrm{p}) \times H^1(\Omega_\mathrm{p}) \to \mathbb{R}$ are defined as

$$\begin{aligned} a_\mathrm{f}^R(\mathbf{u}, \boldsymbol{v}) &= (2\nu\mathsf{D}(\mathbf{u}), \mathsf{D}(\boldsymbol{v}))_{0,\Omega_\mathrm{f}} + \frac{1}{\alpha}\sum_{i=1}^{d-1}(\mathbf{u}\cdot\boldsymbol{\tau}_i, \boldsymbol{v}\cdot\boldsymbol{\tau}_i)_{0,\Gamma_\mathrm{I}} + \gamma_\mathrm{f}(\mathbf{u}\cdot\boldsymbol{n}, \boldsymbol{v}\cdot\boldsymbol{n})_{0,\Gamma_\mathrm{I}} \\ &= a_\mathrm{f}(\mathbf{u}, \boldsymbol{v}) + \gamma_\mathrm{f}(\mathbf{u}\cdot\boldsymbol{n}, \boldsymbol{v}\cdot\boldsymbol{n})_{0,\Gamma_\mathrm{I}}, \\ b_\mathrm{f}(\boldsymbol{v}, p) &= -(\nabla\cdot\boldsymbol{v}, p)_{0,\Omega_\mathrm{f}}, \\ a_\mathrm{p}^R(\varphi, \psi) &= (\mathsf{K}\nabla\varphi, \nabla\psi)_{0,\Omega_\mathrm{p}} + \frac{1}{\gamma_\mathrm{p}}(\varphi, \psi)_{0,\Gamma_\mathrm{I}} \\ &= a_\mathrm{p}(\varphi, \psi) + \frac{1}{\gamma_\mathrm{p}}(\varphi, \psi)_{0,\Gamma_\mathrm{I}}, \end{aligned}$$

while the right-hand sides $\ell_\mathrm{f} \in \boldsymbol{V}_\mathrm{f}^*$ and $\ell_\mathrm{p} \in Q_\mathrm{p}^*$ are unchanged:

$$\begin{aligned} \ell_\mathrm{f}(\boldsymbol{v}) &= (\mathbf{f}_\mathrm{f}, \boldsymbol{v})_0 + (\mathbf{g}_\mathrm{f}, \boldsymbol{v})_{0,\Gamma_\mathrm{f,N}}, \\ \ell_\mathrm{p}(\psi) &= \big(f_\mathrm{p}, \psi\big)_0 - \big(g_\mathrm{p}, \psi\big)_{0,\Gamma_\mathrm{p,N}}. \end{aligned}$$

In Eq. (6.30) the terms involving γ_f and γ_p only make sense if the Robin data $\varphi_\mathrm{p} + \gamma_\mathrm{f}\mathsf{K}\nabla\varphi_\mathrm{p}\cdot\boldsymbol{n}$ and $-\boldsymbol{u}_\mathrm{f}\cdot\boldsymbol{n} + {\gamma_\mathrm{p}}^{-1}\boldsymbol{n}\cdot\mathsf{T}(\boldsymbol{u}_\mathrm{f}, p_\mathrm{f})\cdot\boldsymbol{n}$ are in $L^2(\Gamma_\mathrm{I})$ which is in general not true. Note that the Neumann data T_i^N extends the notion of $\mathsf{K}\nabla\varphi_\mathrm{p}\cdot\boldsymbol{n}$ and $\boldsymbol{n}\cdot\mathsf{T}(\boldsymbol{u}_\mathrm{f}, p_\mathrm{f})\cdot\boldsymbol{n}$ to be in Λ_i^*, see Definitions 6.4.5 and 6.4.6. However, the terms $\big(\mathsf{K}\nabla\varphi_\mathrm{p}\cdot\boldsymbol{n}, \boldsymbol{v}\cdot\boldsymbol{n}\big)_{0,\Gamma_\mathrm{I}}$ and $(\boldsymbol{n}\cdot\mathsf{T}(\boldsymbol{u}_\mathrm{f}, p_\mathrm{f})\cdot\boldsymbol{n}, \psi)_{0,\Gamma_\mathrm{I}}$ pair objects in $\Lambda_\mathrm{p}^* \times \Lambda_\mathrm{f}$ and $\Lambda_\mathrm{f}^* \times \Lambda_\mathrm{p}$, respectively which can only be meaningful if

$$\Lambda_\mathrm{p} = \Lambda_\mathrm{f}, \tag{6.31}$$

which is therefore now implicitly assumed. Then the weak form can be reformulated to be

$$\begin{cases} a_{\mathrm{f}}^{\mathrm{R}}(\boldsymbol{u}_{\mathrm{f}}, \boldsymbol{v}) + b_{\mathrm{f}}(\boldsymbol{v}, p_{\mathrm{f}}) + c_{\mathrm{f}}(\varphi_{\mathrm{p}}, \boldsymbol{v}) = \ell_{\mathrm{f}}(\boldsymbol{v}) + \ell_{\mathrm{f,p}}(\boldsymbol{v}), \\ b_{\mathrm{f}}(\boldsymbol{u}_{\mathrm{f}}, q) = 0, \\ a_{\mathrm{p}}^{\mathrm{R}}(\varphi_{\mathrm{p}}, \psi) + c_{\mathrm{p}}((\boldsymbol{u}_{\mathrm{f}}, p_{\mathrm{f}}), \psi) = \ell_{\mathrm{p}}(\psi) + \ell_{\mathrm{p,f}}(\psi), \end{cases} \tag{6.32}$$

with the coupling terms and additional right-hand sides

$$\begin{aligned} c_{\mathrm{f}}(\varphi, \boldsymbol{v}) &= (\varphi, \boldsymbol{v}\cdot\boldsymbol{n})_{0,\Gamma_{\mathrm{I}}} - \gamma_{\mathrm{f}} a_{\mathrm{p}}(\varphi, \psi_{\boldsymbol{v}}), \\ c_{\mathrm{p}}((\mathbf{u}, p), \psi) &= -(\mathbf{u}\cdot\boldsymbol{n}, \psi)_{0,\Gamma_{\mathrm{I}}} + \frac{1}{\gamma_{\mathrm{p}}} a_{\mathrm{f}}(\mathbf{u}, \boldsymbol{v}_{\psi}) + \frac{1}{\gamma_{\mathrm{p}}} b_{\mathrm{f}}(\boldsymbol{v}_{\psi}, p), \\ \ell_{\mathrm{f,p}}(\boldsymbol{v}) &= -\gamma_{\mathrm{f}} \ell_{\mathrm{p}}(\psi_{\boldsymbol{v}}), \\ \ell_{\mathrm{p,f}}(\psi) &= \frac{1}{\gamma_{\mathrm{p}}} \ell_{\mathrm{f}}(\boldsymbol{v}_{\psi}), \end{aligned}$$

where $\boldsymbol{v}_{\psi}$ is an extension of ψ to $\boldsymbol{V}_{\mathrm{f}}$ (with $\boldsymbol{v}_{\psi}\cdot\boldsymbol{n} = \psi$) and $\psi_{\boldsymbol{v}}$ extends $\boldsymbol{v}\cdot\boldsymbol{n}$ to Q_{p}. Note that these definitions do not depend on the particular choice of extensions as long as the same ones are used for both c_{f} and $\ell_{\mathrm{f,p}}$ as well as for c_{p} and $\ell_{\mathrm{p,f}}$, compare with Definitions 5.1.5 and 5.3.9.

Remark 6.6.2 Another way to obtain the Robin–Robin weak form (6.32) is to add multiples of the terms $(\mathbf{u}\cdot\boldsymbol{n}, \boldsymbol{v}\cdot\boldsymbol{n})_{0,\Gamma_{\mathrm{I}}} + \left(\mathsf{K}\nabla\varphi_{\mathrm{p}}\cdot\boldsymbol{n}, \boldsymbol{v}\cdot\boldsymbol{n}\right)_{0,\Gamma_{\mathrm{I}}}$ to the first and $(\varphi, \psi)_{0,\Gamma_{\mathrm{I}}} + (\boldsymbol{n}\cdot\mathsf{T}(\boldsymbol{u}_{\mathrm{f}}, p_{\mathrm{f}})\cdot\boldsymbol{n}, \psi)_{0,\Gamma_{\mathrm{I}}}$ to the last equation in the Neumann–Neumann weak formulation (6.7). These terms are consistent in the sense that they vanish for the (unique) solution of the Neumann–Neumann problem. This leads to a strategy for the proof of the equivalence to the Neumann–Neumann formulation in Eq. (6.6.3).

6.6.2 Remarks on the Condition $\Lambda_{\mathrm{p}} = \Lambda_{\mathrm{f}}$

Starting with the formulation (6.29) where the interface conditions are of Robin type, a meaningful weak formulation is derived under the condition $\Lambda_{\mathrm{p}} = \Lambda_{\mathrm{f}}$, i.e., Eq. (6.31). There are two common situations where this condition is violated. One is concerned with the boundary at the interface Γ_{I}. Note that Λ_{p} is the subspace of functions $\lambda \in H^{1/2}(\Gamma_{\mathrm{I}})$ which can be extended by zero to $H^{1/2}(\Gamma_{\mathrm{I}} \cup \Gamma_{\mathrm{p,D}})$, so λ must vanish at $\overline{\Gamma_{\mathrm{I}}} \cap \overline{\Gamma_{\mathrm{p,D}}}$ fast enough, see Theorem 4.3.4, the subsequent discussion, and Proposition 4.3.6. Similar statements are true for Λ_{f}. This means, condition (6.31) can only hold if the essential boundary parts in the respective subdomains touch the interface at the same points, i.e., if $\overline{\Gamma_{\mathrm{p,D}}} \cap \overline{\Gamma_{\mathrm{I}}} = \overline{\Gamma_{\mathrm{f,D}}} \cap \overline{\Gamma_{\mathrm{I}}}$. A special situation is one

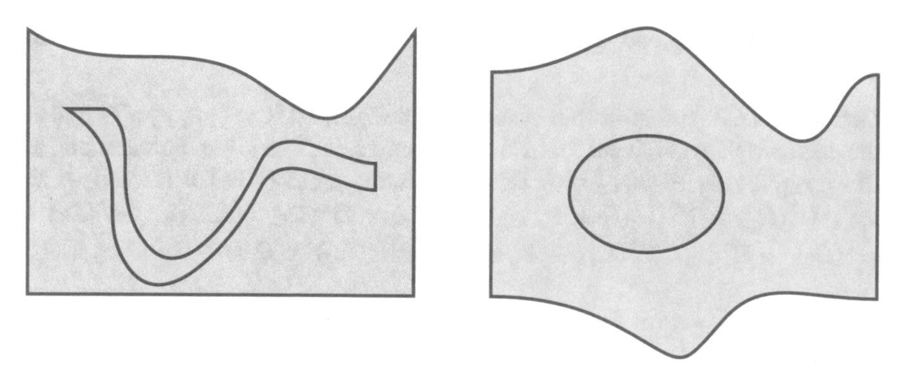

Fig. 6.3 Sketch of two examples where one subdomain is completely enclosed by the other so that the interface Γ_{I} has no boundary

subdomain being completely inside the other, for example a porous obstacle in a free flow or a crack inside a porous medium, see Fig. 6.3. In this case the boundary of Γ_{I} is empty and therefore $\overline{\Gamma_{\mathrm{I}}} \cap \Gamma_{i,\mathrm{D}} = \emptyset$, $i \in \{\mathrm{p}, \mathrm{f}\}$. Hence, the first common situation where the condition $\Lambda_{\mathrm{p}} = \Lambda_{\mathrm{f}}$ is violated does not apply here. Some analysis is devoted to this special case in [GOS11a, GOS11b]. The other possible situation where the condition $\Lambda_{\mathrm{p}} = \Lambda_{\mathrm{f}}$ is violated is the presence of a nonsmooth interface. While the trace of a function $\boldsymbol{v} \in \mathbf{H}^1(\Omega_{\mathrm{f}})$ is in the vector-valued space $\mathbf{H}^{1/2}(\Gamma_{\mathrm{I}})$, its normal component $\boldsymbol{v} \cdot \boldsymbol{n}$ on the interface need not be in $H^{1/2}(\Gamma_{\mathrm{I}})$ because the normal might have jumps, see also Remark 6.2.2. Then Λ_{f} contains $H^{1/2}(\Gamma_{\mathrm{I}})$ but is not equal, while Λ_{p} is a subspace. In the left picture of Fig. 6.3 the interface is only Lipschitz continuous and therefore $\Lambda_{\mathrm{p}} \neq \Lambda_{\mathrm{f}}$, while in the right picture the interface is smooth and hence $\Lambda_{\mathrm{p}} = \Lambda_{\mathrm{f}}$.

In many applications the interface is modeled to be Lipschitz continuous, for example piecewise linear. Also the condition $\overline{\Gamma_{\mathrm{p,D}}} \cap \overline{\Gamma_{\mathrm{I}}} = \overline{\Gamma_{\mathrm{f,D}}} \cap \overline{\Gamma_{\mathrm{I}}}$ is not in general satisfied such that it is quite common that the interface spaces do not coincide. The Robin–Robin formulation can then be viewed as the Neumann–Neumann formulation (6.7) but with additional terms, which have to be suitably adapted whenever the extensions $\psi_{\boldsymbol{v}} \in Q_{\mathrm{p}}$ of $\boldsymbol{v} \cdot \boldsymbol{n} \in \Lambda_{\mathrm{f}}$ and $\boldsymbol{v}_\psi \in \boldsymbol{V}_{\mathrm{f}}$ of $\psi \in \Lambda_{\mathrm{p}}$ do not exist. If the interface is piecewise smooth, with only finitely many kinks, then in some sense $(\Lambda_{\mathrm{p}} \cup \Lambda_{\mathrm{f}}) \setminus (\Lambda_{\mathrm{p}} \cap \Lambda_{\mathrm{f}})$ is small and these extensions mostly do exist.

The above considerations therefore lead to a definition of a Robin–Robin weak formulation even if $\Lambda_{\mathrm{p}} \neq \Lambda_{\mathrm{f}}$. However in such a case the conditions on the interface are fulfilled only in some weaker sense. To this end, define $P_i : \Lambda_j \to \Lambda_i$, $i \neq j$, such that $P_i(\xi) = \xi$ for all $\xi \in \Lambda_j \cap \Lambda_i$ and $P_i(\xi) = 0$ if $\xi \in \Lambda_j \setminus \Lambda_i$, i.e., $P_i = \chi_{\Lambda_i}\,\mathrm{id}$, where χ_M is the indicator function of the set M and id the identity. The functions P_i have the property $P_i \circ P_i = P_i$ but are no projections, because the spaces Λ_i on the interface Γ_{I} are not closed subspaces. Also both P_i are not continuous, however, they are bounded. With the help of the functions P_i, the bilinear forms $a_{\mathrm{f}}^{\mathrm{R}}$ and $a_{\mathrm{p}}^{\mathrm{R}}$, the coupling terms c_{f} and c_{f}, as well as additional right-hand sides $\ell_{\mathrm{f,p}}$ and $\ell_{\mathrm{p,f}}$, which are used in the weak formulation (6.32), are extended:

Definition 6.6.3 (Robin–Robin Weak Formulation) Let $\gamma_\mathrm{f} \geq 0$, $\gamma_\mathrm{p} > 0$, and all the data as well as (bi)linear forms from Definition 6.2.1 be given. Furthermore, let $P_i = \chi_{\Lambda_i}\,\mathrm{id} : \Lambda_j \to \Lambda_i$, $j \neq i$. Define the forms $a_\mathrm{f}^\mathrm{R} : \boldsymbol{H}^1(\Omega_\mathrm{f}) \times \boldsymbol{H}^1(\Omega_\mathrm{f}) \to \mathbb{R}$, $a_\mathrm{p}^\mathrm{R} : H^1(\Omega_\mathrm{p}) \times H^1(\Omega_\mathrm{p}) \to \mathbb{R}$, $c_\mathrm{f} : H^1(\Omega_\mathrm{p}) \times \boldsymbol{H}^1(\Omega_\mathrm{f}) \to \mathbb{R}$, $c_\mathrm{p} : \boldsymbol{H}^1(\Omega_\mathrm{f}) \times H^1(\Omega_\mathrm{p}) \to \mathbb{R}$, $\ell_{\mathrm{f,p}} : \boldsymbol{H}^1(\Omega_\mathrm{f}) \to \mathbb{R}$, and $\ell_{\mathrm{p,f}} : H^1(\Omega_\mathrm{p}) \to \mathbb{R}$ as

$$
\begin{aligned}
a_\mathrm{f}^\mathrm{R}(\mathbf{u}, \boldsymbol{v}) &= a_\mathrm{f}(\mathbf{u}, \boldsymbol{v}) + \gamma_\mathrm{f}\big(\mathbf{u}\cdot\boldsymbol{n}, P_\mathrm{p}(\boldsymbol{v}\cdot\boldsymbol{n})\big)_{0,\Gamma_\mathrm{I}},\\
a_\mathrm{p}^\mathrm{R}(\varphi, \psi) &= a_\mathrm{p}(\varphi, \psi) + \frac{1}{\gamma_\mathrm{p}}(\varphi, P_\mathrm{f}(\psi))_{0,\Gamma_\mathrm{I}},\\
c_\mathrm{f}(\varphi, \boldsymbol{v}) &= (\varphi, \boldsymbol{v}\cdot\boldsymbol{n})_{0,\Gamma_\mathrm{I}} - \gamma_\mathrm{f} a_\mathrm{p}\big(\varphi, \psi_{P_\mathrm{p}(\boldsymbol{v}\cdot\boldsymbol{n})}\big),\\
c_\mathrm{p}((\mathbf{u}, p), \psi) &= -(\mathbf{u}\cdot\boldsymbol{n}, \psi)_{0,\Gamma_\mathrm{I}} + \frac{1}{\gamma_\mathrm{p}} a_\mathrm{f}\big(\mathbf{u}, \boldsymbol{v}_{P_\mathrm{f}(\psi)}\big) + \frac{1}{\gamma_\mathrm{p}} b_\mathrm{f}\big(\boldsymbol{v}_{P_\mathrm{f}(\psi)}, p\big),\\
\ell_{\mathrm{f,p}}(\boldsymbol{v}) &= -\gamma_\mathrm{f}\ell_\mathrm{p}\big(\psi_{P_\mathrm{p}(\boldsymbol{v}\cdot\boldsymbol{n})}\big),\\
\ell_{\mathrm{p,f}}(\psi) &= \frac{1}{\gamma_\mathrm{p}}\ell_\mathrm{f}\big(\boldsymbol{v}_{P_\mathrm{f}(\psi)}\big).
\end{aligned}
\tag{6.33}
$$

Then the Robin–Robin weak formulation of the Stokes–Darcy coupled problem is: Find $(\boldsymbol{u}_\mathrm{f}, p_\mathrm{f}, \varphi_\mathrm{p}) \in \mathbf{H}^1(\Omega_\mathrm{f}) \times L^2(\Omega_\mathrm{f}) \times H^1(\Omega_\mathrm{p})$, such that for all $\boldsymbol{v} \in \boldsymbol{V}_\mathrm{f}$, $q \in Q_\mathrm{f}$, and all $\psi \in Q_\mathrm{p}$ it is

$$
\begin{cases}
a_\mathrm{f}^\mathrm{R}(\boldsymbol{u}_\mathrm{f}, \boldsymbol{v}) + b_\mathrm{f}(\boldsymbol{v}, p_\mathrm{f}) + c_f(\varphi_\mathrm{p}, \boldsymbol{v}) = \ell_\mathrm{f}(\boldsymbol{v}) + \ell_{\mathrm{f,p}}(\boldsymbol{v})\\
b_\mathrm{f}(\boldsymbol{u}_\mathrm{f}, q) = 0\\
a_\mathrm{p}^\mathrm{R}(\varphi_\mathrm{p}, \psi) + c_\mathrm{p}((\boldsymbol{u}_\mathrm{f}, p_\mathrm{f}), \psi) = \ell_\mathrm{p}(\psi) + \ell_{\mathrm{p,f}}(\psi),
\end{cases}
\tag{6.34}
$$

together with $\boldsymbol{u}_\mathrm{f} = \mathbf{u}_\mathrm{b}$ on $\Gamma_{\mathrm{f,D}}$ and $\varphi_\mathrm{p} = \varphi_\mathrm{b}$ on $\Gamma_{\mathrm{p,D}}$.

Essentially the above extensions (6.34) can be summarized as follows: Whenever a trace of a test function ($\boldsymbol{v}$ or ψ) is not in $\Lambda_\mathrm{f} \cap \Lambda_\mathrm{p}$ it is set to zero in some terms and Eq. (6.34) reduce to the Neumann–Neumann system (6.7). Note that the forms above indeed are not (bi)linear, because the functions P_p and P_f are not. However, if $\Lambda_\mathrm{f} = \Lambda_\mathrm{p}$ this extended weak formulation reduces to that in Eq. (6.32) and all the defined forms are (bi)linear.

6.6.3 Equivalence to the Neumann–Neumann Formulation

It is shown in Sect. 6.3 that the Neumann–Neumann weak formulation (6.7) admits a unique solution. Furthermore, the difference to the Robin–Robin weak formulation is the addition of terms which are consistent. This is made more precise in the following theorem:

Theorem 6.6.4 (Equivalence of Weak Forms) *Let* $(\boldsymbol{u}_{\mathrm{f}}, p_{\mathrm{f}}, \varphi_{\mathrm{p}})$ *be the unique solution to the Neumann–Neumann weak formulation* (6.7)*, Definition 6.2.1. Then this solution also solves the Robin–Robin formulation* (6.34)*, Definition 6.6.3. Conversely, a solution of* (6.34) *also solves* (6.7)*.*

Proof Let $(\boldsymbol{v}, q, \psi) \in \boldsymbol{V}_{\mathrm{f}} \times Q_{\mathrm{f}} \times Q_{\mathrm{p}}$. If the function $P_{\mathrm{p}}(T_{\mathrm{f}}(\boldsymbol{v}, q)) = P_{\mathrm{p}}(\boldsymbol{v} \cdot \boldsymbol{n})$ onto Λ_{p} vanishes, the first equation in (6.34) reduces to that of (6.7), similarly for the last equation if $P_{\mathrm{f}}(T_{\mathrm{p}}\psi) = 0$. Now assume $P_{\mathrm{p}}(T_{\mathrm{f}}(\boldsymbol{v}, q)) \neq 0$ and denote its extension to Q_{p} by $\psi_{\boldsymbol{v}} = \psi_{P_{\mathrm{p}}(\boldsymbol{v}\cdot\boldsymbol{n})}$. Then the additional terms in the first equation of (6.34) are

$$\gamma_{\mathrm{f}}(\boldsymbol{u}_{\mathrm{f}} \cdot \boldsymbol{n}, \psi_{\boldsymbol{v}})_{0,\Gamma_{\mathrm{I}}} - \gamma_{\mathrm{f}} a_{\mathrm{p}}(\varphi, \psi_{\boldsymbol{v}}) = \gamma_{\mathrm{f}} \ell_{\mathrm{p}}(\psi_{\boldsymbol{v}}).$$

This identity is the last equation in (6.7) with $\psi = \psi_{\boldsymbol{v}}$ and therefore holds for all such $\boldsymbol{v}$. Now consider the case where $P_{\mathrm{f}}(T_{\mathrm{p}}\psi) \neq 0$. Then the additional terms in the last equation in the Robin–Robin formulation (6.34) read

$$\frac{1}{\gamma_{\mathrm{p}}}\big(\varphi_{\mathrm{p}}, \psi\big)_{0,\Gamma_{\mathrm{I}}} + \frac{1}{\gamma_{\mathrm{p}}} a_{\mathrm{f}}(\boldsymbol{u}_{\mathrm{f}}, \boldsymbol{v}_{\psi}) + \frac{1}{\gamma_{\mathrm{p}}} b_{\mathrm{f}}(\boldsymbol{v}_{\psi}, p_{\mathrm{f}}) = \frac{1}{\gamma_{\mathrm{p}}} \ell_{\mathrm{f}}(\boldsymbol{v}_{\psi}),$$

where $\boldsymbol{v}_{\psi}$ denotes the extension of $P_{\mathrm{f}}(T_{\mathrm{p}}\psi)$ to $\boldsymbol{V}_{\mathrm{f}}$, i.e., $\boldsymbol{v}_{\psi} \cdot \boldsymbol{n} = P_{\mathrm{f}}(T_{\mathrm{p}}\psi)$. Therefore, this identity holds because of the first equation in the Neumann–Neumann formulation (6.7).

Now let $(\boldsymbol{u}_{\mathrm{f}}, p_{\mathrm{f}}, \varphi_{\mathrm{p}})$ be a solution to the Robin–Robin formulation (6.34). As a reminder the relevant two equations of (6.34) are rewritten here:

$$\begin{aligned} a_{\mathrm{f}}(\boldsymbol{u}_{\mathrm{f}}, \boldsymbol{v}) + b_{\mathrm{f}}(\boldsymbol{v}, p_{\mathrm{f}}) + \gamma_{\mathrm{f}}\big(\boldsymbol{u}_{\mathrm{f}} \cdot \boldsymbol{n}, P_{\mathrm{p}}(\boldsymbol{v} \cdot \boldsymbol{n})\big)_{0,\Gamma_{\mathrm{I}}} + \big(\varphi_{\mathrm{p}}, \boldsymbol{v} \cdot \boldsymbol{n}\big)_{0,\Gamma_{\mathrm{I}}} \\ - \gamma_{\mathrm{f}} a_{\mathrm{p}}(\varphi_{\mathrm{p}}, \psi_{\boldsymbol{v}}) = \ell_{\mathrm{f}}(\boldsymbol{v}) - \gamma_{\mathrm{f}} \ell_{\mathrm{p}}(\psi_{\boldsymbol{v}}), \end{aligned}$$

$$\begin{aligned} a_{\mathrm{p}}\big(\varphi_{\mathrm{p}}, \psi\big) + \frac{1}{\gamma_{\mathrm{p}}}\big(\varphi_{\mathrm{p}}, P_{\mathrm{f}}(\psi)\big)_{0,\Gamma_{\mathrm{I}}} - (\boldsymbol{u}_{\mathrm{f}} \cdot \boldsymbol{n}, \psi)_{0,\Gamma_{\mathrm{I}}} \\ + \frac{1}{\gamma_{\mathrm{p}}} a_{\mathrm{f}}(\boldsymbol{u}_{\mathrm{f}}, \boldsymbol{v}_{\psi}) + \frac{1}{\gamma_{\mathrm{p}}} b_{\mathrm{f}}(\boldsymbol{v}_{\psi}, p_{\mathrm{f}}) = \ell_{\mathrm{p}}(\psi) + \frac{1}{\gamma_{\mathrm{p}}} \ell_{\mathrm{f}}(\boldsymbol{v}_{\psi}). \end{aligned}$$

The extensions $\psi_{\boldsymbol{v}} \in Q_{\mathrm{p}}$ and $\boldsymbol{v}_{\psi} \in \boldsymbol{V}_{\mathrm{f}}$ of $P_{\mathrm{p}}(\boldsymbol{v}\cdot\boldsymbol{n})$ and $P_{\mathrm{f}}(\psi)$ on the interface Γ_{I} can be inserted into these two equations. In particular, the second equation with $\psi = \psi_{\boldsymbol{v}}$ and multiplied by γ_{f} can be added to the first, and the first equation with $\boldsymbol{v} = \boldsymbol{v}_{\psi}$

multiplied by $-\frac{1}{\gamma_{\mathrm{p}}}$ can be added to the second, yielding

$$a_{\mathrm{f}}\left(\boldsymbol{u}_{\mathrm{f}}, \boldsymbol{v} + \frac{\gamma_{\mathrm{f}}}{\gamma_{\mathrm{p}}}\boldsymbol{v}_{\psi_{\boldsymbol{v}}}\right) + b_{\mathrm{f}}\left(\boldsymbol{v} + \frac{\gamma_{\mathrm{f}}}{\gamma_{\mathrm{p}}}\boldsymbol{v}_{\psi_{\boldsymbol{v}}}, p_{\mathrm{f}}\right) + \left(\varphi_{\mathrm{p}}, \left(\boldsymbol{v} + \frac{\gamma_{\mathrm{f}}}{\gamma_{\mathrm{p}}}\boldsymbol{v}_{\psi_{\boldsymbol{v}}}\right)\cdot \boldsymbol{n}\right)_{0,\Gamma_{\mathrm{I}}}$$

$$= \ell_{\mathrm{f}}\left(\boldsymbol{v} + \frac{\gamma_{\mathrm{f}}}{\gamma_{\mathrm{p}}}\boldsymbol{v}_{\psi_{\boldsymbol{v}}}\right),$$

$$a_{\mathrm{p}}\left(\varphi_{\mathrm{p}}, \psi + \frac{\gamma_{\mathrm{f}}}{\gamma_{\mathrm{p}}}\psi_{\boldsymbol{v}_\psi}\right) - \left(\boldsymbol{u}_{\mathrm{f}}\cdot \boldsymbol{n}, \psi + \frac{\gamma_{\mathrm{f}}}{\gamma_{\mathrm{p}}}\psi_{\boldsymbol{v}_\psi}\right)_{0,\Gamma_{\mathrm{I}}} = \ell_{\mathrm{p}}\left(\psi + \frac{\gamma_{\mathrm{f}}}{\gamma_{\mathrm{p}}}\psi_{\boldsymbol{v}_\psi}\right).$$

These two equations are the first and last equation in the Neumann–Neumann weak formulation (6.7) with $\boldsymbol{v} + \frac{\gamma_{\mathrm{f}}}{\gamma_{\mathrm{p}}}\boldsymbol{v}_{\psi_{\boldsymbol{v}}}$ and $\psi + \frac{\gamma_{\mathrm{f}}}{\gamma_{\mathrm{p}}}\psi_{\boldsymbol{v}_\psi}$ instead of $\boldsymbol{v}$ and ψ. The function $\boldsymbol{v}_{\psi_{\boldsymbol{v}}} \in \boldsymbol{V}_{\mathrm{f}}$ has the trace $T_{\mathrm{f}}(\boldsymbol{v}_{\psi_{\boldsymbol{v}}}, \cdot) = P_{\mathrm{f}}(P_{\mathrm{p}}(T_{\mathrm{f}}(\boldsymbol{v}, \cdot)))$ and similarly it is $T_{\mathrm{p}}(\psi_{\boldsymbol{v}_\psi}) = P_{\mathrm{p}}(P_{\mathrm{f}}(T_{\mathrm{p}}\psi))$. Finally, any $\tilde{\boldsymbol{v}} \in \boldsymbol{V}_{\mathrm{f}}$ and $\tilde{\psi} \in Q_{\mathrm{p}}$ can be written as $\tilde{\boldsymbol{v}} = \boldsymbol{v} + \frac{\gamma_{\mathrm{f}}}{\gamma_{\mathrm{p}}}\boldsymbol{v}_{\psi_{\boldsymbol{v}}}$ and $\tilde{\psi} = \psi + \frac{\gamma_{\mathrm{f}}}{\gamma_{\mathrm{p}}}\psi_{\boldsymbol{v}_\psi}$ with $\boldsymbol{v} \in \boldsymbol{V}_{\mathrm{f}}$ and $\psi \in Q_{\mathrm{p}}$. Indeed, let $\alpha = \left(1 + \gamma_{\mathrm{f}}/\gamma_{\mathrm{p}}\right)^{-1}$, define $\psi := \tilde{\psi} - \frac{\gamma_{\mathrm{f}}}{\gamma_{\mathrm{p}}}\psi_{\boldsymbol{v}_{\alpha\tilde{\psi}}}$, and note that

$$\begin{aligned} P_{\mathrm{f}}(T_{\mathrm{p}}\psi) &= P_{\mathrm{f}}\left(T_{\mathrm{p}}\tilde{\psi} - \frac{\gamma_{\mathrm{f}}}{\gamma_{\mathrm{p}}}T_{\mathrm{p}}\psi_{\boldsymbol{v}_{\alpha\tilde{\psi}}}\right) \\ &= P_{\mathrm{f}}\left(T_{\mathrm{p}}\tilde{\psi} - \alpha\frac{\gamma_{\mathrm{f}}}{\gamma_{\mathrm{p}}}P_{\mathrm{p}}(P_{\mathrm{f}}(T_{\mathrm{p}}\tilde{\psi}))\right) \\ &= P_{\mathrm{f}}\left(\left(1 - \alpha\frac{\gamma_{\mathrm{f}}}{\gamma_{\mathrm{p}}}\right)T_{\mathrm{p}}\tilde{\psi}\right) \\ &= P_{\mathrm{f}}\left(T_{\mathrm{p}}(\alpha\tilde{\psi})\right), \end{aligned}$$

i.e., $\boldsymbol{v}_\psi = \boldsymbol{v}_{\alpha\tilde{\psi}}$ and therefore $\psi_{\boldsymbol{v}_\psi} = \psi_{\boldsymbol{v}_{\alpha\tilde{\psi}}}$ and hence $\tilde{\psi} = \psi + \frac{\gamma_{\mathrm{f}}}{\gamma_{\mathrm{p}}}\psi_{\boldsymbol{v}_\psi}$. Here it is used that $T_{\mathrm{p}}\psi_{\boldsymbol{v}_\eta} = P_{\mathrm{f}}(T_{\mathrm{p}}\eta)$ for all $\eta \in Q_{\mathrm{p}}$. Using the same ideas lead to a construction of $\boldsymbol{v}$ for a given $\tilde{\boldsymbol{v}}$. Thus, it is shown that the Neumann–Neumann formulation (6.7) holds if the Robin–Robin one does. □

Remark 6.6.5 The previous result, Theorem 6.6.4, shows existence of a solution to the Robin–Robin formulation (6.34), Definition 6.6.3. Furthermore, its second part assures that this solution is unique. In other words, unique solvability of the Neumann–Neumann problem (6.7) carries over to (6.34).

6.6.4 Subdomain Operators

The Neumann–Neumann formulation (6.7) could be rewritten using operators on the interface, see Definitions 6.4.11 and 6.4.12 in Sect. 6.4.4. Analogously, such operators solving Robin instead of Dirichlet or Neumann problems are introduced.

Definition 6.6.6 (Solution Operators $K_i^{\gamma_i}$) With the spaces from Definitions 6.4.1 and 6.4.2, define for $\gamma_{\mathrm{p}} > 0$ and $\gamma_{\mathrm{f}} \geq 0$ the Robin solution operators $K_{\mathrm{p}}^{\gamma_{\mathrm{p}}} : \Lambda_{\mathrm{p}}^* \to \widetilde{Q}_{\mathrm{p}}$ and $K_{\mathrm{f}}^{\gamma_{\mathrm{f}}} : \Lambda_{\mathrm{f}}^* \to \widetilde{\boldsymbol{V}}_{\mathrm{f}} \times Q_{\mathrm{f}}$ which map given Robin data μ on the interface Γ_{I} to a Darcy and Stokes solution, respectively:

$$K_{\mathrm{p}}^{\gamma_{\mathrm{p}}}(\mu) = \varphi \text{ solves}$$

$$\begin{cases} -\nabla \cdot (\mathsf{K}\nabla\varphi) = f_{\mathrm{p}} & \text{in } \Omega_{\mathrm{p}}, \\ -\mathsf{K}\nabla\varphi \cdot \boldsymbol{n} = g_{\mathrm{p}} & \text{on } \Gamma_{\mathrm{p,N}}, \\ \varphi = \varphi_{\mathrm{b}} & \text{on } \Gamma_{\mathrm{p,D}}, \\ -\gamma_{\mathrm{p}}\mathsf{K}\nabla\varphi \cdot \boldsymbol{n} + \varphi = \mu & \text{on } \Gamma_{\mathrm{I}}. \end{cases}$$

$$K_{\mathrm{f}}^{\gamma_{\mathrm{f}}}(\mu) = (\mathbf{u}, p) \text{ solves}$$

$$\begin{cases} -\nabla \cdot \mathsf{T}(\mathbf{u}, p) = \mathbf{f}_{\mathrm{f}} & \text{in } \Omega_{\mathrm{f}}, \\ \nabla \cdot \mathbf{u} = 0 & \text{in } \Omega_{\mathrm{f}}, \\ \mathsf{T}(\mathbf{u}, p) \cdot \boldsymbol{n} = \mathbf{f}_{\mathrm{f}}^N & \text{on } \Gamma_{\mathrm{f,N}}, \\ \mathbf{u} = \mathbf{f}_{\mathrm{f}}^D & \text{on } \Gamma_{\mathrm{f,D}}, \\ \mathbf{u} \cdot \boldsymbol{\tau} + \alpha\boldsymbol{\tau} \cdot \mathsf{T}(\mathbf{u}, p) \cdot \boldsymbol{n} = 0 & \text{on } \Gamma_{\mathrm{I}}, \\ -\gamma_{\mathrm{f}}\mathbf{u} \cdot \boldsymbol{n} - \boldsymbol{n} \cdot \mathsf{T}(\mathbf{u}, p) \cdot \boldsymbol{n} = \mu & \text{on } \Gamma_{\mathrm{I}}. \end{cases}$$

The respective weak formulations are obtained as before, see Sects. 5.1.1 and 5.3.1. In particular for all $\psi \in Q_{\mathrm{p}}$ and all $(\boldsymbol{v}, q) \in \boldsymbol{V}_{\mathrm{f}} \times Q_{\mathrm{f}}$ it is

$$a_{\mathrm{p}}^{\mathrm{R}}(\varphi, \psi) = \ell_{\mathrm{p}}(\psi) + \frac{1}{\gamma_{\mathrm{p}}}\big(\mu, T_{\mathrm{p}}\psi\big)_{\Lambda_{\mathrm{p}}{}^* \times \Lambda_{\mathrm{p}}}$$

and

$$\begin{aligned} a_{\mathrm{f}}^{\mathrm{R}}(\mathbf{u}, \boldsymbol{v}) + b_{\mathrm{f}}(\boldsymbol{v}, p) &= \ell_{\mathrm{f}}(\boldsymbol{v}) - (\mu, T_{\mathrm{f}}\boldsymbol{v})_{\Lambda_{\mathrm{f}}{}^* \times \Lambda_{\mathrm{f}}}, \\ b_f(\mathbf{u}, q) &= 0. \end{aligned}$$

Note that in Sect. 6.6.2 functions P_p and P_f are introduced which slightly modified the bilinear forms a_p^R and a_f^R whenever $T_\mathrm{p}\psi \notin \Lambda_\mathrm{f}$ and $T_\mathrm{f}(\boldsymbol{v}, \cdot) \notin \Lambda_\mathrm{p}$ respectively. Without this modification the above operators $K_i^{\gamma_i}$ are of the type introduced in the Sects. 5.1.3 and 5.3.3. With it however, the Robin operators $K_i^{\gamma_i}$ have the form of the Neumann operators $K_{i,\mathrm{N}}$ whenever $P_\mathrm{f}(T_\mathrm{p}\psi) = 0$ and $P_\mathrm{p}(T_\mathrm{f}(\boldsymbol{v}, \cdot)) = 0$. Despite this difficulty these operators are still referred to as of Robin type. Furthermore, the image spaces of $K_i^{\gamma_i}$ coincide with those of the respective Neumann and Dirichlet operators, i.e., $\operatorname{Im} K_{i,\mathrm{N}} = \operatorname{Im} K_{i,\mathrm{D}} = \operatorname{Im} K_i^\gamma$ for all $\gamma > 0$. Also, it is $K_\mathrm{f}^0 = K_{\mathrm{f,N}}$ while the other operators $K_{i,j}$ can not be identified with the K_i^γ with a suitable γ.

It is possible to recover Robin data similar to the Neumann case in Definitions 6.4.5 and 6.4.6:

Definition 6.6.7 Let $\varphi \in H^1(\Omega_\mathrm{p})$ be given such that for all $\psi \in Q_\mathrm{p} \cap \ker T_\mathrm{p} = H^1_{\Gamma_{\mathrm{p,D}} \boldsymbol{u}_\mathrm{p} \Gamma_\mathrm{I}}(\Omega_\mathrm{p})$, it is[2]

$$a_\mathrm{p}(\varphi, \psi) = \ell_p(\psi).$$

Furthermore, let $(\mathbf{u}, p) \in \boldsymbol{H}^1(\Omega_\mathrm{f}) \times L^2(\Omega_\mathrm{f})$ be given such that for all $(\boldsymbol{v}, q) \in (\boldsymbol{V}_\mathrm{f} \times Q_\mathrm{f}) \cap \ker T_\mathrm{f}$ it is[3]

$$\begin{cases} a_\mathrm{f}(\mathbf{u}, \boldsymbol{v}) + b_\mathrm{f}(\boldsymbol{v}, p) = \ell_\mathrm{f}(\boldsymbol{v}), \\ \qquad\qquad b_\mathrm{f}(\mathbf{u}, q) = 0. \end{cases}$$

Then the Robin data $T_{\mathrm{p},\gamma}^\mathrm{R} : H^1(\Omega_\mathrm{p}) \to \Lambda_\mathrm{p}^*$ and $T_{\mathrm{f},\gamma}^\mathrm{R} : \mathbf{H}^1(\Omega_\mathrm{f}) \times L^2(\Omega_\mathrm{f}) \to \Lambda_\mathrm{f}^*$ is defined as

$$\begin{aligned} \left(T_{\mathrm{p},\gamma}^\mathrm{R}\varphi, \lambda\right)_{\Lambda_\mathrm{p}^* \times \Lambda_\mathrm{p}} &= \gamma\left(\ell_\mathrm{p}(\psi_\lambda) - a_\mathrm{p}(\varphi, \psi_\lambda)\right) + (\varphi, \lambda)_{0,\Gamma_\mathrm{I}}, \\ \left(T_{\mathrm{f},\gamma}^\mathrm{R}(\mathbf{u}, p), \lambda\right)_{\Lambda_\mathrm{f}^* \times \Lambda_\mathrm{f}} &= \ell_\mathrm{f}(\boldsymbol{v}_\lambda) - a_\mathrm{f}(\mathbf{u}, \boldsymbol{v}_\lambda) - b_\mathrm{f}(\boldsymbol{v}_\lambda, p) + \gamma(\mathbf{u} \cdot \boldsymbol{n}, \lambda)_{0,\Gamma_\mathrm{I}}, \end{aligned} \tag{6.35}$$

Compare this also with Definitions 5.1.5 and 5.3.9. This is equivalent to adding the Neumann and Dirichlet data with appropriate scaling, i.e.,

$$T_{\mathrm{p},\gamma}^\mathrm{R} = -\gamma T_\mathrm{p}^\mathrm{N} + T_\mathrm{p} \quad \text{and} \quad T_{\mathrm{f},\gamma}^\mathrm{R} = -T_\mathrm{f}^\mathrm{N} + \gamma T_\mathrm{f}.$$

Identities similar to the ones for the operators $K_{i,\mathrm{N}}$, $i \in \{\mathrm{p}, \mathrm{f}\}$ in Eqs. (6.9) hold whenever $\Lambda_\mathrm{f} = \Lambda_\mathrm{p}$:

$$\begin{aligned} T_{i,-\gamma_i}^\mathrm{R} \circ K_i^{\gamma_i} &= \mathrm{id} \quad \text{in } \Lambda_i^*, \\ K_i^{\gamma_i} \circ T_{i,-\gamma_i}^\mathrm{R} &= \mathrm{id} \quad \text{in } \operatorname{Im} K_{i,\mathrm{N}} = \operatorname{Im} K_{i,\mathrm{D}} = \operatorname{Im} K_i^{\gamma_i}. \end{aligned}$$

[2]With this ψ it is $a_\mathrm{p}(\varphi, \psi) = a_\mathrm{p}^R(\varphi, \psi)$.

[3]With this $\boldsymbol{v}$ it is $a_\mathrm{f}^R(\mathbf{u}, \boldsymbol{v}) = a_\mathrm{f}(\mathbf{u}, \boldsymbol{v})$.

If however the interface spaces Λ_f and Λ_p do not coincide, such identities do not hold because terms involving $\mathrm{id} - P_i$ enter the equation. In Sect. 6.6.6 the operators $T^\mathrm{R}_{i,\gamma}$ are concatenated with K^γ_j, $j \in \{\mathrm{p}, \mathrm{f}\}$, $j \neq i$, which requires that the image space of one is in the preimage space of the other, i.e., that $T^\mathrm{R}_{\mathrm{p},\gamma}$ maps to Λ^*_f and $T^\mathrm{R}_{\mathrm{f},\gamma}$ to Λ^*_p. The following extension is suitable:

Definition 6.6.8 The Robin data operators $T^\mathrm{R}_{i,\gamma}$ from Definition 6.6.7 are extended via the functions $P_i : L^2(\Gamma_\mathrm{I}) \to \Lambda_i$ from Definition 6.6.3 to $T^\mathrm{R}_{\mathrm{p},\gamma} : H^1(\Omega_\mathrm{p}) \to \Lambda^*_\mathrm{f}$ and $T^\mathrm{R}_{\mathrm{f},\gamma} : \boldsymbol{H}^1(\Omega_\mathrm{f}) \times L^2(\Omega_\mathrm{f}) \to \Lambda^*_\mathrm{p}$ as follows

$$\left(T^\mathrm{R}_{\mathrm{p},\gamma}(\varphi), \eta\right)_{\Lambda_\mathrm{f}{}^* \times \Lambda_\mathrm{f}} = \gamma\left(\ell_\mathrm{p}(\psi_{P_\mathrm{p}(\eta)}) - a_\mathrm{p}(\varphi, \psi_{P_\mathrm{p}(\eta)})\right) + (\varphi, \eta)_{0,\Gamma_\mathrm{I}}, \tag{6.36a}$$

$$\begin{aligned}\left(T^\mathrm{R}_{\mathrm{f},\gamma}(\mathbf{u}, p), \eta\right)_{\Lambda_\mathrm{p}{}^* \times \Lambda_\mathrm{p}} &= \ell_\mathrm{f}(\boldsymbol{v}_{P_\mathrm{f}(\eta)}) - a_\mathrm{f}(\mathbf{u}, \boldsymbol{v}_{P_\mathrm{f}(\eta)}) - b_\mathrm{f}(\boldsymbol{v}_{P_\mathrm{f}(\eta)}, p)\\ &\quad + \gamma(\mathbf{u} \cdot \boldsymbol{n}, \lambda)_{0,\Gamma_\mathrm{I}}.\end{aligned} \tag{6.36b}$$

This means η is replaced by zero whenever the earlier definitions are not applicable; in that case the Robin data operators above reduce to the trace operator T_p and γT_f, respectively.

6.6.5 Operators Acting Only on the Interface

In Sect. 6.4.3 the operators $H_{i,j}$ are defined which map Dirichlet/Neumann data to Neumann/Dirichlet data solving one of the two subproblems. Here, similarly, operators $H_i^{\gamma_1,\gamma_2}$ which solve Robin problems using γ_1 and return other Robin data (using γ_2) are introduced. With their help the Robin–Robin weak formulation (6.34) can be rewritten.

Definition 6.6.9 (Robin Operators on the Interface) Using the Robin solution and data operators, Definitions 6.6.6 and 6.6.7, define for all $\gamma_\mathrm{p} > 0$ and $\gamma_\mathrm{f} \geq 0$

$$\begin{aligned}H_\mathrm{p}^{\gamma_\mathrm{p},\gamma_\mathrm{f}} &= T^\mathrm{R}_{\mathrm{p},\gamma_\mathrm{f}} \circ K_\mathrm{p}^{\gamma_\mathrm{p}},\\ H_\mathrm{f}^{\gamma_\mathrm{f},\gamma_\mathrm{p}} &= T^\mathrm{R}_{\mathrm{f},\gamma_\mathrm{p}} \circ K_\mathrm{f}^{\gamma_\mathrm{f}}.\end{aligned}$$

Note that the conditions on γ_p and γ_f can be slightly weakened without sacrificing the well-posedness of $K_i^{\gamma_i}$. For negative values of γ_i the respective bilinear form a_i^R is still positive definite if $|\gamma_i|$ is small compared to K or ν respectively, see also the proofs of the coercivity of a_i in Sects. 5.1.2 and 5.3.2. This is however a strong limitation, because small values of K and ν are of interest. So even though the following inverse identities hold formally, they have little practical relevance:

$$H_i^{\gamma_1,-\gamma_2} \circ H_i^{\gamma_2,-\gamma_1} = \mathrm{id} \qquad \text{in } \Lambda_i^*. \tag{6.37}$$

6.6.6 Weak Formulation Rewritten

Analogously to the fixed point equations (6.12) and (6.13) and Definition 6.4.11, for the Neumann–Neumann problem consider the following

Definition 6.6.10 (Robin Fixed Point Equations) Using the Robin operators on the interface from Definition 6.6.9, define the two fixed point equations

$$H_{\mathrm{f}}^{\gamma_{\mathrm{f}},\gamma_{\mathrm{p}}}\left(H_{\mathrm{p}}^{\gamma_{\mathrm{p}},\gamma_{\mathrm{f}}}(\lambda)\right) = \lambda, \tag{6.38}$$

$$H_{\mathrm{p}}^{\gamma_{\mathrm{p}},\gamma_{\mathrm{f}}}\left(H_{\mathrm{f}}^{\gamma_{\mathrm{f}},\gamma_{\mathrm{p}}}(\mu)\right) = \mu. \tag{6.39}$$

According to the extensions of the Robin data operators in Definition 6.6.8, these fixed point equations are well defined. Corollary 6.4.14 shows that the fixed point equations (6.12) and (6.13) are equivalent. Its counterpart with respect to the equations above is the following.

Corollary 6.6.11 (Equivalence of Fixed Point Equations) *If* λ *is a solution of* (6.38)*, then* $\mu = H_{\mathrm{p}}^{\gamma_{\mathrm{p}},\gamma_{\mathrm{f}}}(\lambda)$ *solves* (6.39)*. Conversely if* μ *solves* (6.39)*, then* $\lambda = H_{\mathrm{f}}^{\gamma_{\mathrm{f}},\gamma_{\mathrm{p}}}(\mu)$ *solves* (6.38)*.*

Proof Apply $H_{\mathrm{p}}^{\gamma_{\mathrm{p}},\gamma_{\mathrm{f}}}$ to (6.38) and $H_{\mathrm{f}}^{\gamma_{\mathrm{f}},\gamma_{\mathrm{p}}}$ to (6.39). □

The fixed point formulations furthermore allow for a connection to the coupled Robin–Robin problem (6.34). The following theorem is an analogue to Theorem 6.4.15.

Theorem 6.6.12 *Let* λ *solve the fixed point equation* (6.38)*. Then*

$$(\mathbf{u}, p) := K_{\mathrm{f}}^{\gamma_{\mathrm{f}}}\left(H_{\mathrm{p}}^{\gamma_{\mathrm{p}},\gamma_{\mathrm{f}}}(\lambda)\right) \qquad \textit{and} \qquad \varphi := K_{\mathrm{p}}^{\gamma_{\mathrm{p}}}(\lambda)$$

solve (6.34)*. If conversely* $(\mathbf{u}, p, \varphi)$ *solve* (6.34)*, then* $\lambda = T^{\mathrm{R}}_{\mathrm{f},\gamma_{\mathrm{p}}}(\mathbf{u}, p)$ *solves* (6.38)*.*

Proof Let λ solve Eq. (6.38), i.e., $\lambda = H_{\mathrm{f}}^{\gamma_{\mathrm{f}},\gamma_{\mathrm{p}}}\left(H_{\mathrm{p}}^{\gamma_{\mathrm{p}},\gamma_{\mathrm{f}}}(\lambda)\right)$. By definition the solution $(\mathbf{u}, p, \varphi)$ satisfies the correct Dirichlet data away from the interface, $\mathbf{u} = \mathbf{u}_{\mathrm{b}}$ on $\Gamma_{\mathrm{f,D}}$ and $\varphi = \varphi_{\mathrm{b}}$ on $\Gamma_{\mathrm{p,D}}$. Now let $\boldsymbol{v} \in \boldsymbol{V}_{\mathrm{f}}$, $q \in Q_{\mathrm{f}}$, and $\psi \in Q_{\mathrm{p}}$ be given test functions and $\mu = H_{\mathrm{p}}^{\gamma_{\mathrm{p}},\gamma_{\mathrm{f}}}(\lambda)$. Then according to the definition of $K_{\mathrm{f}}^{\gamma_{\mathrm{f}}}$ it is

$$\begin{aligned} a_{\mathrm{f}}^{\mathrm{R}}(\mathbf{u}, \boldsymbol{v}) + b_{\mathrm{f}}(\boldsymbol{v}, p) &= \ell_{\mathrm{f}}(\boldsymbol{v}) - (\mu, T_{\mathrm{f}}\boldsymbol{v})_{\Lambda_{\mathrm{f}}{}^* \times \Lambda_{\mathrm{f}}}, \\ b_{\mathrm{f}}(\mathbf{u}, q) &= 0. \end{aligned}$$

The definition of $T^{\mathrm{R}}_{\mathrm{p},\gamma_\mathrm{f}}$ yields

$$\begin{aligned}(\mu, \boldsymbol{v}\cdot\boldsymbol{n})_{\Lambda_\mathrm{f}^*\times\Lambda_\mathrm{f}} &= \left(T^{\mathrm{R}}_{\mathrm{p},\gamma_\mathrm{f}}(\varphi), \boldsymbol{v}\cdot\boldsymbol{n}\right)_{\Lambda_\mathrm{p}^*\times\Lambda_\mathrm{p}} \\ &= \gamma_\mathrm{f}\big(\ell_\mathrm{p}(\psi_{P_\mathrm{p}(\boldsymbol{v}\cdot\boldsymbol{n})}) - a_\mathrm{p}(\varphi, \psi_{P_\mathrm{p}(\boldsymbol{v}\cdot\boldsymbol{n})})\big) + (\varphi, \boldsymbol{v}\cdot\boldsymbol{n})_{0,\Gamma_\mathrm{I}} \\ &= c_\mathrm{f}(\varphi, \boldsymbol{v}) - \ell_{\mathrm{f,p}}(\boldsymbol{v})\end{aligned}$$

with the function P_p introduced in Sect. 6.6.2, and hence the first equation in (6.34). Similarly, the definition of $K_\mathrm{p}^{\gamma_\mathrm{p}}$ gives

$$a_\mathrm{p}^\mathrm{R}(\varphi, \psi) = \ell_\mathrm{p}(\psi) + \frac{1}{\gamma_\mathrm{p}}(\lambda, \psi)_{\Lambda_\mathrm{p}^*\times\Lambda_\mathrm{p}}.$$

Using the fixed point equation (6.38), it is $\lambda = H_\mathrm{f}^{\gamma_\mathrm{f},\gamma_\mathrm{p}}(\mu)$, i.e., $\lambda = T^{\mathrm{R}}_{\mathrm{f},\gamma_\mathrm{p}}(\mathbf{u}, p)$, and therefore

$$\begin{aligned}\frac{1}{\gamma_\mathrm{p}}(\lambda, \psi)_{\Lambda_\mathrm{p}^*\times\Lambda_\mathrm{p}} &= \frac{1}{\gamma_\mathrm{p}}\left(T^{\mathrm{R}}_{\mathrm{f},\gamma_\mathrm{p}}(\mathbf{u}, p), \psi\right)_{\Lambda_\mathrm{p}^*\times\Lambda_\mathrm{p}} \\ &= \frac{1}{\gamma_\mathrm{p}}\Big(\ell_f\big(\boldsymbol{v}_{P_\mathrm{f}(\psi)}\big) - a_\mathrm{f}\big(\mathbf{u}, \boldsymbol{v}_{P_\mathrm{f}(\psi)}\big) - b_f\big(\boldsymbol{v}_{P_\mathrm{f}(\psi)}, p\big)\Big) \\ &\qquad + (\mathbf{u}\cdot\boldsymbol{n}, \psi)_{0,\Gamma_\mathrm{I}} \\ &= \ell_{\mathrm{p,f}}(\psi) - c_\mathrm{p}((\mathbf{u}, p), \psi),\end{aligned}$$

which is the last equation in (6.34). □

With a very similar proof one can also show that a solution of the other fixed point equation (6.39) on the interface leads to a solution of the weak formulation (6.34). However, Corollary 6.6.11 together with the previous result, Theorem 6.6.12, provide an easier proof of the following analogue to Theorem 6.4.16:

Theorem 6.6.13 *Let μ solve the interface condition* (6.39). *Then*

$$(\mathbf{u}, p) := K_\mathrm{f}^{\gamma_\mathrm{f}}(\mu) \qquad \textit{and} \qquad \varphi := K_\mathrm{p}^{\gamma_\mathrm{p}}\left(H_\mathrm{f}^{\gamma_\mathrm{f},\gamma_\mathrm{p}}(\mu)\right)$$

solve (6.34). *If conversely* $(\mathbf{u}, p, \varphi)$ *solve* (6.34), *then* $\mu = T^{\mathrm{R}}_{\mathrm{p},\gamma_\mathrm{f}}(\varphi)$ *solves* (6.39).

Proof With μ solving (6.39), also $\lambda = H_\mathrm{f}^{\gamma_\mathrm{f},\gamma_\mathrm{p}}(\mu)$ solves (6.38), see Corollary 6.6.11. Then note that $(\mathbf{u}, p) = K_\mathrm{f}^{\gamma_\mathrm{f}}(\mu) = K_\mathrm{f}^{\gamma_\mathrm{f}}\left(H_\mathrm{p}^{\gamma_\mathrm{p},\gamma_\mathrm{f}}(\lambda)\right)$ and $\varphi = K_\mathrm{p}^{\gamma_\mathrm{p}}\left(H_\mathrm{f}^{\gamma_\mathrm{f},\gamma_\mathrm{p}}(\mu)\right) = K_\mathrm{p}^{\gamma_\mathrm{p}}(\lambda)$, so that $(\mathbf{u}, p, \varphi)$ solves the Robin–Robin weak formulation (6.34), according to Theorem 6.6.12. If conversely $(\mathbf{u}, p, \varphi)$ solves (6.34), then the same theorem assures that $\lambda = T^{\mathrm{R}}_{\mathrm{f},\gamma_{\mathrm{p}}}(\mathbf{u}, p)$ solves (6.38), and hence $\mu = H_{\mathrm{p}}^{\gamma_{\mathrm{p}},\gamma_{\mathrm{f}}}(\lambda) = T^{\mathrm{R}}_{\mathrm{p},\gamma_{\mathrm{f}}}(\varphi)$ solves (6.39), see Corollary 6.6.11. □

Remark 6.6.14 Together with the identities (6.37) it can be seen that there are equations of Steklov–Poincaré type which are equivalent to the fixed point problems (6.38) and (6.39):

$$H_{\mathrm{f}}^{\gamma_{\mathrm{f}},\gamma_{\mathrm{p}}}\left(H_{\mathrm{p}}^{\gamma_{\mathrm{p}},\gamma_{\mathrm{f}}}(\lambda)\right) = \lambda \quad\Longleftrightarrow\quad H_{\mathrm{p}}^{\gamma_{\mathrm{p}},\gamma_{\mathrm{f}}}(\lambda) - H_{\mathrm{f}}^{-\gamma_{\mathrm{p}},-\gamma_{\mathrm{f}}}(\lambda) = 0,$$

$$H_{\mathrm{p}}^{\gamma_{\mathrm{p}},\gamma_{\mathrm{f}}}\left(H_{\mathrm{f}}^{\gamma_{\mathrm{f}},\gamma_{\mathrm{p}}}(\mu)\right) = \mu \quad\Longleftrightarrow\quad H_{\mathrm{f}}^{\gamma_{\mathrm{f}},\gamma_{\mathrm{p}}}(\mu) - H_{\mathrm{p}}^{-\gamma_{\mathrm{f}},-\gamma_{\mathrm{p}}}(\mu) = 0.$$

Note that while these equations are formally correct, they suffer the same deficiencies as the identities (6.37). This is why the Steklov–Poincaré equations with Robin operators are not further studied.

Remark 6.6.15 Similar to Sect. 6.5 one can show that the Robin operators above are affine linear and derive a linear equation on the interface similar to (6.24) and (6.25). No particular problems arise in this approach but for ease of notation and presentation it is avoided here.

6.6.7 Decoupling the Equations Further

In this subsection another view on the Robin–Robin weak formulation (6.34) is developed, very similar to the decoupling in Sect. 6.4.6. For all $\boldsymbol{v} \in \boldsymbol{V}_{\mathrm{f}}$, $q \in Q_{\mathrm{f}}$, $\psi \in Q_{\mathrm{p}}$, $\xi_{\mathrm{f}} \in \Lambda_{\mathrm{f}}$, and $\xi_{\mathrm{p}} \in \Lambda_{\mathrm{p}}$ it is

$$\begin{aligned}
a_{\mathrm{f}}^{\mathrm{R}}(\boldsymbol{u}_{\mathrm{f}}, \boldsymbol{v}) + b_{\mathrm{f}}(\boldsymbol{v}, p_{\mathrm{f}}) + (\mu, \boldsymbol{v}\cdot\boldsymbol{n})_{\Lambda_{\mathrm{f}}^*\times\Lambda_{\mathrm{f}}} &= \ell_f(\boldsymbol{v}),\\
b_{\mathrm{f}}(\boldsymbol{u}_{\mathrm{f}}, q) &= 0,\\
\big(\lambda, \xi_{\mathrm{p}}\big)_{\Lambda_{\mathrm{p}}^*\times\Lambda_{\mathrm{p}}} - \gamma_{\mathrm{p}}\big(\boldsymbol{u}_{\mathrm{f}}\cdot\boldsymbol{n}, \xi_{\mathrm{p}}\big)_{0,\Gamma_{\mathrm{I}}} &\\
+a_{\mathrm{f}}(\boldsymbol{u}_{\mathrm{f}}, \boldsymbol{v}_{P_{\mathrm{f}}(\xi_{\mathrm{p}})}) + b_{\mathrm{f}}(\boldsymbol{v}_{P_{\mathrm{f}}(\xi_{\mathrm{p}})}, p_{\mathrm{f}}) - \ell_{\mathrm{f}}(\boldsymbol{v}_{P_{\mathrm{f}}(\xi_{\mathrm{p}})}) &= 0,\\
a_{\mathrm{p}}^{\mathrm{R}}(\varphi_{\mathrm{p}}, \psi) - \frac{1}{\gamma_{\mathrm{p}}}(\lambda, \psi)_{\Lambda_{\mathrm{p}}^*\times\Lambda_{\mathrm{p}}} &= \ell_{\mathrm{p}}(\psi),\\
(\mu, \xi_{\mathrm{f}})_{\Lambda_{\mathrm{f}}^*\times\Lambda_{\mathrm{f}}} - \big(\varphi_{\mathrm{p}}, \xi_{\mathrm{f}}\big)_{0,\Gamma_{\mathrm{I}}} + \gamma_{\mathrm{f}} a_{\mathrm{p}}(\varphi_{\mathrm{p}}, \psi_{P_{\mathrm{p}}(\xi_{\mathrm{f}})}) - \gamma_{\mathrm{f}}\ell_{\mathrm{p}}(\psi_{P_{\mathrm{p}}(\xi_{\mathrm{f}})}) &= 0,
\end{aligned} \tag{6.40}$$

where $\boldsymbol{v}_{\lambda_{\mathrm{p}}}$ and $\psi_{\lambda_{\mathrm{f}}}$ are extensions of λ_{p} and λ_{f} to $\boldsymbol{V}_{\mathrm{f}}$ and Q_{p} such that $\boldsymbol{v}_{\lambda_{\mathrm{p}}}\cdot\boldsymbol{n} = \lambda_{\mathrm{p}}$ and $T_{\mathrm{p}}\psi_{\lambda_{\mathrm{f}}} = \psi_{\lambda_{\mathrm{f}}}\big|_{\Gamma_{\mathrm{I}}} = \lambda_{\mathrm{f}}$. The second and fourth equations above are essentially the application of c_{p} and c_{f} from Eq. (6.33).

6.6.8 Alternative Robin–Robin Method

In [DQV07] and [CGHW11] an alternative weak formulation is introduced which solves Robin problems in each subdomain, see also the C-RR method in [CJW14]. It is assumed that $\Lambda_\mathrm{p} = \Lambda_\mathrm{f}$. Here a solution consists of $\boldsymbol{u}_\mathrm{f}$, p_f, and φ_p together with two variables on the interface $\eta_f, \eta_p \in \Lambda_\mathrm{p} = \Lambda_\mathrm{f}$. Then for each $\boldsymbol{v} \in \boldsymbol{V}_\mathrm{f}$, $q \in Q_\mathrm{f}$ and $\psi \in Q_\mathrm{p}$ it shall hold

$$\begin{aligned} a_\mathrm{f}^\mathrm{R}(\boldsymbol{u}_\mathrm{f}, \boldsymbol{v}) + b_\mathrm{f}(\boldsymbol{v}, p_\mathrm{f}) - (\eta_\mathrm{f}, \boldsymbol{v} \cdot \boldsymbol{n})_{0,\Gamma_\mathrm{I}} &= \ell_\mathrm{f}(\boldsymbol{v}), \\ b(\boldsymbol{u}_\mathrm{f}, q) &= 0, \\ \eta_\mathrm{p} &= d\boldsymbol{u}_\mathrm{f} \cdot \boldsymbol{n} + c\eta_\mathrm{f}, \\ a_\mathrm{p}^\mathrm{R}(\varphi_\mathrm{p}, \psi) - \frac{1}{\gamma_\mathrm{p}}\left(\eta_\mathrm{p}, \psi\right)_{0,\Gamma_\mathrm{I}} &= \ell_\mathrm{p}(\psi), \\ \eta_\mathrm{f} &= b\varphi_\mathrm{p} + a\eta_\mathrm{p}, \end{aligned} \tag{6.41}$$

where a, b, c, and d are constants which are chosen such that the solution to the Neumann–Neumann formulation (6.7) also solves the above.[4] In contrast to Eqs. (6.20) and (6.40) the intermediate variables on the interface now are directly coupled. Subtracting the equations in (6.7) from the first and third equation in (6.41) yields

$$\begin{aligned} \gamma_\mathrm{f}(\boldsymbol{u}_\mathrm{f} \cdot \boldsymbol{n}, \boldsymbol{v} \cdot \boldsymbol{n})_{0,\Gamma_\mathrm{I}} - (\eta_\mathrm{f}, \boldsymbol{v} \cdot \boldsymbol{n})_{0,\Gamma_\mathrm{I}} - \left(\varphi_\mathrm{p}, \boldsymbol{v} \cdot \boldsymbol{n}\right)_{0,\Gamma_\mathrm{I}} &= 0, \\ \eta_\mathrm{p} &= d\boldsymbol{u}_\mathrm{f} \cdot \boldsymbol{n} + c\eta_\mathrm{f}, \\ \frac{1}{\gamma_\mathrm{p}}\left(\varphi_\mathrm{p}, \psi\right)_{0,\Gamma_\mathrm{I}} - \frac{1}{\gamma_\mathrm{p}}\left(\eta_\mathrm{p}, \psi\right)_{0,\Gamma_\mathrm{I}} + (\boldsymbol{u}_\mathrm{f} \cdot \boldsymbol{n}, \psi)_{0,\Gamma_\mathrm{I}} &= 0, \\ \eta_\mathrm{f} &= b\varphi_\mathrm{p} + a\eta_\mathrm{p}. \end{aligned}$$

The first and third equations only include inner products on the interface Γ_I, hence can be simplified to be equations solely in $\Lambda_\mathrm{f} = \Lambda_\mathrm{p}$:

$$\begin{aligned} \eta_\mathrm{f} &= \gamma_\mathrm{f}\boldsymbol{u}_\mathrm{f} \cdot \boldsymbol{n} - \varphi_\mathrm{p}, \\ \eta_\mathrm{p} &= d\boldsymbol{u}_\mathrm{f} \cdot \boldsymbol{n} + c\eta_\mathrm{f}, \\ \eta_\mathrm{p} &= \gamma_\mathrm{p}\boldsymbol{u}_\mathrm{f} \cdot \boldsymbol{n} + \varphi_\mathrm{p}, \\ \eta_\mathrm{f} &= b\varphi_\mathrm{p} + a\eta_\mathrm{p}. \end{aligned}$$

[4] The signs and positions of η_f and η_p are chosen such that the resulting system matches that in [DQV07] and [CGHW11].

These equations can be solved for a, b, c, and d yielding

$$a = \frac{\gamma_f}{\gamma_p}, \qquad b = -1 - a, \qquad c = -1, \qquad d = \gamma_f + \gamma_p, \tag{6.42}$$

see also Lemma 2.2 in [CGHW11]. In comparison to the Robin–Robin weak formulation (6.40) the interface variables in Eq. (6.41) are of higher regularity, namely in $\Lambda_f = \Lambda_p$ instead of its dual. However, η_f and η_p are directly coupled to each other while this is not the case for λ and μ in Eq. (6.40). Both approaches share the advantage that each interface variable only depends on one of the two subdomain solutions directly.

Remark 6.6.16 The notation η_f and η_p is chosen to match that in the literature [DQV07, CGHW11, CJW14]. These interface variables play a similar role to λ and μ in the previous Sect. 6.6.7, but in general are not equal.

6.7 The Finite Element Method for the Stokes–Darcy Problem

A general introduction to the finite element method is not given here, instead it is referred to the classic literature, e.g., [Bra07b, BS08, BF91, Cia02]. This is also where relevant notions such as grid, edge, face, vertex, and finite element space, as well as properties of the linear systems obtained through the finite element method for the subproblems can be found.

Consider one of the weak formulations of the coupled Stokes–Darcy problem (6.7) or (6.34). Furthermore, let $\{\mathcal{T}_h\}$ be a regular family of grids for the entire domain Ω such that the interface Γ_I does not intersect with the interior of any grid cell, i.e., it is the union of a number of edges/faces of the grid. This leads to subgrids $\mathcal{T}_{h,f}$ and $\mathcal{T}_{h,p}$ in each subdomain Ω_f and Ω_p which match at the interface in the sense that each edge/face $F \subset \Gamma_I$ of $\mathcal{T}_{h,f}$ is exactly one edge/face of $\mathcal{T}_{h,p}$. In general it is possible to use non-matching grids, but for simplicity here this is not pursued. Each of the spaces $\boldsymbol{H}^1(\Omega_f)$, $L^2(\Omega_f)$, and $H^1(\Omega_p)$ in which the solution is sought is approximated by suitable conforming finite element spaces $\boldsymbol{V}_{h,f}$, $Q_{h,f}$, and $Q_{h,p}$ leading to a linear system of equations of the following form:

$$\begin{pmatrix} S & C_1 \\ C_2 & D \end{pmatrix} \begin{pmatrix} (\mathbf{u}_{h,f}, p_{h,f}) \\ \varphi_{h,p} \end{pmatrix} = b. \tag{6.43}$$

Here S is a saddle point matrix and D is the discretization of the bilinear form a_p or a_p^R, respectively. The matrices C_1 and C_2 represent the coupling terms. In the case of the Neumann–Neumann weak form (6.7) it is $C_1^\top = -C_2$ while in the case of the Robin–Robin form (6.34) no such relation holds. To be precise, let $\{\boldsymbol{v}_i\}$, $\{q_i\}$, and $\{\psi_i\}$ be a finite element basis of $\boldsymbol{V}_{h,f}$, $Q_{h,f}$, and $Q_{h,p}$, respectively. Then the system to be solved is

$$\begin{pmatrix} A & B^\top & C_\mathrm{f} \\ B & 0 & 0 \\ C_{\mathrm{p},1} & C_{\mathrm{p},2} & D \end{pmatrix} \begin{pmatrix} \mathbf{u}_{h,\mathrm{f}} \\ p_{h,\mathrm{f}} \\ \varphi_{h,\mathrm{p}} \end{pmatrix} = \begin{pmatrix} b_1 \\ 0 \\ b_2 \end{pmatrix}.$$

The matrices and right-hand side have the following entries for the two considered weak formulations:

Neumann–Neumann (6.7)	Robin–Robin (6.34)
$A_{ij} = a_\mathrm{f}(\boldsymbol{v}_j, \boldsymbol{v}_i),$	$A_{ij} = a_\mathrm{f}^\mathrm{R}(\boldsymbol{v}_j, \boldsymbol{v}_i),$
$B_{ij} = b_\mathrm{f}(\boldsymbol{v}_j, q_i),$	$B_{ij} = b_\mathrm{f}(\boldsymbol{v}_j, q_i),$
$(C_\mathrm{f})_{ij} = \left(\psi_j, \boldsymbol{v}_i \cdot \boldsymbol{n}\right)_{0,\Gamma_\mathrm{I}},$	$(C_\mathrm{f})_{ij} = \left(\psi_j, \boldsymbol{v}_i \cdot \boldsymbol{n}\right)_{0,\Gamma_\mathrm{I}} - \gamma_\mathrm{f} a_\mathrm{p}(\psi_j, \psi_{P_\mathrm{p}(\boldsymbol{v}_i\cdot\boldsymbol{n})}),$
$(C_{\mathrm{p},1})_{ij} = -\left(\psi_i, \boldsymbol{v}_j \cdot \boldsymbol{n}\right)_{0,\Gamma_\mathrm{I}},$	$(C_{\mathrm{p},1})_{ij} = -\left(\psi_i, \boldsymbol{v}_j \cdot \boldsymbol{n}\right)_{0,\Gamma_\mathrm{I}} + \dfrac{1}{\gamma_\mathrm{p}} a_\mathrm{f}(\boldsymbol{v}_j, \boldsymbol{v}_{P_\mathrm{f}(\psi_i)}),$
$(C_{\mathrm{p},2})_{ij} = 0,$	$(C_{\mathrm{p},2})_{ij} = \dfrac{1}{\gamma_\mathrm{p}} b_\mathrm{f}(\boldsymbol{v}_{P_\mathrm{f}(\psi_i)}, q_j),$
$D_{ij} = a_\mathrm{p}(\psi_j, \psi_j),$	$D_{ij} = a_\mathrm{p}^\mathrm{R}(\psi_j, \psi_j),$
$(b_1)_i = \ell_\mathrm{f}(\boldsymbol{v}_i),$	$(b_1)_i = \ell_\mathrm{f}(\boldsymbol{v}_i) + \ell_{\mathrm{f,p}}(\boldsymbol{v}_i),$
$(b_2)_i = \ell_\mathrm{f}(\psi_i),$	$(b_2)_i = \ell_\mathrm{f}(\psi_i) + \ell_{\mathrm{p,f}}(\psi_i).$

Specific algorithms to solve these type of linear systems are discussed in Chap. 7.

In the Sects. 6.4.4 and 6.6.6 several equations on the interface Γ_I are introduced which are equivalent to the Neumann–Neumann and Robin–Robin weak formulations (6.7) and (6.34), respectively, namely the fixed point equations (6.12), (6.13), (6.38), and (6.39) as well as the Steklov–Poincaré equations (6.16) and (6.17), see also Theorems 6.4.15, 6.4.16, 6.6.12, and 6.6.13. Additionally, in Sects. 6.4.6 and 6.6.7 the respective linear system is rewritten with the help of the interface variables λ and μ. This can be discretized in a straightforward way leading to a system of the form

$$\begin{pmatrix} S & & E_\mathrm{f} & \\ R_\mathrm{f} & -\mathrm{id} & & \\ & E_\mathrm{p} & D & \\ & & R_\mathrm{p} & -\mathrm{id} \end{pmatrix} \begin{pmatrix} (\mathbf{u}_{h,\mathrm{f}}, p_{h,\mathrm{f}}) \\ \lambda_h \\ \varphi_{h,\mathrm{p}} \\ \mu_h \end{pmatrix} = \begin{pmatrix} (b_1, 0) \\ 0 \\ b_2 \\ 0 \end{pmatrix}, \tag{6.44}$$

where $E_\mathrm{f} R_\mathrm{p} = C_1$ and $E_\mathrm{p} R_\mathrm{f} = C_2$, see also Section 2 in [CJW14]. For the Neumann–Neumann problem the spaces for the interface variables are the restrictions of the respective finite element spaces in the subdomains. Hence, μ_h is continuous because only conforming finite elements are used for $\varphi_{h,\mathrm{p}}$ and λ_h is in general only continuous if the interface Γ_I does not contain jumps in the normal, i.e., is flat, compare with Remark 6.2.2. Therefore, one of the discretized interface spaces might have to be discontinuous. Implementing the decoupled system (6.44) rather than the coupled one (6.43) may be helpful because the solver for one subdomain does not need any information on the other subdomain which allows greater modularization. Furthermore, the discretization of the Steklov–Poincaré operators is just the application of E_i, $i \in \{\mathrm{f}, \mathrm{p}\}$, followed by a solving step (inverting S or D, respectively) and an application of the restriction R_i.

Also the alternative Robin–Robin system (6.41) can be written in the form of Eq. (6.44). Here the variables on the interface, η_f and η_p, are continuous whenever the finite element spaces for the Stokes velocity and the pressure in the Darcy subdomain are discretized with continuous, i.e., conforming, finite elements.[5] This is why this approach is denoted C-RR (continuous Robin–Robin) in [CJW14]. Note that in the C-RR method, due to the direct coupling of the interface variables, the system to be solved cannot be written in the form of Eq. (6.43).

6.7.1 The D-RR Method

In [CJW14], another method is proposed and called D-RR. It directly discretizes the preliminary weak form (6.30), i.e., derivatives of functions in the subdomain are evaluated at the interface. The resulting linear system of equations has the same form as the other approaches, and the matrices A, B, and D have the same entries. The coupling terms however differ, for the D-RR method these are

$$
\begin{aligned}
(C_\mathrm{f})_{ij} &= \left(\psi_j, \boldsymbol{v}_i \cdot \boldsymbol{n}\right)_{0,\Gamma_\mathrm{I}} + \gamma_\mathrm{f}\left(\mathsf{K}\nabla\psi_j \cdot \boldsymbol{n}, \boldsymbol{v}_i \cdot \boldsymbol{n}\right)_{0,\Gamma_\mathrm{I}},\\
(C_{\mathrm{p},1})_{ij} &= -\left(\psi_i, \boldsymbol{v}_j \cdot \boldsymbol{n}\right)_{0,\Gamma_\mathrm{I}} + \frac{1}{\gamma_\mathrm{p}}\left(\boldsymbol{n} \cdot 2\nu\mathsf{D}(\boldsymbol{v}_j) \cdot \boldsymbol{n}, \psi_i\right)_{0,\Gamma_\mathrm{I}},\\
(C_{\mathrm{p},2})_{ij} &= -\frac{1}{\gamma_\mathrm{p}}\left(q_j, \psi_i\right)_{0,\Gamma_\mathrm{I}},\\
(b_1)_i &= \ell_\mathrm{f}(\boldsymbol{v}_i),\\
(b_2)_i &= \ell_\mathrm{f}(\psi_i).
\end{aligned}
$$

[5] For this approach it is assumed that $\Lambda_\mathrm{f} = \Lambda_\mathrm{p}$, see Sect. 6.6.8.

In other words, compared with the Neumann–Neumann discretization, the terms

$$\gamma_{\mathrm{f}}\big(\mathbf{u}_{h,\mathrm{f}} \cdot \boldsymbol{n}, \boldsymbol{v}_i \cdot \boldsymbol{n}\big)_{0,\Gamma_{\mathrm{I}}} + \gamma_{\mathrm{f}}\big(\mathsf{K}\nabla\varphi_{h,\mathrm{p}} \cdot \boldsymbol{n}, \boldsymbol{v}_i \cdot \boldsymbol{n}\big)_{0,\Gamma_{\mathrm{I}}}$$

and

$$\frac{1}{\gamma_{\mathrm{p}}}\big(\varphi_{h,\mathrm{p}}, \psi_i\big)_{0,\Gamma_{\mathrm{I}}} + \frac{1}{\gamma_{\mathrm{p}}}\big(\boldsymbol{n} \cdot \mathsf{T}(\mathbf{u}_{h,\mathrm{f}}, p_{h,\mathrm{f}}) \cdot \boldsymbol{n}, \psi_i\big)_{0,\Gamma_{\mathrm{I}}}$$

are added to the system, i.e., the two interface equations (6.3a) and (6.3b) tested with the appropriate traces of the respective test functions. In this sense, the D-RR method is a conforming discretization. However, these additions involve lower order terms such that the accuracy of the resulting discrete solution might be worse compared with the Neumann–Neumann one. On the other hand, the implementation cost is comparable and much smaller than that of the Robin–Robin approach.

Since the used finite element functions are piecewise smooth this is possible on each cell separately yielding a discontinuous interface function even for straight interfaces, hence the name discontinuous Robin–Robin (D-RR). This approach can also be viewed as an approximation to the discretization of Eq. (6.34), after all the Robin traces $T^R_{\mathrm{p},\gamma}$ and $T^R_{\mathrm{f},\gamma}$, see their definitions (6.35), are extensions of suitable combinations of derivatives and values, compare also with Definitions 5.1.5 and 5.3.9.

Chapter 7
Algorithms

The Neumann–Neumann as well as the Robin–Robin systems (6.7) and (6.34) along with their decoupled variants (6.20) and (6.40) can be solved iteratively. The first two can be represented as systems

$$\begin{pmatrix} S & C_1 \\ C_2 & D \end{pmatrix} \begin{pmatrix} (\boldsymbol{u}_\mathrm{f}, p_\mathrm{f}) \\ \varphi_\mathrm{p} \end{pmatrix} = \begin{pmatrix} b_1 \\ b_2 \end{pmatrix}. \tag{7.1}$$

Compare this with the discrete system (6.43) where for simplicity the same symbols in the matrix and right-hand side are used.[1] The decoupled approaches (6.20) and (6.40) lead to systems of the form

$$\begin{pmatrix} S & & & E_\mathrm{f} \\ R_\mathrm{f} & -\mathrm{id} & & \\ & E_\mathrm{p} & D & \\ & & R_\mathrm{p} & -\mathrm{id} \end{pmatrix} \begin{pmatrix} (\boldsymbol{u}_\mathrm{f}, p_\mathrm{f}) \\ \lambda \\ \varphi_\mathrm{p} \\ \mu \end{pmatrix} = \begin{pmatrix} b_1 \\ 0 \\ b_2 \\ 0 \end{pmatrix}, \tag{7.2}$$

with $C_1 = E_\mathrm{f} R_\mathrm{p}$ and $C_2 = E_\mathrm{p} R_\mathrm{f}$ in correspondence to the discrete version in Eq. (6.44). In the case of the Neumann–Neumann approach the interface operators are recovered as $H_{\mathrm{f},N}(\mu) = R_\mathrm{f}\big(S^{-1}(b_1 - E_\mathrm{f}\mu)\big)$ and $H_{\mathrm{p},N}(\lambda) = R_\mathrm{p}\big(D^{-1}\big(b_2 - E_\mathrm{p}(\lambda)\big)\big)$, while in the Robin–Robin case $H_{\mathrm{f},N}$ and $H_{\mathrm{p},N}$ have to be replaced with $H_\mathrm{f}^{\gamma_\mathrm{f},\gamma_\mathrm{p}}$ and $H_\mathrm{p}^{\gamma_\mathrm{p},\gamma_\mathrm{f}}$, respectively. Note that the entire system (including S, D, b_1, b_2, E_f, E_p, R_f, R_p, λ, and μ) differs compared with the Neumann–Neumann case, but the structure of the equations remains.

[1] There is a slight notational inconsistency here. For simplicity, the vector $(b_1, 0)$ in Eqs. (6.43) and (6.44) is denoted only b_1 here.

U. Wilbrandt, *Stokes–Darcy Equations*, Advances in Mathematical Fluid Mechanics,
https://doi.org/10.1007/978-3-030-02904-3_7

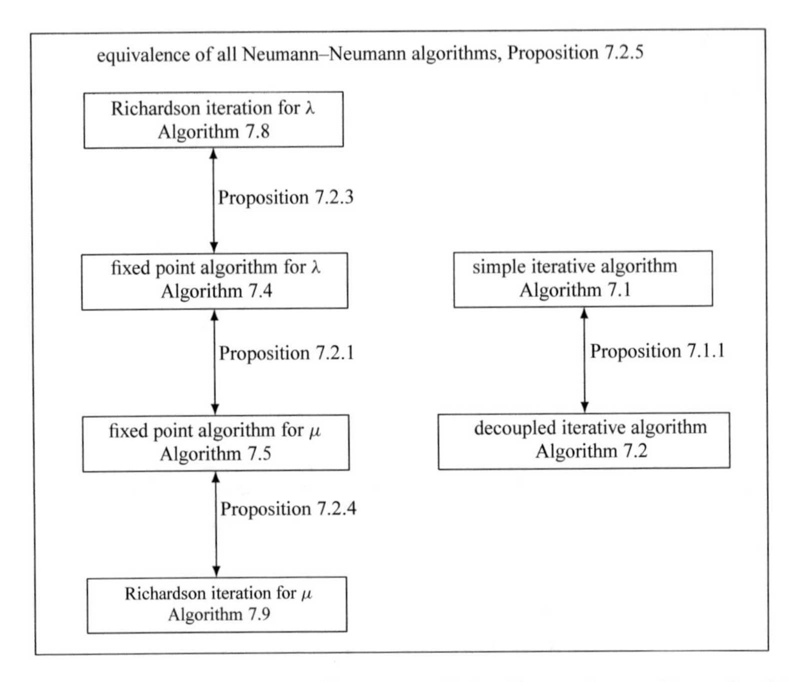

Fig. 7.1 A graph showing the relationships among all algorithms and propositions related to the Neumann–Neumann approach. The expected convergence behavior is shown in Proposition 7.3.1

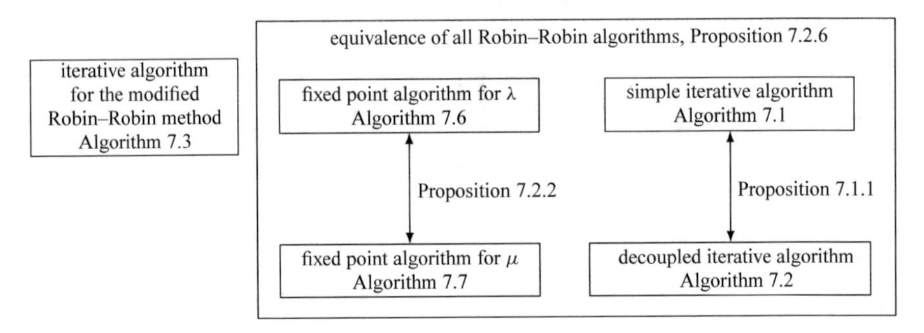

Fig. 7.2 A graph showing the relationships among all algorithms and propositions related to the Robin–Robin approaches. The expected convergence behavior is shown in Proposition 7.3.3 and in 7.3.2 for Algorithm 7.3

7.1 Classical Iterative Subdomain Methods

In this section several algorithms are presented which solve the Stokes–Darcy problem (7.1) and (7.2). In the finite element setting it is possible to set up the entire matrix at once. However, this typically requires a lot of coding and direct access to the underlying finite element spaces and matrices. Instead, all the presented

Algorithm 7.1

choose φ_p^0
for $k = 0, 1, \ldots$ **do**
$\left(\mathbf{u}_f^{k+1}, p_f^{k+1}\right) = S^{-1}\left(b_1 - C_1\varphi_p^k\right)$
$\varphi_p^{k+1} = D^{-1}\left(b_2 - C_2\left(\mathbf{u}_f^{k+1}, p_f^{k+1}\right)\right)$
end for

A simple iterative subdomain algorithm to solve the discrete Stokes–Darcy problem (7.1)

algorithms are based on solving the individual subproblems separately. The coupled system does not need to exist in memory at all. This enables the use of possibly highly tailored codes for each subproblem which only need to be able to solve Dirichlet, Neumann, and/or Robin problems and return data on the interface. In principle, the two subproblems could be solved using different codes.

A first rather simple approach to the Stokes–Darcy problem (7.1) is shown in Algorithm 7.1. Here each iteration consists of one solving step in each subdomain. Using the k-th iterate instead of the $(k + 1)$-th in the second step yields a version where each solving step can be done in parallel. The main advantage of Algorithm 7.1 is its simplicity and straightforward implementation. To better understand it, a reformulation is meaningful. Let s^k be the pair $(\mathbf{u}_f^k, p_f^k)$. Then Algorithm 7.1 leads to

$$\begin{aligned}
\begin{pmatrix} s^{k+1} \\ \varphi_p^{k+1} \end{pmatrix} &= \begin{pmatrix} S^{-1}(b_1 - C_1\varphi_p^k) \\ D^{-1}(b_2 - C_2 s^{k+1}) \end{pmatrix} \\
&= \begin{pmatrix} S^{-1}(b_1 - C_1\varphi_p^k) \\ D^{-1}b_2 - D^{-1}C_2S^{-1}(b_1 - C_1\varphi_p^k) \end{pmatrix} \\
&= \begin{pmatrix} S^{-1} & 0 \\ -D^{-1}C_2S^{-1} & D^{-1} \end{pmatrix} \begin{pmatrix} b_1 - C_1\varphi_p^k \\ b_2 \end{pmatrix} \\
&= \begin{pmatrix} S^{-1} & 0 \\ -D^{-1}C_2S^{-1} & D^{-1} \end{pmatrix} \left[\begin{pmatrix} b_1 \\ b_2 \end{pmatrix} - \begin{pmatrix} 0 & C_1 \\ 0 & 0 \end{pmatrix} \begin{pmatrix} s^k \\ \varphi_p^k \end{pmatrix} \right] \\
&= \begin{pmatrix} S & 0 \\ C_2 & D \end{pmatrix}^{-1} \left[\begin{pmatrix} b_1 \\ b_2 \end{pmatrix} - \begin{pmatrix} 0 & C_1 \\ 0 & 0 \end{pmatrix} \begin{pmatrix} s^k \\ \varphi_p^k \end{pmatrix} \right] \\
&= \begin{pmatrix} s^k \\ \varphi_p^k \end{pmatrix} + \begin{pmatrix} S & 0 \\ C_2 & D \end{pmatrix}^{-1} \left[\begin{pmatrix} b_1 \\ b_2 \end{pmatrix} - \begin{pmatrix} S & C_1 \\ C_2 & D \end{pmatrix} \begin{pmatrix} s^k \\ \varphi_p^k \end{pmatrix} \right].
\end{aligned}$$

This is the form a block-wise Gauss–Seidel[2] iteration has. Furthermore, a damping can be added via a weighting with the previous iterate:

$$
\begin{aligned}
\begin{pmatrix} s^{k+1} \\ \varphi_{\mathrm{p}}^{k+1} \end{pmatrix} &= (1-\omega)\begin{pmatrix} s^{k} \\ \varphi_{\mathrm{p}}^{k} \end{pmatrix} + \omega \begin{pmatrix} S^{-1}(b_1 - C_1\varphi_{\mathrm{p}}^k) \\ D^{-1}(b_2 - \omega C_2 s^{k+1}) \end{pmatrix} \\
&= \omega \begin{pmatrix} S^{-1}(b_1 + \frac{1-\omega}{\omega} S s^k - C_1\varphi_{\mathrm{p}}^k) \\ D^{-1}(b_2 + \frac{1-\omega}{\omega} D\varphi_{\mathrm{p}}^k - \omega C_2 s^{k+1}) \end{pmatrix} \\
&= \omega \begin{pmatrix} S^{-1}(b_1 + \frac{1-\omega}{\omega} S s^k - C_1\varphi_{\mathrm{p}}^k) \\ D^{-1}\left(b_2 + \frac{1-\omega}{\omega} D\varphi_{\mathrm{p}}^k - \omega C_2 S^{-1}(b_1 + \frac{1-\omega}{\omega} S s^k - C_1\varphi_{\mathrm{p}}^k)\right) \end{pmatrix} \\
&= \omega \begin{pmatrix} S^{-1} & 0 \\ -D^{-1}\omega C_2 S^{-1} & D^{-1} \end{pmatrix} \begin{pmatrix} b1 + \frac{1-\omega}{\omega} S s^k - C_1\varphi_{\mathrm{p}}^k \\ b2 + \frac{1-\omega}{\omega} D\varphi_{\mathrm{p}}^k \end{pmatrix} \\
&= \omega \begin{pmatrix} S & 0 \\ \omega C_2 & D \end{pmatrix}^{-1} \left[\begin{pmatrix} b_1 \\ b_2 \end{pmatrix} - \begin{pmatrix} S & C_1 \\ C_2 & D \end{pmatrix} \begin{pmatrix} s^{k} \\ \varphi_{\mathrm{p}}^{k} \end{pmatrix} + \frac{1}{\omega} \begin{pmatrix} S & 0 \\ \omega C_2 & D \end{pmatrix} \begin{pmatrix} s^{k} \\ \varphi_{\mathrm{p}}^{k} \end{pmatrix} \right] \\
&= \begin{pmatrix} s^{k} \\ \varphi_{\mathrm{p}}^{k} \end{pmatrix} + \omega \begin{pmatrix} S & 0 \\ \omega C_2 & D \end{pmatrix}^{-1} \left[\begin{pmatrix} b_1 \\ b_2 \end{pmatrix} - \begin{pmatrix} S & C_1 \\ C_2 & D \end{pmatrix} \begin{pmatrix} s^{k} \\ \varphi_{\mathrm{p}}^{k} \end{pmatrix} \right].
\end{aligned}
$$

This is then the block-wise SOR method. Also SSOR can be done in a block-wise manner. In essence, Algorithm 7.1 is a fixed point (Richardson) iteration preconditioned with a block-wise Gauss–Seidel method.

It is shown that the decoupled system (7.2) is equivalent to the smaller coupled problem (7.1), see Eqs. (6.7) and (6.20) in the Neumann–Neumann case and Eqs. (6.34) and (6.40) for the Robin–Robin approach. A straightforward recursion to solve (7.2) leads to Algorithm 7.2, which therefore produces the same iterates as Algorithm 7.1, if the initial guess is chosen appropriately:

Proposition 7.1.1 *Let φ_{p}^0 be given. Then the iterates $(\boldsymbol{u}_{\mathrm{f}}^k, p_{\mathrm{f}}^k)$ and φ_{p}^k, $k \in \mathbb{N}$, obtained from Algorithm 7.1, initialized with φ_{p}^0, coincide with those from Algorithm 7.2, initialized with $\mu^0 = R_p\varphi_{\mathrm{p}}^0$. If in contrast μ^0 is given, Algorithm 7.2 yields the same iterates as Algorithm 7.1 with $\varphi_{\mathrm{p}}^0 = E_p\mu^0$ (or any other extension).*

The alternative Robin–Robin method (C-RR) is only representable in the decoupled way, see also Sect. 6.6.8, leading to Algorithm 7.3. Here the updates of the interface variables are modified in comparison with Algorithm 7.2. Both algorithms correspond to blockwise Gauss–Seidel iterations and can be turned into Jacobi type if the indices $^{k+1}$ are replaced by ^{k} in the right-hand sides.

[2]See for example [Saa03] for an introduction to classical iterative solvers such as Gauss–Seidel, Jacobi, SOR (successive overrelaxation), and SSOR (symmetric SOR) as well as for Krylov solvers, such as cg (conjugate gradients) and gmres (generalized minimal residual).

Algorithm 7.2	**Algorithm 7.3**
choose μ^0	choose η_f^0
for $k = 0, 1, \dots$ **do**	**for** $k = 0, 1, \dots$ **do**
$\left(\mathbf{u}_\mathrm{f}^{k+1}, p_\mathrm{f}^{k+1}\right) = S^{-1}(b_1 - E_\mathrm{f}\mu^k)$	$\left(\mathbf{u}_\mathrm{f}^{k+1}, p_\mathrm{f}^{k+1}\right) = S^{-1}(b_1 - E_\mathrm{f}\eta_\mathrm{f}^k)$
$\lambda^{k+1} = R_\mathrm{f}\left(\mathbf{u}_\mathrm{f}^{k+1}, p_\mathrm{f}^{k+1}\right)$	$\eta_\mathrm{p}^{k+1} = (\gamma_\mathrm{f} + \gamma_\mathrm{p})\mathbf{u}_\mathrm{f}^{k+1} \cdot \boldsymbol{n} - \eta_\mathrm{f}^k$
$\varphi_\mathrm{p}^{k+1} = D^{-1}(b_2 - E_\mathrm{p}\lambda^{k+1})$	$\varphi_\mathrm{p}^{k+1} = D^{-1}\left(b_2 - E_\mathrm{p}\eta_\mathrm{p}^{k+1}\right)$
$\mu^{k+1} = R_\mathrm{p}\varphi_\mathrm{p}^{k+1}$	$\eta_\mathrm{f}^{k+1} = -\left(1 + \frac{\gamma_\mathrm{f}}{\gamma_\mathrm{p}}\right)\varphi_\mathrm{p}^{k+1} + \frac{\gamma_\mathrm{f}}{\gamma_\mathrm{p}}\eta_\mathrm{f}^{k+1}$
end for	**end for**

Simple iterative algorithms to solve the decoupled Stokes–Darcy problem (7.2) (left) and its variant for the C-RR method (right)

The main difference between Algorithms 7.1 and 7.2 is that the latter stores the intermediate results λ^{k+1} and μ^{k+1} which are only implicitly given in the former. This idea can be extended to store only one solution variable: If one is not directly interested in, say $(\boldsymbol{u}_\mathrm{f}, p_\mathrm{f})$, one can rewrite Algorithm 7.1 as a recursion for φ_p^{k+1}:

$$\varphi_\mathrm{p}^{k+1} = D^{-1}\left(b_2 - C_2 S^{-1}\left(b_1 - C_1\varphi_\mathrm{p}^k\right)\right). \tag{7.3}$$

The same strategy in Algorithm 7.2 leads to

$$\varphi_\mathrm{p}^{k+1} = D^{-1}\left(b_2 - E_\mathrm{p} R_\mathrm{f} S^{-1}\left(b_1 - E_\mathrm{f} R_\mathrm{p}\varphi_\mathrm{p}^k\right)\right). \tag{7.4}$$

Using the identities $E_\mathrm{f} R_\mathrm{p} = C_1$ and $E_\mathrm{p} R_\mathrm{f} = C_2$ as in Sect. 6.7, this is exactly the recursion in Eq. (7.3), and is the claim of Proposition 7.1.1. The additional coupling between the interface variables in the C-RR method, Algorithm 7.3, does not allow such a simple representation.

7.2 Algorithms for Interface Equations

The simplest algorithm to solve the fixed point equations (6.12) and (6.13) is to define the next iterate by the left-hand side of these equations, see Algorithms 7.4 and 7.5.

As before, a damping strategy can be added via a weighting with the previous iterate and is left out here only for brevity. Since the fixed point equations (6.12) and (6.13) are equivalent, Corollary 6.4.14, so are the above algorithms:

Proposition 7.2.1 *Let λ^k, $k \in \mathbb{N}$, be the iterates of Algorithm 7.4 initialized with λ^0. Then $\mu^k = H_{\mathrm{p,N}}(\lambda^k)$ are the iterates of Algorithm 7.5 initialized with $\mu^0 = H_{\mathrm{p,N}}(\lambda^0)$. Conversely, if μ^k, $k \in \mathbb{N}$, are the iterates of Algorithm 7.5 initialized*

Algorithm 7.4

choose λ^0
for $k = 0, 1, \dots$ **do**
$\lambda^{k+1} = H_{\mathrm{f,N}}\left(H_{\mathrm{p,N}}(\lambda^k)\right)$
end for

Algorithm 7.5

choose μ^0
for $k = 0, 1, \dots$ **do**
$\mu^{k+1} = H_{\mathrm{p,N}}\left(H_{\mathrm{f,N}}(\mu^k)\right)$
end for

Simple fixed point algorithms to solve Eqs. (6.12) (left) and (6.13) (right)

Algorithm 7.6

choose λ^0
for $k = 0, 1, \dots$ **do**
$\lambda^{k+1} = H_{\mathrm{f}}^{\gamma_{\mathrm{f}},\gamma_{\mathrm{p}}}\left(H_{\mathrm{p}}^{\gamma_{\mathrm{p}},\gamma_{\mathrm{f}}}(\lambda^k)\right)$
end for

Algorithm 7.7

choose μ^0
for $k = 0, 1, \dots$ **do**
$\mu^{k+1} = H_{\mathrm{p}}^{\gamma_{\mathrm{p}},\gamma_{\mathrm{f}}}\left(H_{\mathrm{f}}^{\gamma_{\mathrm{f}},\gamma_{\mathrm{p}}}(\mu^k)\right)$
end for

Simple fixed point algorithms to solve Eqs. (6.38) (left) and (6.39) (right)

with μ^0, then $\lambda^k = H_{\mathrm{f,N}}(\mu^k)$ are the iterates of Algorithm 7.4 initialized with $\lambda^0 = H_{\mathrm{f,N}}(\mu^0)$.

The algorithms above can as well be applied to solve the fixed point equations (6.38) and (6.39) for the Robin–Robin problem, see Algorithms 7.6 and 7.7. Similar to Proposition 7.2.1 these two are equivalent:

Proposition 7.2.2 *Let λ^k, $k \in \mathbb{N}$, be the iterates of Algorithm 7.6 initialized with λ^0. Then $\mu^k = H_{\mathrm{p}}^{\gamma_{\mathrm{p}},\gamma_{\mathrm{f}}}(\lambda^k)$ are the iterates of Algorithm 7.7 initialized with $\mu^0 = H_{\mathrm{p}}^{\gamma_{\mathrm{p}},\gamma_{\mathrm{f}}}(\lambda^0)$. Conversely, if μ^k, $k \in \mathbb{N}$, are the iterates of Algorithm 7.7 initialized with μ^0, then $\lambda^k = H_{\mathrm{f}}^{\gamma_{\mathrm{f}},\gamma_{\mathrm{p}}}(\mu^k)$ are the iterates of Algorithm 7.6 initialized with $\lambda^0 = H_{\mathrm{f}}^{\gamma_{\mathrm{f}},\gamma_{\mathrm{p}}}(\mu^0)$.*

A first algorithm for the Steklov–Poincaré equations (6.16) could consist of a recursion of the form

$$\lambda^{k+1} = \lambda^k - \omega\Big(H_{\mathrm{f,D}}(\lambda^k) - H_{\mathrm{p,N}}(\lambda^k)\Big),$$

with a damping $\omega > 0$. However with $\lambda^k \in \widetilde{\Lambda}_{\mathrm{f}}$ (as required by $H_{\mathrm{f,D}}$) it would be $\lambda^{k+1} \in \Lambda_{\mathrm{f}}^*$, because the image space of $H_{\mathrm{f,D}}$ is Λ_{f}^* and includes that of $H_{\mathrm{p,N}}$, see also Eqs. (6.10). Therefore, to allow a meaningful definition of λ^{k+2} in the next step, it is necessary to insert an operator P which maps Λ_{f}^* to $\widetilde{\Lambda}_{\mathrm{f}}$. This leads to the preconditioned Richardson iteration of Algorithm 7.8 with the preconditioner P. A candidate for P is $H_{\mathrm{f,N}}$ as it has the desired domain of definition and image space. However it is not linear, especially it does not map zero to zero, which would imply that the iteration would not stay at the sought solution. The linear part of $H_{\mathrm{f,N}}$, $L_{\mathrm{f,N}} = H_{\mathrm{f,N}} - \lambda_{0,\mathrm{f}} = H_{\mathrm{f,N}} - H_{\mathrm{f,N}}(0)$, does not suffer from this drawback and is the

Algorithm 7.8

choose λ^0
for $k = 0, 1, \ldots$ **do**
$\lambda^{k+1} = \lambda^k - \omega P\big(H_{\mathrm{f,D}}(\lambda^k) - H_{\mathrm{p,N}}(\lambda^k)\big)$
end for

Preconditioned Richardson iteration for the Steklov–Poincaré equation (6.16)

canonical preconditioner P, see also Definition 6.5.1. With this choice the recursion in Algorithm 7.8 is

$$\begin{aligned}
\lambda^{k+1} &= \lambda^k - \omega L_{\mathrm{f,N}}\Big(H_{\mathrm{f,D}}(\lambda^k) - H_{\mathrm{p,N}}(\lambda^k)\Big) \\
&= \lambda^k - \omega\Big(L_{\mathrm{f,N}}\Big(H_{\mathrm{f,D}}(\lambda^k)\Big) - L_{\mathrm{f,N}}\Big(H_{\mathrm{p,N}}(\lambda^k)\Big)\Big) \\
&= \lambda^k - \omega\Big(H_{\mathrm{f,N}}\Big(H_{\mathrm{f,D}}(\lambda^k)\Big) - H_{\mathrm{f,N}}\Big(H_{\mathrm{p,N}}(\lambda^k)\Big)\Big) \\
&= (1-\omega)\lambda^k + \omega H_{\mathrm{f,N}}\Big(H_{\mathrm{p,N}}(\lambda^k)\Big),
\end{aligned}$$

where the linearity of $L_{\mathrm{f,N}}$, Definition 6.5.1, and the identity (6.11c) are used. This is exactly Algorithm 7.4 for the fixed point equation (6.12) with a damping ω which is stated in the following proposition.

Proposition 7.2.3 *Let λ^0 be fixed. Then Algorithm 7.8 with $P = L_{\mathrm{f,N}}$ yields the same iterates as Algorithm 7.4 (or rather a damped version of it, if $\omega \neq 1$).*

Similarly, the preconditioned Richardson iteration for the Steklov–Poincaré equation (6.17) leads to Algorithm 7.9. Here the canonical preconditioner $P = L_{p,N}$ is appropriate and leads to the same iterates as a damped version of Algorithm 7.5.

Proposition 7.2.4 *Let μ^0 be fixed. Then Algorithm 7.9 with $P = L_{\mathrm{p,N}}$ yields same iterates as Algorithm 7.5 (or rather a damped version of it, if $\omega \neq 1$).*

In contrast to both Robin–Robin approaches the Neumann–Neumann method allows (useful) Steklov–Poincaré equations (6.16) and (6.17), see also Remark 6.6.14. Therefore, also other Krylov type iterative methods, such as

Algorithm 7.9

choose μ^0
for $k = 0, 1, \ldots$ **do**
$\mu^{k+1} = \mu^k - \omega P\big(H_{\mathrm{p,D}}(\mu^k) - H_{\mathrm{f,N}}(\mu^k)\big)$
end for

Preconditioned Richardson iteration for the Steklov–Poincaré equations (6.17)

conjugate gradients (cg) or generalized minimal residuals (gmres), can be applied and are converging much faster.

7.2.1 Connection to the Classical Iterative Subdomain Methods

It is noted at the beginning of this chapter that in the case of the Neumann–Neumann coupled system (6.7), the operators on the interface are represented as $H_{\mathrm{f,N}}(\mu) = R_{\mathrm{f}}\big(S^{-1}(b_1 - E_{\mathrm{f}}\mu)\big)$ and $H_{\mathrm{p,N}}(\lambda) = R_{\mathrm{p}}\big(D^{-1}(b_2 - E_{\mathrm{p}}(\lambda))\big)$, while in the case of the Robin–Robin coupled system it is[3] $H_{\mathrm{f}}^{\gamma_{\mathrm{f}},\gamma_{\mathrm{p}}}(\mu) = R_{\mathrm{f}}\big(S^{-1}(b_1 - E_{\mathrm{f}}\mu)\big)$ and $H_{\mathrm{p}}^{\gamma_{\mathrm{p}},\gamma_{\mathrm{f}}}(\lambda) = R_{\mathrm{p}}\big(D^{-1}(b_2 - E_{\mathrm{p}}(\lambda))\big)$, such that the recursions in the Algorithms 7.4 and 7.6 turn into

$$\lambda^{k+1} = R_{\mathrm{f}}S^{-1}\Big(b_1 - E_{\mathrm{f}}R_{\mathrm{p}}D^{-1}\Big(b_2 - E_{\mathrm{p}}(\lambda^k)\Big)\Big)$$

which is exactly what Algorithm 7.2 does. The same approach can be done for the Algorithms 7.5 and 7.7 which solve for the fixed points μ:

$$\mu^{k+1} = R_{\mathrm{p}}D^{-1}\Big(b_2 - E_{\mathrm{p}}R_{\mathrm{f}}S^{-1}\Big(b_1 - E_{\mathrm{f}}\mu^k\Big)\Big). \tag{7.5}$$

In fact, applying R_{p} to Eq. (7.4) leads to the above Eq. (7.5). Hence, assuming suitable initial conditions, all the fixed point Algorithms 7.4–7.7 are equivalent to the simple iterative Algorithms 7.1 and 7.2.

Furthermore, for the Neumann–Neumann problem these are also equivalent to the preconditioned Richardson iterations with preconditioners $P = L_{\mathrm{f,N}}$ and $P = L_{\mathrm{p,N}}$, respectively. This is summarized in the following two propositions.

Proposition 7.2.5 *Considering the Neumann–Neumann approach, the Algorithms 7.1, 7.2, 7.4, 7.5, 7.8, and 7.9 are equivalent.*

Proposition 7.2.6 *For the Robin–Robin approach, Algorithms 7.1, 7.2, 7.6, and 7.7 are equivalent.*

In conclusion, only three distinct algorithms are introduced in this chapter, namely the ones based on the Neumann–Neumann and Robin–Robin formulation as well as the modified Robin–Robin method (C-RR), Algorithm 7.3, which behaves differently compared with the first two approaches.

[3]Reminder: The operators S, D, b_1, b_2, E_{f}, E_{p}, R_{f}, and R_{p} differ compared with the Neumann–Neumann case.

7.3 Convergence Behavior

In this section the convergence behavior of three distinct algorithms is analyzed with an emphasis on its dependence on the kinematic viscosity ν and the hydraulic conductivity tensor K. For simplicity, K is assumed to be a positive constant instead of a symmetric positive definite tensor. The general case however poses no further difficulties, other than notation. First, the Neumann–Neumann approach is studied and it turns out that the iteration diverges whenever ν and K are small. The alternative Robin–Robin method (C-RR) is studied next. Its convergence is slow for small ν and K. Finally, the Robin–Robin approach is analyzed and its convergence is somewhat independent of ν and K.

To keep the analysis as similar as possible among the three methods, the focus is on the decoupled Algorithms 7.2 and 7.3. For the C-RR method this is done in Section 3.2 in [CGHW11] and the proofs here use similar ideas and notation.

7.3.1 *Neumann–Neumann*

Consider Algorithm 7.2 solving the Neumann–Neumann problem (6.20). Writing each step in weak form leads to

- Find $(\boldsymbol{u}_\mathrm{f}^{k+1}, p_\mathrm{f}^{k+1})$ such that for all $\boldsymbol{v} \in \boldsymbol{V}_\mathrm{f}$ and $q \in Q_\mathrm{f}$ it is

$$a_\mathrm{f}(\boldsymbol{u}_\mathrm{f}^{k+1}, \boldsymbol{v}) + b_f(\boldsymbol{v}, p_\mathrm{f}^{k+1}) = \ell_\mathrm{f}(\boldsymbol{v}) - \left(\mu^k, \boldsymbol{v} \cdot \boldsymbol{n}\right)_{0,\Gamma_\mathrm{I}},$$
$$b_f(\boldsymbol{u}_\mathrm{f}^{k+1}, q) = 0,$$

- set $\lambda^{k+1} = \boldsymbol{u}_\mathrm{f}^{k+1} \cdot \boldsymbol{n}$,
- find φ_p^{k+1} such that for all $\psi \in Q_\mathrm{p}$ it is

$$a_\mathrm{p}(\varphi_\mathrm{p}^{k+1}, \psi) = \ell_\mathrm{p}(\psi) + \left(\lambda^{k+1}, \psi\right)_{0,\Gamma_\mathrm{I}},$$

- set $\mu^{k+1} = \varphi_\mathrm{p}^{k+1}|_{\Gamma_\mathrm{I}}$.

Next, define the errors of the iterates:

$$\varepsilon_\lambda^k = \lambda - \lambda^k, \quad \varepsilon_\mu^k = \mu - \mu^k, \quad e_\varphi^k = \varphi_\mathrm{p} - \varphi_\mathrm{p}^k, \quad e_\mathbf{u}^k = \boldsymbol{u}_\mathrm{f} - \boldsymbol{u}_\mathrm{f}^k, \quad e_p^k = p_\mathrm{f} - p_\mathrm{f}^k.$$

Since the constituting equations are linear the errors solve related equations, namely for all $\boldsymbol{v} \in \boldsymbol{V}_{\mathrm{f}}$, $q \in Q_{\mathrm{f}}$, and $\psi \in Q_{\mathrm{p}}$ it is

$$a_{\mathrm{f}}(e_{\mathbf{u}}^{k+1}, \boldsymbol{v}) + b_{\mathrm{f}}(\boldsymbol{v}, e_p^{k+1}) = -\left(\varepsilon_\mu^k, \boldsymbol{v} \cdot \boldsymbol{n}\right)_{0,\Gamma_{\mathrm{I}}}, \tag{7.6}$$

$$b_f(e_{\mathbf{u}}^{k+1}, q) = 0, \tag{7.7}$$

$$\varepsilon_\lambda^{k+1} = e_{\mathbf{u}}^{k+1} \cdot \boldsymbol{n}, \tag{7.8}$$

$$a_{\mathrm{p}}(e_\varphi^{k+1}, \psi) = \left(\varepsilon_\lambda^{k+1}, \psi\right)_{0,\Gamma_{\mathrm{I}}}, \tag{7.9}$$

$$\varepsilon_\mu^{k+1} = e_\varphi^{k+1}|_{\Gamma_{\mathrm{I}}}. \tag{7.10}$$

Using the notation introduced in Sect. 6.5, this means $(e_{\mathbf{u}}^{k+1}, e_p^{k+1}) = \mathscr{L}_{\mathrm{f,N}}(\varepsilon_\mu^k)$ and $e_\varphi^{k+1} = \mathscr{L}_{\mathrm{p,N}}(\varepsilon_\lambda^{k+1})$ as well as $\varepsilon_\lambda^{k+1} = T_{\mathrm{f}}(e_{\mathbf{u}}^{k+1}, e_p^{k+1}) = L_{\mathrm{f,N}}(\varepsilon_\mu^k)$ and $\varepsilon_\mu^{k+1} = T_{\mathrm{p}}(e_\varphi^{k+1}) = L_{\mathrm{p,N}}(\varepsilon_\lambda^{k+1})$. Combining these equations leads to

$$\left\|\varepsilon_\lambda^{k+1}\right\|_{0,\Gamma_{\mathrm{I}}} = \left\|\left(T_{\mathrm{f}} \circ \mathscr{L}_{\mathrm{f,N}} \circ T_{\mathrm{p}} \circ \mathscr{L}_{\mathrm{p,N}}\right)\left(\varepsilon_\lambda^k\right)\right\|_{0,\Gamma_{\mathrm{I}}} \leq c_{\mathrm{NN}} \left\|\varepsilon_\lambda^k\right\|_{0,\Gamma_{\mathrm{I}}},$$
$$\left\|\varepsilon_\mu^{k+1}\right\|_{0,\Gamma_{\mathrm{I}}} = \left\|\left(T_{\mathrm{p}} \circ \mathscr{L}_{\mathrm{p,N}} \circ T_{\mathrm{f}} \circ \mathscr{L}_{\mathrm{f,N}}\right)\left(\varepsilon_\mu^k\right)\right\|_{0,\Gamma_{\mathrm{I}}} \leq c_{\mathrm{NN}} \left\|\varepsilon_\mu^k\right\|_{0,\Gamma_{\mathrm{I}}},$$

with

$$c_{\mathrm{NN}} := \frac{c_{T_{\mathrm{f}}}^2 c_{T_{\mathrm{p}}}^2}{\alpha_{a_{\mathrm{f}}} \alpha_{a_{\mathrm{p}}}}.$$

Here it is used that the continuity constants of the solution operators $\mathscr{L}_{i,\mathrm{N}}$, $i \in \{\mathrm{f}, \mathrm{p}\}$, are c_{T_i}/α_{a_i}, see the discussion after Definition 6.5.3. Similar results are obtained for the errors in the solution variables $e_{\mathbf{u}}^{k+1}$ and e_φ^{k+1} to give

$$\left\|e_{\mathbf{u}}^{k+1}\right\|_{1,\Omega_{\mathrm{f}}} \leq c_{\mathrm{NN}} \left\|e_{\mathbf{u}}^k\right\|_{1,\Omega_{\mathrm{f}}}, \qquad \left\|e_\varphi^{k+1}\right\|_{1,\Omega_{\mathrm{p}}} \leq c_{\mathrm{NN}} \left\|e_\varphi^k\right\|_{1,\Omega_{\mathrm{p}}}.$$

The following proposition summarizes this result.

Proposition 7.3.1 *The reduction of the error in each step of the Algorithms using the Neumann–Neumann approach, see Proposition 7.2.5, is*

$$c_{\mathrm{NN}} = \frac{c_{T_{\mathrm{f}}}^2 c_{T_{\mathrm{p}}}^2}{\alpha_{a_{\mathrm{f}}} \alpha_{a_{\mathrm{p}}}}.$$

The coercivity constants $\alpha_{a_{\mathrm{f}}}$ and $\alpha_{a_{\mathrm{p}}}$ scale like ν and K, respectively, see Sect. 6.3.2 as well as Sects. 5.1.2 and 5.3.2. Therefore, $c_{\mathrm{NN}} \to \infty$ for $\nu \to 0$ and $\mathsf{K} \to 0$ such that the Neumann–Neumann algorithms are only expected to converge for larger ν and K and diverge otherwise. This can be compensated with a damping ω. However, the algorithm then only converges if the damping is very small, which makes it slow.

7.3.2 Alternative Robin–Robin (C-RR)

Consider Algorithm 7.3 which solves the alternative Robin–Robin method introduced in Sect. 6.6.8. The proof presented here is essentially taken from [CGHW11], Section 3.2. There, a slightly modified version is presented which corresponds to a Jacobi type method whereas here a Gauss–Seidel type iteration is analyzed. As in the Neumann–Neumann case each step in the algorithm is written using the weak formulations:

- Find $(\boldsymbol{u}_{\mathrm{f}}^{k+1}, p_{\mathrm{f}}^{k+1}) = K_{\mathrm{f}}^{\gamma_{\mathrm{f}}}(-\eta_{\mathrm{f}})$, i.e., such that for all $\boldsymbol{v} \in \boldsymbol{V}_{\mathrm{f}}$ and $q \in Q_{\mathrm{f}}$ it is

$$a_{\mathrm{f}}^{\mathrm{R}}(\boldsymbol{u}_{\mathrm{f}}^{k+1}, \boldsymbol{v}) + b_f(\boldsymbol{v}, p_{\mathrm{f}}^{k+1}) = \ell_{\mathrm{f}}(\boldsymbol{v}) + \left(\eta_{\mathrm{f}}^k, \boldsymbol{v}\cdot\boldsymbol{n}\right)_{0,\Gamma_{\mathrm{I}}},$$
$$b_f(\boldsymbol{u}_{\mathrm{f}}^{k+1}, q) = 0,$$

- set $\eta_{\mathrm{p}}^{k+1} = c\eta_{\mathrm{f}}^k + d\boldsymbol{u}_{\mathrm{f}}^{k+1}\cdot\boldsymbol{n}$,
- find $\varphi_{\mathrm{p}}^{k+1} = K_{\mathrm{p}}^{\gamma_{\mathrm{p}}}(\gamma_{\mathrm{p}}\eta_{\mathrm{p}}^{k+1})$ such that for all $\psi \in Q_{\mathrm{p}}$ it is

$$a_{\mathrm{p}}^{\mathrm{R}}(\varphi_{\mathrm{p}}^{k+1}, \psi) = \ell_{\mathrm{p}}(\psi) + \left(\eta_{\mathrm{p}}^{k+1}, \psi\right)_{0,\Gamma_{\mathrm{I}}},$$

- set $\eta_{\mathrm{f}}^{k+1} = a\eta_{\mathrm{p}}^{k+1} + b\varphi_{\mathrm{p}}^{k+1}$.

The errors of the iterates

$$\varepsilon_{\mathrm{p}}^k = \eta_{\mathrm{p}} - \eta_{\mathrm{p}}^k,\quad \varepsilon_{\mathrm{f}}^k = \eta_{\mathrm{f}} - \eta_{\mathrm{f}}^k,\quad e_{\varphi}^k = \varphi_{\mathrm{p}} - \varphi_{\mathrm{p}}^k,\quad e_{\mathbf{u}}^k = \boldsymbol{u}_{\mathrm{f}} - \boldsymbol{u}_{\mathrm{f}}^k,\quad e_p^k = p_{\mathrm{f}} - p_{\mathrm{f}}^k,$$

solve the following equations[4] for all $\boldsymbol{v} \in \boldsymbol{V}_{\mathrm{f}}$, $q \in Q_{\mathrm{f}}$, and $\psi \in Q_{\mathrm{p}}$:

$$a_{\mathrm{f}}^{\mathrm{R}}(e_{\mathbf{u}}^{k+1}, \boldsymbol{v}) + b_{\mathrm{f}}(\boldsymbol{v}, e_p^{k+1})$$
$$= a_{\mathrm{f}}(e_{\mathbf{u}}^{k+1}, \boldsymbol{v}) + b_{\mathrm{f}}(\boldsymbol{v}, e_p^{k+1}) + \gamma_{\mathrm{f}}\left(e_{\mathbf{u}}^{k+1}\cdot\boldsymbol{n}, \boldsymbol{v}\cdot\boldsymbol{n}\right)_{0,\Gamma_{\mathrm{I}}} = \left(\varepsilon_{\mathrm{f}}^k, \boldsymbol{v}\cdot\boldsymbol{n}\right)_{0,\Gamma_{\mathrm{I}}}, \tag{7.11}$$

$$b_f(e_{\mathbf{u}}^{k+1}, q) = 0, \tag{7.12}$$

$$a_{\mathrm{p}}^{\mathrm{R}}(e_{\varphi}^{k+1}, \psi) = a_{\mathrm{p}}(e_{\varphi}^{k+1}, \psi) + \frac{1}{\gamma_{\mathrm{p}}}\left(e_{\varphi}^{k+1}, \psi\right)_{0,\Gamma_{\mathrm{I}}} = \left(\varepsilon_{\mathrm{p}}^{k+1}, \psi\right)_{0,\Gamma_{\mathrm{I}}}. \tag{7.13}$$

[4]Note that these equations are the linear parts of the Robin solution operators K_i^{γ}, $i \in \{\mathrm{f}, \mathrm{p}\}$, introduced in Sect. 6.6.4, similar to $\mathscr{L}_{i,N}$ and the respective Neumann operators $K_{i,N}$ in Sect. 6.5.

Additionally, the interface errors satisfy

$$\begin{aligned}
\varepsilon_{\mathrm{p}}^{k+1} &= \eta_{\mathrm{p}} - \eta_{\mathrm{p}}^{k+1} \\
&= \gamma_{\mathrm{p}} \boldsymbol{u}_{\mathrm{f}} \cdot \boldsymbol{n} + \varphi_{\mathrm{p}} - c\eta_{\mathrm{f}}^{k} - d\boldsymbol{u}_{\mathrm{f}}^{k+1} \cdot \boldsymbol{n} \\
&= (d - \gamma_{\mathrm{f}}) \boldsymbol{u}_{\mathrm{f}} \cdot \boldsymbol{n} + \varphi_{\mathrm{p}} - c\eta_{\mathrm{f}}^{k} - d\boldsymbol{u}_{\mathrm{f}}^{k+1} \cdot \boldsymbol{n} \qquad (d = \gamma_{\mathrm{f}} + \gamma_{\mathrm{p}}) \\
&= c\varepsilon_{\mathrm{f}}^{k} + d e_{\mathbf{u}}^{k+1} \cdot \boldsymbol{n}, \qquad (c = -1,\ \eta_{\mathrm{f}} = \gamma_{\mathrm{f}} \boldsymbol{u}_{\mathrm{f}} \cdot \boldsymbol{n} - \varphi_{\mathrm{p}}) \\
\varepsilon_{\mathrm{f}}^{k+1} &= \eta_{\mathrm{f}} - \eta_{\mathrm{f}}^{k+1} \\
&= \gamma_{\mathrm{f}} \boldsymbol{u}_{\mathrm{f}} \cdot \boldsymbol{n} - \varphi_{\mathrm{p}} - a\eta_{\mathrm{p}}^{k+1} - b\varphi_{\mathrm{p}}^{k+1} \\
&= a\gamma_{\mathrm{p}} \boldsymbol{u}_{\mathrm{f}} \cdot \boldsymbol{n} + (a+b)\varphi_{\mathrm{p}} - a\eta_{\mathrm{p}}^{k+1} - b\varphi_{\mathrm{p}}^{k+1} \qquad (-1 = a+b,\ \gamma_{\mathrm{f}} = a\gamma_{\mathrm{p}}) \\
&= a\varepsilon_{\mathrm{p}}^{k+1} + b e_{\varphi}^{k+1}. \qquad (\eta_{\mathrm{p}} = \gamma_{\mathrm{p}} \boldsymbol{u}_{\mathrm{f}} \cdot \boldsymbol{n} + \varphi_{\mathrm{p}})
\end{aligned}$$

Assume the algorithm is initialized with a function on the interface which is in $L^2(\Gamma_{\mathrm{I}})$, then all the iterates and errors on the interface are in $L^2(\Gamma_{\mathrm{I}})$ as well. Then the norm of $\varepsilon_{\mathrm{p}}^{k+1}$ is

$$\left\| \varepsilon_{\mathrm{p}}^{k+1} \right\|_{0,\Gamma_{\mathrm{I}}}^2 = c^2 \left\| \varepsilon_{\mathrm{f}}^{k} \right\|_{0,\Gamma_{\mathrm{I}}}^2 + d^2 \left\| e_{\mathbf{u}}^{k+1} \cdot \boldsymbol{n} \right\|_{0,\Gamma_{\mathrm{I}}}^2 + 2cd \left(\varepsilon_{\mathrm{f}}^{k}, e_{\mathbf{u}}^{k+1} \cdot \boldsymbol{n} \right)_{0,\Gamma_{\mathrm{I}}}.$$

The last term also appears in Eq. (7.11) with $\boldsymbol{v} = e_{\mathbf{u}}^{k+1}$:

$$\left(\varepsilon_{\mathrm{f}}^{k}, e_{\mathbf{u}}^{k+1} \right)_{0,\Gamma_{\mathrm{I}}} = a_f(e_{\mathbf{u}}^{k+1}, e_{\mathbf{u}}^{k+1}) + \gamma_{\mathrm{f}} \left\| e_{\mathbf{u}}^{k+1} \cdot \boldsymbol{n} \right\|_{0,\Gamma_{\mathrm{I}}}^2, \tag{7.14}$$

where also the incompressibility condition, Eq. (7.12), is taken into account. Combining the last two equations gives

$$\left\| \varepsilon_{\mathrm{p}}^{k+1} \right\|_{0,\Gamma_{\mathrm{I}}}^2 = c^2 \left\| \varepsilon_{\mathrm{f}}^{k} \right\|_{0,\Gamma_{\mathrm{I}}}^2 + \underbrace{(d^2 + 2cd\gamma_{\mathrm{f}}) \left\| e_{\mathbf{u}}^{k+1} \cdot \boldsymbol{n} \right\|_{0,\Gamma_{\mathrm{I}}}^2 + 2cd a_f(e_{\mathbf{u}}^{k+1}, e_{\mathbf{u}}^{k+1})}_{=:I_{\mathrm{p}}}. \tag{7.15}$$

Similarly, the norm of $\varepsilon_{\mathrm{f}}^{k+1}$ is given as

$$\left\| \varepsilon_{\mathrm{f}}^{k+1} \right\|_{0,\Gamma_{\mathrm{I}}}^2 = a^2 \left\| \varepsilon_{\mathrm{p}}^{k} \right\|_{0,\Gamma_{\mathrm{I}}}^2 + b^2 \left\| e_{\varphi}^{k+1} \right\|_{0,\Gamma_{\mathrm{I}}}^2 + 2ab \left(\varepsilon_{\mathrm{p}}^{k}, e_{\varphi}^{k+1} \right)_{0,\Gamma_{\mathrm{I}}}$$

and inserting $\psi = e_{\varphi}^{k+1}$ in Eq. (7.13) yields

$$\left(\varepsilon_{\mathrm{p}}^{k}, e_{\varphi}^{k+1} \right)_{0,\Gamma_{\mathrm{I}}} = \gamma_{\mathrm{p}} a_{\mathrm{p}}(e_{\varphi}^{k+1}, e_{\varphi}^{k+1}) + \left\| e_{\varphi}^{k+1} \right\|_{0,\Gamma_{\mathrm{I}}}^2.$$

Again, combining these two equations leads to

$$\left\|\varepsilon_{\mathrm{f}}^{k+1}\right\|_{0,\Gamma_{\mathrm{I}}}^{2}=a^{2}\left\|\varepsilon_{\mathrm{p}}^{k}\right\|_{0,\Gamma_{\mathrm{I}}}^{2}+\underbrace{(b^{2}+2ab)\left\|e_{\varphi}^{k+1}\right\|_{0,\Gamma_{\mathrm{I}}}^{2}+2ab\gamma_{\mathrm{p}}a_{\mathrm{p}}(e_{\varphi}^{k+1},e_{\varphi}^{k+1})}_{=:I_{\mathrm{f}}}. \tag{7.16}$$

In [CGHW11] it is shown that for $\gamma_{\mathrm{f}}=\gamma_{\mathrm{p}}$ these errors converge to zero, however no particular convergence speed is shown. In case the parameters $0<\gamma_{\mathrm{f}}<\gamma_{\mathrm{p}}$ furthermore satisfy

$$\gamma_{\mathrm{p}}-\gamma_{\mathrm{f}}<\frac{2\alpha_{a_{\mathrm{f}}}}{C_{T_{\mathrm{f}}}^{2}}\qquad\text{and}\qquad\frac{1}{\gamma_{\mathrm{f}}}-\frac{1}{\gamma_{\mathrm{p}}}<\frac{2\alpha_{a_{\mathrm{p}}}}{C_{T_{\mathrm{p}}}^{2}}, \tag{7.17}$$

with the coercivity constants α_{a_i}, $i\in\{\mathrm{f},\mathrm{p}\}$, and the continuity constants C_{T_i} of the trace operators T_i, it can be shown that the remainders I_{p} and I_{f} in the right-hand sides of Eqs. (7.15) and (7.16) are bounded by zero. First, consider Eq. (7.15). The continuity of the trace operator T_{f} and the coercivity of a_{f} give

$$\left\|e_{\mathbf{u}}^{k+1}\cdot\boldsymbol{n}\right\|_{0,\Gamma_{\mathrm{I}}}^{2}\leq C_{T_{\mathrm{f}}}^{2}\left\|e_{\mathbf{u}}^{k+1}\right\|_{1,\Omega_{\mathrm{f}}}^{2}\leq\frac{C_{T_{\mathrm{f}}}^{2}}{\alpha_{a_{\mathrm{f}}}}a_{\mathrm{f}}(e_{\mathbf{u}}^{k+1},e_{\mathbf{u}}^{k+1}). \tag{7.18}$$

Using $d=\gamma_{\mathrm{f}}+\gamma_{\mathrm{p}}$ and $c=-1$ the factor $(d^{2}+2cd\gamma_{\mathrm{f}})$ occurring in I_{p} is ${\gamma_{\mathrm{p}}}^{2}-{\gamma_{\mathrm{f}}}^{2}$, i.e., positive, and therefore the above estimate can be used to bound I_{p}:

$$\begin{aligned}I_{\mathrm{p}}&\leq\left(\frac{C_{T_{\mathrm{f}}}^{2}}{\alpha_{a_{\mathrm{f}}}}\left({\gamma_{\mathrm{p}}}^{2}-{\gamma_{\mathrm{f}}}^{2}\right)-2\gamma_{\mathrm{f}}-2\gamma_{\mathrm{p}}\right)a_{\mathrm{f}}(e_{\mathbf{u}}^{k+1},e_{\mathbf{u}}^{k+1})\\&\leq(\gamma_{\mathrm{f}}+\gamma_{\mathrm{p}})\left(\frac{C_{T_{\mathrm{f}}}^{2}}{\alpha_{a_{\mathrm{f}}}}(\gamma_{\mathrm{p}}-\gamma_{\mathrm{f}})-2\right)a_{\mathrm{f}}(e_{\mathbf{u}}^{k+1},e_{\mathbf{u}}^{k+1})\\&\leq 0,\end{aligned}$$

where the first condition in (7.17) is applied, together with the facts $\gamma_{\mathrm{f}}+\gamma_{\mathrm{p}}>0$ and $a_{\mathrm{f}}(e_{\mathbf{u}}^{k+1},e_{\mathbf{u}}^{k+1})\geq 0$. Next, consider Eq. (7.16). Note that using $b=-1-a$ and $a=\gamma_{\mathrm{f}}/\gamma_{\mathrm{p}}$ it is $b^{2}+2ab=b(a-1)=1-a^{2}>0$ and hence

$$\left\|e_{\varphi}^{k+1}\right\|_{0,\Gamma_{\mathrm{I}}}^{2}\leq C_{T_{\mathrm{p}}}^{2}\left\|e_{\varphi}^{k+1}\right\|_{1,\Omega_{\mathrm{p}}}^{2}\leq\frac{C_{T_{\mathrm{p}}}^{2}}{\alpha_{a_{\mathrm{p}}}}a_{\mathrm{p}}(e_{\varphi}^{k+1},e_{\varphi}^{k+1}) \tag{7.19}$$

implies

$$\begin{aligned}
I_{\mathrm{f}} &\leq b\left(\frac{C_{T_{\mathrm{p}}}^2}{\alpha_{a_{\mathrm{p}}}}(b+2a)+2a\gamma_{\mathrm{p}}\right)a_{\mathrm{p}}(e_{\varphi}^{k+1},e_{\varphi}^{k+1})\\
&= b\left(\frac{C_{T_{\mathrm{p}}}^2}{\alpha_{a_{\mathrm{p}}}}\left(\frac{\gamma_{\mathrm{f}}}{\gamma_{\mathrm{p}}}-1\right)+2\gamma_{\mathrm{f}}\right)a_{\mathrm{p}}(e_{\varphi}^{k+1},e_{\varphi}^{k+1})\\
&= -b\gamma_{\mathrm{f}}\left(\frac{C_{T_{\mathrm{p}}}^2}{\alpha_{a_{\mathrm{p}}}}\left(\frac{1}{\gamma_{\mathrm{f}}}-\frac{1}{\gamma_{\mathrm{p}}}\right)-2\right)a_{\mathrm{p}}(e_{\varphi}^{k+1},e_{\varphi}^{k+1})\\
&\leq 0,
\end{aligned}$$

where the last inequality holds because $-\gamma_{\mathrm{f}}b > 0$, the term involving a_{p} is positive, and the second condition in (7.17) implies that the factor in parentheses above is negative.

Therefore, it is shown that $I_{\mathrm{p}} \leq 0$ and $I_{\mathrm{f}} \leq 0$ so that Eqs. (7.15) and (7.16) simplify to

$$\left\|\varepsilon_{\mathrm{p}}^{k+1}\right\|_{0,\Gamma_{\mathrm{I}}} \leq \left\|\varepsilon_{\mathrm{f}}^{k}\right\|_{0,\Gamma_{\mathrm{I}}} \qquad \text{and} \qquad \left\|\varepsilon_{\mathrm{f}}^{k+1}\right\|_{0,\Gamma_{\mathrm{I}}} \leq \frac{\gamma_{\mathrm{f}}}{\gamma_{\mathrm{p}}}\left\|\varepsilon_{\mathrm{p}}^{k}\right\|_{0,\Gamma_{\mathrm{I}}}$$

which in turn gives

$$\left\|\varepsilon_{\mathrm{p}}^{k+1}\right\|_{0,\Gamma_{\mathrm{I}}} \leq \frac{\gamma_{\mathrm{f}}}{\gamma_{\mathrm{p}}}\left\|\varepsilon_{\mathrm{p}}^{k-1}\right\|_{0,\Gamma_{\mathrm{I}}} \qquad \text{and} \qquad \left\|\varepsilon_{\mathrm{f}}^{k+1}\right\|_{0,\Gamma_{\mathrm{I}}} \leq \frac{\gamma_{\mathrm{f}}}{\gamma_{\mathrm{p}}}\left\|\varepsilon_{\mathrm{f}}^{k-1}\right\|_{0,\Gamma_{\mathrm{I}}},$$

i.e., an expected error reduction by $\gamma_{\mathrm{f}}/\gamma_{\mathrm{p}}$ during every second iteration step. Hence, it is desirable to choose γ_{p} much larger than γ_{f}, however, the conditions (7.17) imply that $\gamma_{\mathrm{p}} \approx \gamma_{\mathrm{f}}$ at least for small viscosity ν and conductivity K, which determine the coercivity constants $\alpha_{a_{\mathrm{f}}}$ and $\alpha_{a_{\mathrm{p}}}$. A fast convergence can therefore only be expected for larger ν and K. Unlike in the Neumann–Neumann case, a divergence does not occur in a C-RR iteration as long as the Robin parameters γ_{p} and γ_{f} are suitably chosen. The following proposition serves as a summary of the expected convergence properties of the modified Robin–Robin method.

Proposition 7.3.2 *Solving the alternative Robin–Robin method via the Algorithm 7.3 reduces the error at every second step with a factor $\gamma_{\mathrm{f}}/\gamma_{\mathrm{p}}$ whenever the following restrictions hold*

$$\gamma_{\mathrm{p}} - \gamma_{\mathrm{f}} < \frac{2\alpha_{a_{\mathrm{f}}}}{C_{T_{\mathrm{f}}}^2} \qquad \textit{and} \qquad \frac{1}{\gamma_{\mathrm{f}}} - \frac{1}{\gamma_{\mathrm{p}}} < \frac{2\alpha_{a_{\mathrm{p}}}}{C_{T_{\mathrm{p}}}^2}. \qquad \text{(7.17 revisited)}$$

7.3.3 Robin–Robin

In this subsection consider again Algorithm 7.2 but solving the Robin–Robin problem (6.40). In contrast to the C-RR method here the variables on the interface are in general not in $L^2(\Gamma_\mathrm{I})$ but in the respective duals of Λ_p and Λ_f. As before, writing each step in weak form leads to[5]

- Find $(\boldsymbol{u}_\mathrm{f}^{k+1}, p_\mathrm{f}^{k+1}) = K_\mathrm{f}^{\gamma_\mathrm{f}}(\mu^k)$, i.e., for all $\boldsymbol{v} \in \boldsymbol{V}_\mathrm{f}$ and $q \in Q_\mathrm{f}$ it is

$$a_\mathrm{f}^\mathrm{R}(\boldsymbol{u}_\mathrm{f}^{k+1}, \boldsymbol{v}) + b_f(\boldsymbol{v}, p_\mathrm{f}^{k+1}) = \ell_\mathrm{f}(\boldsymbol{v}) - \left(\mu^k, \boldsymbol{v}\cdot\boldsymbol{n}\right)_{\Lambda_\mathrm{f}^*\times\Lambda_\mathrm{f}},$$

$$b_f(\boldsymbol{u}_\mathrm{f}^{k+1}, q) = 0,$$

- find $\lambda^{k+1} = T_{\mathrm{f},\gamma_\mathrm{p}}^\mathrm{R}(\boldsymbol{u}_\mathrm{f}^{k+1}, p_\mathrm{f}^{k+1})$, i.e., for all $\xi \in \Lambda_\mathrm{p}$ it is

$$\begin{aligned}\left(\lambda^{k+1}, \xi\right)_{\Lambda_\mathrm{p}^*\times\Lambda_\mathrm{p}} = \gamma_\mathrm{p}\left(\boldsymbol{u}_\mathrm{f}^{k+1}\cdot\boldsymbol{n}, \xi\right)_{0,\Gamma_\mathrm{I}} &- a_\mathrm{f}(\boldsymbol{u}_\mathrm{f}^{k+1}, \boldsymbol{v}_{P_\mathrm{f}(\xi)}) \\ &- b_\mathrm{f}(\boldsymbol{v}_{P_\mathrm{f}(\xi)}, p_\mathrm{f}^{k+1}) + \ell_\mathrm{f}(\boldsymbol{v}_{P_\mathrm{f}(\xi)}),\end{aligned}$$

- find $\varphi_\mathrm{p}^{k+1} = K_\mathrm{p}^{\gamma_\mathrm{p}}(\lambda^{k+1})$, i.e., for all $\psi \in Q_\mathrm{p}$ it is

$$a_\mathrm{p}^\mathrm{R}(\varphi_\mathrm{p}^{k+1}, \psi) = \ell_\mathrm{p}(\psi) + \frac{1}{\gamma_\mathrm{p}}\left(\lambda^{k+1}, \psi\right)_{\Lambda_\mathrm{p}^*\times\Lambda_\mathrm{p}},$$

- find $\mu^{k+1} = T_{\mathrm{p},\gamma_\mathrm{f}}^\mathrm{R}(\varphi_\mathrm{p}^{k+1})$, i.e., for all $\xi \in \Lambda_\mathrm{f}$ it is

$$\left(\mu^{k+1}, \xi\right)_{\Lambda_\mathrm{f}^*\times\Lambda_\mathrm{f}} = \left(\varphi_\mathrm{p}^{k+1}, \xi\right)_{0,\Gamma_\mathrm{I}} - \gamma_\mathrm{f} a_\mathrm{p}(\varphi_\mathrm{p}^{k+1}, \psi_{P_\mathrm{p}(\xi)}) + \gamma_\mathrm{f}\ell_\mathrm{p}(\psi_{P_\mathrm{p}(\xi)}).$$

The errors of the iterates are once more denoted by

$$\varepsilon_\lambda^k = \lambda - \lambda^k, \quad \varepsilon_\mu^k = \mu - \mu^k, \quad e_\varphi^k = \varphi_\mathrm{p} - \varphi_\mathrm{p}^k, \quad \mathbf{e}_\mathbf{u}^k = \boldsymbol{u}_\mathrm{f} - \boldsymbol{u}_\mathrm{f}^k, \quad e_p^k = p_\mathrm{f} - p_\mathrm{f}^k,$$

and solve the following equations for all $\boldsymbol{v} \in \boldsymbol{V}_\mathrm{f}$, $q \in Q_\mathrm{f}$, and $\psi \in Q_\mathrm{p}$:

$$a_\mathrm{f}^\mathrm{R}(\mathbf{e}_\mathbf{u}^{k+1}, \boldsymbol{v}) + b_\mathrm{f}(\boldsymbol{v}, e_p^{k+1}) = -\left(\varepsilon_\mu^k, \boldsymbol{v}\cdot\boldsymbol{n}\right)_{\Lambda_\mathrm{f}^*\times\Lambda_\mathrm{f}}, \tag{7.20}$$

$$b_f(\mathbf{e}_\mathbf{u}^{k+1}, q) = 0, \tag{7.21}$$

$$a_\mathrm{p}^\mathrm{R}(e_\varphi^{k+1}, \psi) = \frac{1}{\gamma_\mathrm{p}}\left(\varepsilon_\lambda^{k+1}, \psi\right)_{\Lambda_\mathrm{p}^*\times\Lambda_\mathrm{p}}. \tag{7.22}$$

[5]The solution operators $K_\mathrm{f}^{\gamma_\mathrm{f}}$ and $K_\mathrm{p}^{\gamma_\mathrm{p}}$ are introduced at the beginning of Sect. 6.6.4, the Robin traces $T_{\mathrm{f},\gamma_\mathrm{p}}^R$ and $T_{\mathrm{p},\gamma_\mathrm{f}}^R$ are introduced in Definition 6.6.8.

Additionally, on the interface the errors $\varepsilon_{\lambda}^{k+1} = T^{\mathrm{R}}_{\mathrm{f},\gamma_{\mathrm{p}}}(\mathbf{e}_{\mathbf{u}}^{k+1}, e_p^{k+1})$ and $\varepsilon_{\mu}^{k+1} = T^{\mathrm{R}}_{\mathrm{p},\gamma_{\mathrm{f}}}(e_{\varphi}^{k+1})$ satisfy for all $\xi \in \Lambda_{\mathrm{f}}$ and Λ_{p}, respectively:

$$\left(\varepsilon_{\lambda}^{k+1}, \xi\right)_{\Lambda_{\mathrm{p}}{}^{*}\times\Lambda_{\mathrm{p}}} = \gamma_{\mathrm{p}}\left(\mathbf{e}_{\mathbf{u}}^{k+1}\cdot \boldsymbol{n}, \xi\right)_{0,\Gamma_{\mathrm{I}}} - a_{\mathrm{f}}(\mathbf{e}_{\mathbf{u}}^{k+1}, \boldsymbol{v}_{P_{\mathrm{f}}(\xi)}) - b_{\mathrm{f}}(\boldsymbol{v}_{P_{\mathrm{f}}(\xi)}, e_p^{k+1}), \tag{7.23}$$

$$\left(\varepsilon_{\mu}^{k+1}, \xi\right)_{\Lambda_{\mathrm{f}}{}^{*}\times\Lambda_{\mathrm{f}}} = \left(e_{\varphi}^{k+1}, \xi\right)_{0,\Gamma_{\mathrm{I}}} - \gamma_{\mathrm{f}} a_{\mathrm{p}}(e_{\varphi}^{k+1}, \psi_{P_{\mathrm{p}}(\xi)}). \tag{7.24}$$

As an intermediate result it holds that the error in the Stokes pressure can be bounded by the error in the Stokes velocity: In fact, according to Remark 5.3.6 let $\boldsymbol{v} \in \boldsymbol{V}_{\mathrm{f}} \subset \boldsymbol{H}^1(\Omega_{\mathrm{f}})$ be given with $T_{\mathrm{f}}(\boldsymbol{v}, \cdot) = \boldsymbol{v}\cdot\boldsymbol{n}\big|_{\Gamma_{\mathrm{I}}} = 0$, $\nabla\cdot\boldsymbol{v} = e_p^{k+1}$, and $\|\boldsymbol{v}\| \le \frac{C}{\beta}\left\|e_p^{k+1}\right\|_{0,\Omega_{\mathrm{f}}}$. Then it is

$$\begin{aligned}
\left\|e_p^{k+1}\right\|_{0,\Omega_{\mathrm{f}}}^2 &= \left(\nabla\cdot\boldsymbol{v}, e_p^{k+1}\right)_{0,\Omega_{\mathrm{f}}} \\
&= -b_{\mathrm{f}}(\boldsymbol{v}, e_p^{k+1}) \\
&= a_{\mathrm{f}}(e_{\mathbf{u}}^{k+1}, \boldsymbol{v}) \qquad\qquad \text{using Eq. (7.20) and } \boldsymbol{v}\cdot\boldsymbol{n} = 0 \\
&\le c_{a_{\mathrm{f}}}\left\|e_{\mathbf{u}}^{k+1}\right\|_{1,\Omega_{\mathrm{f}}} \|\boldsymbol{v}\|_{1,\Omega_{\mathrm{f}}} \\
&\le \frac{Cc_{a_{\mathrm{f}}}}{\beta}\left\|\mathbf{e}_{\mathbf{u}}^{k+1}\right\|_{1,\Omega_{\mathrm{f}}}\left\|e_p^{k+1}\right\|_{1,\Omega_{\mathrm{f}}},
\end{aligned}$$

and hence $\left\|e_p^{k+1}\right\|_{0,\Omega_{\mathrm{f}}} \le \frac{Cc_{a_{\mathrm{f}}}}{\beta}\|\mathbf{e}_{\mathbf{u}}^{k+1}\|_{1,\Omega_{\mathrm{f}}}$.

With this result the norms of the errors on the interface can be bounded using Eqs. (7.23) and (7.24) together with the linearity of the involved bilinear forms and the Cauchy–Schwarz inequality (2.1) as follows:

$$\begin{aligned}
\left\|\varepsilon_{\lambda}^{k+1}\right\|_{\Lambda_{\mathrm{p}}^{*}} &= \sup_{\xi\in\Lambda_{\mathrm{p}}} \frac{\left(\varepsilon_{\lambda}^{k+1}, \xi\right)_{\Lambda_{\mathrm{p}}{}^{*}\times\Lambda_{\mathrm{p}}}}{\|\xi\|_{\Lambda_{\mathrm{p}}}} \\
&\le \sup_{\xi\in\Lambda_{\mathrm{p}}} \frac{1}{\|\xi\|_{\Lambda_{\mathrm{p}}}}\Bigg(\gamma_{\mathrm{p}}\left\|\mathbf{e}_{\mathbf{u}}^{k+1}\cdot\boldsymbol{n}\right\|_{0,\Gamma_{\mathrm{I}}}\|\xi\|_{0,\Gamma_{\mathrm{I}}} \\
&\qquad + c_{E_{\mathrm{f}}}\left(c_{a_{\mathrm{f}}}\left\|\mathbf{e}_{\mathbf{u}}^{k+1}\right\|_{1,\Omega_{\mathrm{f}}} + c_{b_{\mathrm{f}}}\left\|e_p^{k+1}\right\|_{0,\Omega_{\mathrm{f}}}\right)\|\xi\|_{\Lambda_{\mathrm{p}}}\Bigg) \\
&\le \left(\gamma_{\mathrm{p}} c_{T_{\mathrm{f}}} + c_{E_{\mathrm{f}}} c_{a_{\mathrm{f}}}\left(1 + Cc_{b_{\mathrm{f}}}\beta^{-1}\right)\right)\left\|\mathbf{e}_{\mathbf{u}}^{k+1}\right\|_{1,\Omega_{\mathrm{f}}},
\end{aligned} \tag{7.25}$$

$$\begin{aligned}\left\|\varepsilon_\mu^{k+1}\right\|_{\Lambda_\mathrm{f}^*} &= \sup_{\xi\in\Lambda_\mathrm{f}} \frac{\left(\varepsilon_\mu^{k+1},\xi\right)_{\Lambda_\mathrm{f}^*\times\Lambda_\mathrm{f}}}{\|\xi\|_{\Lambda_\mathrm{f}}}\\ &\le \sup_{\xi\in\Lambda_\mathrm{f}} \frac{1}{\|\xi\|_{\Lambda_\mathrm{f}}}\left(\left\|e_\varphi^{k+1}\right\|_{0,\Gamma_\mathrm{I}}\|\xi\|_{0,\Gamma_\mathrm{I}} + \gamma_\mathrm{f} c_{a_\mathrm{p}} c_{E_\mathrm{p}}\left\|e_\varphi^{k+1}\right\|_{1,\Omega_\mathrm{p}}\|\xi\|_{\Lambda_\mathrm{f}}\right)\\ &\le \left(c_{T_\mathrm{p}} + \gamma_\mathrm{f} c_{a_\mathrm{p}} c_{E_\mathrm{p}}\right)\left\|e_\varphi^{k+1}\right\|_{1,\Omega_\mathrm{p}},\end{aligned} \tag{7.26}$$

where the continuity constants of the extension operators $\xi \mapsto \boldsymbol{v}_{P_\mathrm{f}(\xi)}$ and $\xi \mapsto \psi_{P_\mathrm{p}(\xi)}$ are denoted by c_{E_f} and c_{E_p}, respectively,[6] see also Definitions 5.1.5 and 5.3.9. Focusing on the constants which depend on ν and K therefore leads to

$$\left\|\varepsilon_\lambda^{k+1}\right\|_{\Lambda_\mathrm{p}^*} \le C\left(\gamma_\mathrm{p}+\nu\right)\left\|\mathbf{e}_\mathbf{u}^{k+1}\right\|_{1,\Omega_\mathrm{f}} \quad \text{and} \quad \left\|\varepsilon_\mu^{k+1}\right\|_{\Lambda_\mathrm{f}^*} \le C(1+\gamma_\mathrm{f}\mathsf{K})\left\|e_\varphi^{k+1}\right\|_{1,\Omega_\mathrm{p}} \tag{7.27}$$

with some constant C.

Next, bounds on $e_\mathbf{u}^{k+1}$ and e_φ^{k+1} in terms of ε_μ^k and $\varepsilon_\lambda^{k+1}$ are needed. Inserting $\psi = e_\varphi^{k+1}$ into (7.22) yields

$$a_\mathrm{p}^\mathrm{R}(e_\varphi^{k+1}, e_\varphi^{k+1}) = \frac{1}{\gamma_\mathrm{p}}\left(\varepsilon_\lambda^{k+1}, e_\varphi^{k+1}\right)_{\Lambda_\mathrm{p}^*\times\Lambda_\mathrm{p}} \tag{7.28}$$

and the coercivity of a_p together with the continuity of the trace T_p leads to

$$\begin{aligned}\alpha_{a_\mathrm{p}}\left\|e_\varphi^{k+1}\right\|_{1,\Omega_\mathrm{p}}^2 &\le a_\mathrm{p}^\mathrm{R}(e_\varphi^{k+1}, e_\varphi^{k+1})\\ &\overset{(7.28)}{\le} \frac{1}{\gamma_\mathrm{p}}\left\|\varepsilon_\lambda^{k+1}\right\|_{\Lambda_\mathrm{p}^*}\left\|e_\varphi^{k+1}\right\|_{\Lambda_\mathrm{p}}\\ &\le \frac{c_{T_\mathrm{p}}}{\gamma_\mathrm{p}}\left\|\varepsilon_\lambda^{k+1}\right\|_{\Lambda_\mathrm{p}^*}\left\|e_\varphi^{k+1}\right\|_{1,\Omega_\mathrm{p}}\end{aligned} \tag{7.29}$$

$$\Longrightarrow \left\|e_\varphi^{k+1}\right\|_{1,\Omega_\mathrm{p}} \le \frac{c_{T_\mathrm{p}}}{\gamma_\mathrm{p}\alpha_{a_\mathrm{p}}}\left\|\varepsilon_\lambda^{k+1}\right\|_{\Lambda_\mathrm{p}^*}. \tag{7.30}$$

Similarly in the Stokes subdomain inserting $\boldsymbol{v} = \mathbf{e}_\mathbf{u}^{k+1}$ into Eq. (7.20) taking (7.21) with $q = e_p^{k+1}$ into account yields

$$a_\mathrm{f}^\mathrm{R}(\mathbf{e}_\mathbf{u}^{k+1}, \mathbf{e}_\mathbf{u}^{k+1}) = -\left(\varepsilon_\mu^k, \mathbf{e}_\mathbf{u}^{k+1}\cdot\boldsymbol{n}\right)_{\Lambda_\mathrm{f}^*\times\Lambda_\mathrm{f}}. \tag{7.31}$$

[6]Since the functions P_f and P_p are not continuous (unless $\Lambda_\mathrm{f} = \Lambda_\mathrm{p}$), these constants are technically not *continuity* constants but rather bounds of the respective operators.

Coercivity of a_{f} and the continuity of the normal trace operator T_{f} imply

$$\begin{aligned}\alpha_{a_{\mathrm{f}}}\left\|\mathbf{e}_{\mathbf{u}}^{k+1}\right\|_{1,\Omega_{\mathrm{f}}}^{2} &\leq a_{\mathrm{f}}^{\mathrm{R}}(\mathbf{e}_{\mathbf{u}}^{k+1},\mathbf{e}_{\mathbf{u}}^{k+1}) \\ &\overset{(7.31)}{\leq} \left\|\varepsilon_{\mu}^{k}\right\|_{\Lambda_{\mathrm{f}}^{*}}\left\|\mathbf{e}_{\mathbf{u}}^{k+1}\cdot\boldsymbol{n}\right\|_{\Lambda_{\mathrm{f}}} \\ &\leq c_{T_{\mathrm{f}}}\left\|\varepsilon_{\mu}^{k}\right\|_{\Lambda_{\mathrm{f}}^{*}}\left\|\mathbf{e}_{\mathbf{u}}^{k+1}\right\|_{1,\Omega_{\mathrm{f}}} \\ \Longrightarrow \quad \left\|\mathbf{e}_{\mathbf{u}}^{k+1}\right\|_{1,\Omega_{\mathrm{f}}} &\leq \frac{c_{T_{\mathrm{p}}}}{\alpha_{a_{\mathrm{f}}}}\left\|\varepsilon_{\mu}^{k}\right\|_{\Lambda_{\mathrm{f}}^{*}}. \end{aligned} \tag{7.32}$$

As in Eq. (7.27), focusing on the constants depending on ν and K gives

$$\left\|e_{\varphi}^{k+1}\right\|_{1,\Omega_{\mathrm{p}}} \leq C\frac{1}{\gamma_{\mathrm{p}}\mathsf{K}}\left\|\varepsilon_{\lambda}^{k+1}\right\|_{\Lambda_{\mathrm{p}}^{*}} \quad \text{and} \quad \left\|\mathbf{e}_{\mathbf{u}}^{k+1}\right\|_{1,\Omega_{\mathrm{f}}} \leq C\frac{1}{\nu}\left\|\varepsilon_{\mu}^{k}\right\|_{\Lambda_{\mathrm{f}}^{*}}. \tag{7.33}$$

Combining Eqs. (7.27) and (7.33) leads to

$$\left\|\varepsilon_{\lambda}^{k+1}\right\|_{\Lambda_{\mathrm{p}}^{*}} \leq c_{\mathrm{RR}}\left\|\varepsilon_{\mu}^{k}\right\|_{\Lambda_{\mathrm{f}}^{*}}, \tag{7.34}$$

$$\left\|\varepsilon_{\lambda}^{k+1}\right\|_{\Lambda_{\mathrm{p}}^{*}} \leq c_{\mathrm{RR}}\left\|\varepsilon_{\lambda}^{k}\right\|_{\Lambda_{\mathrm{p}}^{*}}, \tag{7.35}$$

$$\left\|e_{\varphi}^{k+1}\right\|_{1,\Omega_{\mathrm{p}}} \leq c_{\mathrm{RR}}\left\|e_{\varphi}^{k}\right\|_{1,\Omega_{\mathrm{p}}}, \tag{7.36}$$

$$\left\|\mathbf{e}_{\mathbf{u}}^{k+1}\right\|_{1,\Omega_{\mathrm{f}}} \leq c_{\mathrm{RR}}\left\|\mathbf{e}_{\mathbf{u}}^{k}\right\|_{1,\Omega_{\mathrm{f}}}, \tag{7.37}$$

with[7]

$$c_{\mathrm{RR}} := C\frac{(\gamma_{\mathrm{p}}+\nu)(1+\gamma_{\mathrm{f}}\mathsf{K})}{\gamma_{\mathrm{p}}\mathsf{K}\,\nu}.$$

For small values of viscosity ν and hydraulic conductivity K the constant c_{RR} essentially scales as the corresponding one from the Neumann–Neumann approach, see also Proposition 7.3.1. Even though the bilinear forms $a_{\mathrm{f}}^{\mathrm{R}}$ and $a_{\mathrm{p}}^{\mathrm{R}}$ both include an additional positive semi-definite term, the coercivity constants are the same as for a_{f} and a_{p} used in the Neumann–Neumann approach. The following summarizes these findings:

[7] Here C is a constant independent of ν, K, γ_{f}, and γ_{p}. Note that it is not the same one used earlier.

Proposition 7.3.3 *The reduction of the error in each step of the Algorithms using the Robin–Robin approach, see Proposition 7.2.6, is*

$$c_{\mathrm{RR}} = C \frac{(\gamma_{\mathrm{p}} + \nu)(1 + \gamma_{\mathrm{f}} \mathsf{K})}{\gamma_{\mathrm{p}} \mathsf{K}\, \nu}.$$

7.3.3.1 Towards a Better Estimate

Note that e_{φ}^{k+1} solves a Dirichlet problem with data $T_{\mathrm{p}} e_{\varphi}^{k+1} \in \Lambda_{\mathrm{p}}$, i.e., $e_{\varphi}^{k+1} = K_{\mathrm{D}}(T_{\mathrm{p}} e_{\varphi}^{k+1})$ with the Dirichlet operator K_{D} as in Proposition 5.1.3 (here $\Gamma_{\mathrm{R}} = \emptyset$). The continuity constant of K_{D} is $\big(c_{a_{\mathrm{p}}}/\alpha_{a_{\mathrm{p}}} + 1\big) c_{\mathrm{p},E}$, with the continuity constant $c_{\mathrm{p},E}$ of a right inverse of T_{p}, i.e., an extension operator,[8] see the proof of Proposition 5.1.3, therefore it is

$$\left\| e_{\varphi}^{k+1} \right\|_{1,\Omega_{\mathrm{p}}} \leq c_{\mathrm{p},E} \left(\frac{c_{a_{\mathrm{p}}}}{\alpha_{a_{\mathrm{p}}}} + 1 \right) \left\| e_{\varphi}^{k+1} \right\|_{\Lambda_{\mathrm{p}}} =: \tilde{c}_{\mathrm{p}} \left\| e_{\varphi}^{k+1} \right\|_{\Lambda_{\mathrm{p}}}. \tag{7.38}$$

Note that $\tilde{c}_{\mathrm{p}}$ scales like $\mathcal{O}(1)$ in terms of K. Now assume the Sobolev–Slobodeckij semi-norm on the interface can be bounded by the respective L^2-norm, which, for presentation purposes, is expressed as

$$\|\xi\|_{1/2,\Gamma_{\mathrm{I}}}^2 \leq \varrho(\xi) \|\xi\|_{0,\Gamma_{\mathrm{I}}}^2, \tag{7.39}$$

where $\varrho(\xi)$ depends on $\xi \in H^{1/2}(\Gamma_{\mathrm{I}})$. Then the estimate (7.30) can be improved to

$$\begin{aligned}
\left(\alpha_{a_{\mathrm{p}}} + \frac{1}{\gamma_{\mathrm{p}} \tilde{c}_{\mathrm{p}} \varrho\big(e_{\varphi}^{k+1}\big)} \right) \left\| e_{\varphi}^{k+1} \right\|_{1,\Omega_{\mathrm{p}}}^2 &\leq a_{\mathrm{p}}(e_{\varphi}^{k+1}, e_{\varphi}^{k+1}) + \frac{1}{\gamma_{\mathrm{p}} \varrho\big(e_{\varphi}^{k+1}\big)} \left\| e_{\varphi}^{k+1} \right\|_{\Lambda_{\mathrm{p}}}^2 \\
&\leq a_{\mathrm{p}}(e_{\varphi}^{k+1}, e_{\varphi}^{k+1}) + \frac{1}{\gamma_{\mathrm{p}}} \left\| e_{\varphi}^{k+1} \right\|_{0,\Gamma_{\mathrm{I}}}^2 \\
&= a_{\mathrm{p}}^{\mathrm{R}}(e_{\varphi}^{k+1}, e_{\varphi}^{k+1}) \\
&\overset{(7.28)}{\leq} \frac{1}{\gamma_{\mathrm{p}}} \left\| \varepsilon_{\lambda}^{k+1} \right\|_{\Lambda_{\mathrm{p}}^*} \left\| e_{\varphi}^{k+1} \right\|_{\Lambda_{\mathrm{p}}} \\
\Longrightarrow \left\| e_{\varphi}^{k+1} \right\|_{1,\Omega_{\mathrm{p}}} &\leq \frac{c_{T_{\mathrm{p}}}}{\alpha_{a_{\mathrm{p}}} \gamma_{\mathrm{p}} + \Big(\tilde{c}_{\mathrm{p}} \varrho\big(e_{\varphi}^{k+1}\big) \Big)^{-1}} \left\| \varepsilon_{\lambda}^{k+1} \right\|_{\Lambda_{\mathrm{p}}^*}.
\end{aligned}$$

[8] Note that this extension can (but need not) be the same as $\xi \mapsto \psi_{\xi}$ which is used to define the weak formulation, where additionally the function P_{p} enters, i.e., it might be $c_{\mathrm{p},E} = c_{E_{\mathrm{p}}}$.

Combining all terms independent of K, γ_{p}, and e_φ^{k+1} into a generic constant C leads to

$$\left\| e_\varphi^{k+1} \right\|_{1,\Omega_{\mathrm{p}}} \leq C \frac{1}{\mathsf{K}\gamma_{\mathrm{p}} + 1/\varrho\left(e_\varphi^{k+1}\right)} \left\| \varepsilon_\lambda^{k+1} \right\|_{\Lambda_{\mathrm{p}}^*}. \tag{7.40}$$

A result similar to Eq. (7.38) for the Stokes subdomain can be found as follows. The pair $(\mathbf{e}_{\mathbf{u}}^{k+1}, e_p^{k+1})$ solves a Dirichlet problem with given data $T_{\mathrm{f}}(\mathbf{e}_{\mathbf{u}}^{k+1}, e_p^{k+1})$ using the Dirichlet operator K_{D} as in Proposition 5.3.7 (here $\Gamma_{\mathrm{R}} = \emptyset$). The continuity constant of K_{D} is $(1 + c_{a_{\mathrm{f}}}/\alpha_{a_{\mathrm{f}}})(1 + 1/\beta)c_{\mathrm{f},E}$, where $c_{\mathrm{f},E}$ is the continuity constant of a right inverse of T_{f}, see the proof of Proposition 5.3.7 and in particular Eq. (5.28) with $\tilde{c}_L = 0$. In essence it is

$$\left\| \mathbf{e}_{\mathbf{u}}^{k+1} \right\|_{1,\Omega_{\mathrm{f}}} \leq \tilde{c}_{\mathrm{f}} \left\| \mathbf{e}_{\mathbf{u}}^{k+1} \right\|_{\Lambda_{\mathrm{f}}} \tag{7.41}$$

with a constant $\tilde{c}_{\mathrm{f}}$ independent of ν. Again using the assumption (7.39) the estimate (7.32) can be improved as follows:

$$\begin{aligned}
\left(\alpha_{a_{\mathrm{f}}} + \frac{\gamma_{\mathrm{f}}}{\tilde{c}_{\mathrm{f}}\varrho\left(\mathbf{e}_{\mathbf{u}}^{k+1} \cdot \boldsymbol{n}\right)}\right) \left\| \mathbf{e}_{\mathbf{u}}^{k+1} \right\|_{1,\Omega_{\mathrm{f}}}^2 &\leq a_{\mathrm{f}}(\mathbf{e}_{\mathbf{u}}^{k+1}, \mathbf{e}_{\mathbf{u}}^{k+1}) \\
&\quad + \frac{\gamma_{\mathrm{f}}}{\varrho\left(\mathbf{e}_{\mathbf{u}}^{k+1} \cdot \boldsymbol{n}\right)} \left\| \mathbf{e}_{\mathbf{u}}^{k+1} \cdot \boldsymbol{n} \right\|_{\Lambda_{\mathrm{f}}}^2 \\
&\leq a_{\mathrm{f}}(\mathbf{e}_{\mathbf{u}}^{k+1}, \mathbf{e}_{\mathbf{u}}^{k+1}) + \gamma_{\mathrm{f}} \left\| \mathbf{e}_{\mathbf{u}}^{k+1} \cdot \boldsymbol{n} \right\|_{0,\Gamma_{\mathrm{I}}}^2 \\
&= a_{\mathrm{f}}^{\mathrm{R}}(\mathbf{e}_{\mathbf{u}}^{k+1}, \mathbf{e}_{\mathbf{u}}^{k+1}) \\
&\overset{(7.31)}{\leq} \left\| \varepsilon_\mu^k \right\|_{\Lambda_{\mathrm{f}}^*} \left\| \mathbf{e}_{\mathbf{u}}^{k+1} \cdot \boldsymbol{n} \right\|_{\Lambda_{\mathrm{f}}} \\
\Longrightarrow \quad \left\| \mathbf{e}_{\mathbf{u}}^{k+1} \right\|_{1,\Omega_{\mathrm{f}}} &\leq \frac{c_{T_{\mathrm{f}}}}{\alpha_{a_{\mathrm{f}}} + \gamma_{\mathrm{f}}\left(\tilde{c}_{\mathrm{f}}\varrho\left(\mathbf{e}_{\mathbf{u}}^{k+1} \cdot \boldsymbol{n}\right)\right)^{-1}} \left\| \varepsilon_\mu^k \right\|_{\Lambda_{\mathrm{f}}^*}.
\end{aligned}$$

As before combining all terms which do not depend on ν, γ_{f}, or $\mathbf{e}_{\mathbf{u}}^{k+1}$ leads to

$$\left\| \mathbf{e}_{\mathbf{u}}^{k+1} \right\|_{1,\Omega_{\mathrm{f}}} \leq C \frac{1}{\nu + \gamma_{\mathrm{f}}/\varrho\left(\mathbf{e}_{\mathbf{u}}^{k+1} \cdot \boldsymbol{n}\right)} \left\| \varepsilon_\mu^k \right\|_{\Lambda_{\mathrm{f}}^*}. \tag{7.42}$$

While the derivation of the constant c_{RR} uses Eqs. (7.27) and (7.33), the latter can be replaced with (7.40) and (7.42) to yield

$$\left\| \varepsilon_\lambda^{k+1} \right\|_{\Lambda_\mathrm{p}^*} \leq \tilde{c}_{\mathrm{RR}} \left\| \varepsilon_\mu^k \right\|_{\Lambda_\mathrm{f}^*}, \tag{7.43}$$

$$\left\| \varepsilon_\lambda^{k+1} \right\|_{\Lambda_\mathrm{p}^*} \leq \tilde{c}_{\mathrm{RR}} \left\| \varepsilon_\lambda^k \right\|_{\Lambda_\mathrm{p}^*}, \tag{7.44}$$

$$\left\| e_\varphi^{k+1} \right\|_{1,\Omega_\mathrm{p}} \leq \tilde{c}_{\mathrm{RR}} \left\| e_\varphi^k \right\|_{1,\Omega_\mathrm{p}}, \tag{7.45}$$

$$\left\| \mathbf{e}_\mathbf{u}^{k+1} \right\|_{1,\Omega_\mathrm{f}} \leq \tilde{c}_{\mathrm{RR}} \left\| \mathbf{e}_\mathbf{u}^k \right\|_{1,\Omega_\mathrm{f}}, \tag{7.46}$$

with

$$\tilde{c}_{\mathrm{RR}} := C \frac{(\gamma_\mathrm{p} + \nu)(1 + \gamma_\mathrm{f} \mathsf{K})}{\left(\mathsf{K}\gamma_\mathrm{p} + 1/\varrho\left(e_\varphi^{k+1}\right)\right)\left(\nu + \gamma_\mathrm{f}/\varrho\left(\mathbf{e}_\mathbf{u}^{k+1} \cdot \boldsymbol{n}\right)\right)}.$$

The modified constant $\tilde{c}_{\mathrm{RR}}$ has a much better scaling behavior for small viscosity ν and hydraulic conductivity K. In fact, formally setting $\nu = \mathsf{K} = 0$ yields

$$\tilde{c}_{\mathrm{RR}} \approx C \frac{\gamma_\mathrm{p}}{\gamma_\mathrm{f}} \varrho\left(e_\varphi^{k+1}\right) \varrho\left(\mathbf{e}_\mathbf{u}^{k+1} \cdot \boldsymbol{n}\right) \tag{7.47}$$

and indicates a fixed reduction of the errors in each step of Algorithm 7.2 whenever $\varrho\left(e_\varphi^{k+1}\right)$ and $\varrho\left(\mathbf{e}_\mathbf{u}^{k+1} \cdot \boldsymbol{n}\right)$ do not deteriorate.

In general there is no constant ϱ as in Eq. (7.39) which does not depend on $\xi \in H^{1/2}(\Gamma_\mathrm{I})$, however in practice one expects e_φ^{k+1} and $\mathbf{e}_\mathbf{u}^{k+1} \cdot \boldsymbol{n}$ to be well-behaved so that the terms $\varrho\left(e_\varphi^{k+1}\right)$ and $\varrho\left(\mathbf{e}_\mathbf{u}^{k+1} \cdot \boldsymbol{n}\right)$ remain bounded. In particular, in the finite element framework so-called inverse estimates provide such inequalities with ϱ depending on the grid size h.

7.4 Remarks on the Implementation and Cost

This section describes some experience gained during the implementation of the proposed algorithms in the finite element code `ParMooN`, [WBA+17], followed by a general discussion on their cost per iteration.

The starting point is a code which is able to solve Stokes and Darcy systems already independently, i.e., assemble the discrete versions of S and D in Eqs. (7.1) and (7.2) and apply an approximation of its inverse. It turns out to be most sensible to implement all the remaining operators in Eq. (7.2), namely E_f, R_f, E_p, and R_p. With those at hand, the definition of $H_{\mathrm{f,N}}$, $H_{\mathrm{p,N}}$ for the Neumann–Neumann and $H_\mathrm{f}^{\gamma_\mathrm{f},\gamma_\mathrm{p}}$

and $H_{\text{p}}^{\gamma_{\text{p}},\gamma_{\text{f}}}$ for the Robin–Robin system is straightforward. The Dirichlet operators $H_{\text{f,D}}$ and $H_{\text{p,D}}$ are similar in structure and typically not difficult to implement as well, so that Algorithms 7.8 and 7.9 solving the Steklov–Poincaré equations (6.16) and (6.17) are then available. If Krylov type iterative solvers are implemented such that only the action of the linear operator is required, i.e., without an assembled matrix at hand, then those can be used at no additional implementation cost. Another advantage of implementing the decoupled system (7.2) rather than Eq. (7.1) is that the extension to the modified Robin–Robin method, C-RR, Algorithm 7.3, is uncomplicated. The D-RR method redefines all the operators in the system (7.2) but the structure remains, so that the implementation is only slightly more demanding, for example on the interface Γ_{I} the finite element space has to be discontinuous. The Robin–Robin method is by far the most involved one as the evaluation of the restriction operators R_{f} and R_{p} requires assembling in the respective subdomains to evaluate a_{f}, b_{f} and a_{p} with the correct test functions. Such operators are somewhat unusual and hence necessitate much specific code which had to be written from scratch.

In general it can be a complex task to couple two separate parts of a code to solve the coupled Stokes–Darcy problem. If the C-RR method and the Steklov–Poincaré equations are not of interest, implementing the coupled discrete system (7.1) is sufficient and does not involve spaces on the interface, thus simplifying the software development. It is highly advisable to take advantage of examples with known and very simple solutions during the implementation, see Sects. 8.1.1 and 8.2.

Concerning the cost per iteration, the Neumann–Neumann and Robin–Robin approaches are equal once the involved operators are assembled and stored: The differences such as

- for the Robin–Robin method and D-RR the restrictions R_{f} and R_{p} have a slightly larger sparsity pattern, so that their application is more expensive and
- Algorithms 7.2 and 7.3 require more storage and matrix-vector products compared with Algorithm 7.1

are negligible compared with the solving step and therefore not taken into account. Compared with the Neumann–Neumann or one of the Robin–Robin algorithms, the Krylov type methods include an extra solving step in the preconditioner. This disadvantage can be weakened: If one subproblem dominates the other in terms of computing time (e.g., one subdomain much larger than the other), it is sensible to choose the Steklov–Poincaré equation which is preconditioned with the smaller subproblem, so that one Krylov iteration step is essentially as expensive as solving the dominating subproblem. In such a case the cost per iteration is comparable to the Neumann–Neumann approach. If the two subproblems are of similar size, the Steklov–Poincaré equation can be evaluated in parallel, so that one iteration has a comparable computing time to the Neumann–Neumann method. In both cases the extra implementation effort is manageable.

Chapter 8
Numerical Results

The algorithms described in Chap. 7 are implemented in the C++ finite element code ParMooN, [WBA+17]. In this chapter several examples from the literature are introduced and numerical results shown. To begin with, a more general discussion on numerical examples is given.

8.1 General Remarks on Numerical Examples

Any implementation of an algorithm has to be tested before it can be considered usable. Despite the need of unit tests as in every computer program exceeding a few lines of code, serious finite element software packages additionally contain or easily integrate examples which are of importance to the specific partial differential equation, algorithm or method used. As with scientific work in general, numerical software has to be validated and results have to be reproducible and comparable. This is what standard numerical examples help to achieve.

Defining such numerical examples is a common difficulty in applied mathematics, as somewhat contradicting properties are desirable. First of all examples need to be simple. Advanced methods and algorithms typically require more complicated examples, but unnecessary complexity impedes the implementation in most codes. One can group numerical examples into three categories of increasing complexity. The first consists of trivial examples where the solution is very simple, for example constant or linear. Typically such examples are not found in the literature as their main purpose is to validate a code during the first phase of implementation. Since the given solution is in fact in the discrete ansatz space, any suitable program has to be able to recover this solution up to rounding errors. Here, very coarse grids are employed to simplify debugging.

U. Wilbrandt, *Stokes–Darcy Equations*, Advances in Mathematical Fluid Mechanics,
https://doi.org/10.1007/978-3-030-02904-3_8

In the second category of examples the solutions are also known analytically, but more complex, such as higher order polynomials or trigonometric functions. As in the first category, the computational domain is very simple, typically composed of a few basic geometries. The by far most common domains are the unit square for $d = 2$ and the unit cube for $d = 3$. The chosen boundary conditions are mostly of Dirichlet or Neumann type. These examples can be found in the literature and are, among other things, used to study convergence behavior under uniform grid refinement, i.e., one computes solutions on successively refined grids and compares the $L^2(\Omega)$- and $H^1(\Omega)$-errors with theoretical postulates. In fact, many other results from numerical analysis can be verified with these examples, e.g.,

- a posteriori error estimates,
- optimality of a preconditioner, i.e., independence of the number of iterations with respect to the grid parameter h,
- optimality with respect to a parameter of the underlying partial differential equation,
- parallel efficiency on multi core systems, and
- respecting maximum principles.

Additionally, various comparisons are often conducted with these examples, for example, different solvers, preconditioners, and discretizations.

The third category consists of examples with solutions which do not have a known analytical form, but posses some interesting features which are observable in practice, e.g., a vortex in a flow, boundary layers, or singularities. If feasible, a solution is computed on a very fine grid once[1] and serves as a reference to simulations on coarser grids. Sometimes also real world experiments are possible for simplified cases to replace an analytical solution as reference. To compare results with those of other codes one additionally often computes so-called quantities of interest, which are specific to the example, such as drag and lift in flow simulations or over- and undershoots in convection–diffusion problems. Examples of the third category are used to study methods, discretizations, and algorithms in situations which are closer to practice.

Simulations for real-world problems are in neither of these categories. They typically require more work to set up and the evaluation of models, algorithms and discretizations is not their primary goal. Even though these can be found in the literature, they are often not easily reproducible due to the large computing power needed or lack of detailed description (e.g., complicated domain).

Not all examples fit perfectly in one of the categories, because the simplicity of a problem at hand is highly subjective, but the pattern remains. Defining examples of the first two categories for the Laplace problem (5.1) is often easy, as any sufficiently smooth function can be inserted into the equation, thus defining the right-hand side. Boundary conditions can be set as desired and the corresponding boundary data is then given by the solution calling the respective trace operators. In

[1]This is occasionally called direct numerical simulation (DNS).

the case of the Stokes problem (5.25) the same strategy is not directly possible, as the vector field $\mathbf{u}$ needs to be divergence-free to be a solution.[2] The conditions on the interface Γ_I of the Stokes–Darcy coupled system (6.5) pose further restrictions on the solution components $\boldsymbol{u}_f$, p_f, and φ_p, making them non-obvious to find. With an appropriate ansatz it is often still possible to find suitable examples of the first and second category. In addition, due to linearity, sums and scalar multiples of solutions $(\boldsymbol{u}_f, p_f, \varphi_p)$ solve the Stokes–Darcy coupled system (6.5), too.

8.1.1 *Two Approaches to Find Examples with Known Solutions*

One common approach is to assume all solution components are in Q_k (or P_k) for some k, i.e., for $d = 2$,

$$\begin{aligned} \boldsymbol{u}_f(x, y) &= \begin{pmatrix} \sum_{i=0}^{k} \sum_{j=0}^{k} b_{ij} x^i y^j \\ \sum_{i=0}^{k} \sum_{j=0}^{k} c_{ij} x^i y^j \end{pmatrix}, \\ p_f &= \sum_{i=0}^{k} \sum_{j=0}^{k} d_{ij} x^i y^j, \\ \varphi_p &= \sum_{i=0}^{k} \sum_{j=0}^{k} e_{ij} x^i y^j, \end{aligned} \tag{8.1}$$

with $4(k+1)^d$ unknown coefficients b_{ij}, c_{ij}, d_{ij}, and e_{ij}. Considering the domains $\Omega_f = (0, 1) \times (1, 2)$ and $\Omega_p = (0, 1)^2$, the solutions are then inserted into the equations in the subdomains as well as on the interface to yield several restrictions on the coefficients. If successful, this approach leads to a parametrized solution, i.e., a family of solutions. A `Python` implementation specifically for the Stokes–Darcy coupled problem is provided in Sect. A.2, in particular in Listing 2.

A second common approach is a separation of variables, where each solution component is assumed to be a product of d functions, each depending on exactly one space variable, i.e., for $d = 2$ the ansatz is

$$\begin{aligned} \boldsymbol{u}_f(x, y) &= \begin{pmatrix} f_1(x) f_2(y) \\ g_1(x) g_2(y) \end{pmatrix}, \\ p_f &= 0, \\ \varphi_p &= h_1(x) h_2(y), \end{aligned} \tag{8.2}$$

[2] In fact it is also possible to have a right-hand side $g \in L^2(\Omega)$ in the equation $\nabla \cdot \mathbf{u} = g$. While the saddle point theory covers this case as well, fluids are divergence-free and so examples are expected to respect this.

with smooth, scalar functions f_i, g_i, and h_i, $i = 1, 2$. The pressure is assumed to vanish here only for simplicity and the domain is chosen as before, $\overline{\Omega_{\mathrm{f}} \cup \Omega_{\mathrm{p}}} = [0, 1] \times [0, 2]$, $\Gamma_{\mathrm{I}} = (0, 1) \times \{1\}$. Again, substitution into the equations of the coupled Stokes–Darcy problem yields several restrictions on the scalar functions: The equation $\nabla \cdot \boldsymbol{u}_{\mathrm{f}} = 0$ leads to

$$f_1'(x) f_2(y) + g_1(x) g_2'(y) = 0,$$

which, if f_2 and g_1 are assumed to be nonzero, can be rearranged to

$$\frac{f_1'(x)}{g_1(x)} = \frac{-g_2'(y)}{f_2(y)}.$$

These two fractions need to be constant, say ζ, because the equation holds for all $(x, y) \in \Omega_{\mathrm{f}}$. Hence, it must be

$$g_1 = \frac{1}{\zeta} f_1' \qquad \text{and} \qquad f_2 = -\frac{1}{\zeta} g_2'.$$

The mass conservation (6.3a) and the balance of normal forces (6.3b) then imply

$$0 = \boldsymbol{u}_{\mathrm{f}} \cdot \boldsymbol{n} + \mathsf{K} \nabla \varphi_{\mathrm{p}} \cdot \boldsymbol{n} = -\frac{1}{\zeta} f_1'(x)\, g_2(1) - \mathsf{K}\, h_1(x)\, h_2'(1)$$

$$\Longrightarrow h_1(x) = -\frac{1}{\mathsf{K}\zeta} \frac{g_2(1)}{h_2'(1)} f_1'(x) \qquad \text{and}$$

$$0 = \boldsymbol{n} \cdot \mathsf{T}(\boldsymbol{u}_{\mathrm{f}}, p_{\mathrm{f}}) \cdot \boldsymbol{n} + \varphi_{\mathrm{p}} = \frac{2\nu}{\zeta} f_1'(x) g_2'(1) + h_1(x) h_2(1)$$

$$\Longrightarrow h_1(x) = -\frac{2\nu}{\zeta} \frac{g_2'(1)}{h_2(1)} f_1'(x).$$

Therefore, h_1 is a multiple of f_1' with a constant

$$\vartheta = -\frac{1}{\mathsf{K}\zeta} \frac{g_2(1)}{h_2'(1)} = -\frac{2\nu}{\zeta} \frac{g_2'(1)}{h_2(1)}. \tag{8.3}$$

The Beavers–Joseph–Saffman condition (6.3c) leads to

$$\begin{aligned} 0 &= \boldsymbol{u}_{\mathrm{f}} \cdot \tau + \alpha \tau \cdot \mathsf{T}(\boldsymbol{u}_{\mathrm{f}}, p_{\mathrm{f}}) \cdot \boldsymbol{n} \\ &= -\frac{1}{\zeta} f_1(x) g_2'(1) + \frac{\alpha\nu}{\zeta} \big(f_1(x) g_2''(1) - f_1''(x) g_2(1) \big) \\ \Longrightarrow 0 &= f_1''(x) + \underbrace{\frac{g_2'(1)/(\alpha\nu) - g_2''(1)}{g_2(1)}}_{=:\sigma} f_1(x). \end{aligned}$$

Solutions for positive σ of this ordinary differential equation in f_1 are of the form $f_1(x) = c_1 \cos(\sqrt{\sigma} x) + c_2 \sin(\sqrt{\sigma} x)$ for some constants c_1 and c_2. Thus, one can choose c_1, c_2, g_2, ζ, and h_2 such that σ is positive and the second equality in Eq. (8.3) holds.

Remark 8.1.1 The two presented approaches to construct analytic solutions to the Stokes–Darcy coupled problems are applicable to any Lipschitz domain with a straight interface $\Gamma_\mathrm{I} = \{(x, y_\mathrm{I}) \mid x \in [a, b]\}$ for some fixed y_I, a, and b, as well.

8.2 Numerical Examples

In this section, several examples for the Stokes–Darcy coupling from the literature are introduced. While Example 8.2.1 is of the first category and the riverbed example, Example 8.2.6, is of the third, all other examples are of the second category. For simplicity the hydraulic conductivity tensor K is assumed to be a scalar in the following. In, for example, [DQ09] the simplified condition $\boldsymbol{u}_\mathrm{f} \cdot \tau_i = 0$, $i = 1, \ldots, d-1$, instead of the Beavers–Joseph–Saffman condition (6.3c) is used for small ν. While this seems reasonable physically, it remains unclear how to meaningfully enforce this condition for non-straight interfaces Γ_I. Therefore, here all but one example fulfill the Beavers–Joseph–Saffman condition.

Example 8.2.1 Let $d = 2$, $\Omega_\mathrm{p} = (0,1)^2$ and $\Omega_\mathrm{f} = (0,1) \times (1,2)$, i.e., $\Gamma_\mathrm{I} = (0,1) \times \{1\}$, see Fig. 8.1. Assume that each solution component (the Stokes velocity components as well as p_f and φ_p) belongs to Q_2 as in the ansatz (8.1). Then the interface conditions (6.3) together with the mass conservation $\nabla \cdot \boldsymbol{u}_\mathrm{f} = 0$ (see (6.1d)), and $\mathbf{f}_\mathrm{f} = \mathbf{0}$ in (6.1a) yield a nine-dimensional solution space. Hence, for parameters $a_1, \ldots, a_9 \in \mathbb{R}$ the solution is given by

$$\boldsymbol{u}_\mathrm{f}(x,y) = \begin{pmatrix} \boldsymbol{u}_{\mathrm{f}1} \\ \boldsymbol{u}_{\mathrm{f}2} \end{pmatrix},$$

$$\begin{aligned} \boldsymbol{u}_{\mathrm{f}1} &= \frac{1}{2\nu}\Big(a_9 y^2 - a_9\Big) + a_5\alpha\nu + a_5 y - a_5 - 2a_6 xy + a_7\alpha\nu + a_9\alpha \\ &\quad - 2x(a_6(\alpha\nu - 1) - a_8\alpha\nu) \\ \boldsymbol{u}_{\mathrm{f}2} &= 2\mathsf{K}a_1 + \mathsf{K}a_2 + \mathsf{K}a_9 + 2\mathsf{K}\nu(a_6 + a_8) + a_6 y^2 + a_6(8\mathsf{K}\nu - 1) + a_7 x \\ &\quad + a_8 x^2 + 2y(a_6(\alpha\nu - 1) - a_8\alpha\nu) + 2(4\mathsf{K}\nu - 1)(a_6(\alpha\nu - 1) - a_8\alpha\nu) \end{aligned}$$

$$p_\mathrm{f}(x,y) = a_9(x - 1/2) + \nu(2y - 3)(a_6 + a_8),$$

$$\begin{aligned} \varphi_\mathrm{p}(x,y) &= a_1 + a_2 y + a_3 x + a_4 x^2 y^2 - x^2 y\Big(2a_4 + \frac{a_8}{\mathsf{K}}\Big) + x^2\Big(a_4 + \frac{a_8}{\mathsf{K}}\Big) \\ &\quad - xy^2\Big(-a_3 + a_9 + \frac{a_7}{\mathsf{K}}\Big) + xy\Big(-2a_3 + 2a_9 + \frac{a_7}{\mathsf{K}}\Big) \\ &\quad - y^2\Big(a_1 + a_2 + 4a_6\alpha\nu^2 + a_6\nu - 4a_8\alpha\nu^2 + a_8\nu + \frac{a_9}{2}\Big), \end{aligned}$$

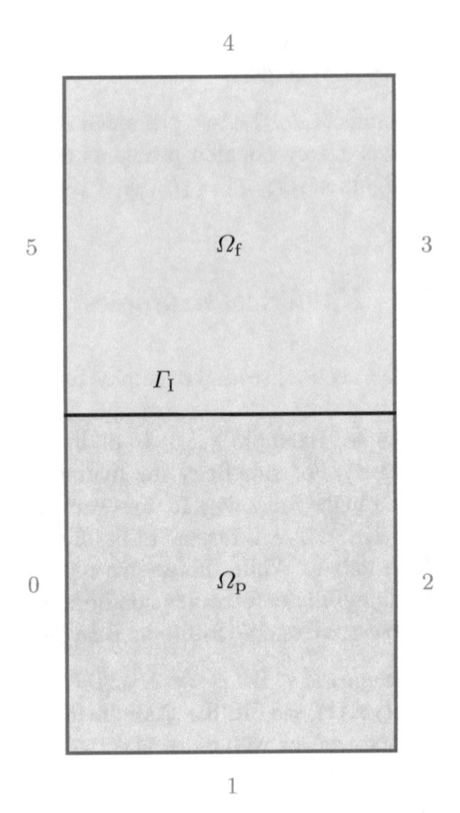

Fig. 8.1 Sketch of the computational domain for Examples 8.2.1–8.2.3. The numbers outside indicate boundary components for easier reference

see also Fig. 8.2 on the left. This solution can be obtained with the help of the function `parametrized_polynomial` implemented in Listing 2. Note that the solution is composed of parts which are scaled with ν, ν^{-1}, K, and K^{-1} at the same time. Furthermore, the parameters a_3 and a_4 only enter φ_p while a_5 only appears in the first component of the Stokes velocity $\boldsymbol{u}_f$, i.e., they do not add terms to either one of the two coupling conditions on the interface Γ_I. The boundary conditions on the outer boundaries can be chosen almost arbitrarily: Since the C-RR method is derived only in the case of $\Lambda_f = \Lambda_p$, a matching condition holds, in particular, on the components 0 and 5 as well as 2 and 3 the same boundary conditions (Dirichlet, Neumann) are imposed.

Remark 8.2.2 In the construction of Example 8.2.1 it is possible to relax the condition $\mathbf{f}_f = \mathbf{0}$ to yield an even larger dimensional solution space. Another particularly simple solution obtained this way is

$$\boldsymbol{u}_f(x, y) = \mathbf{0}, \qquad p_f(x, y) = \varphi_p(x, y) = 0.5 - x. \tag{8.4}$$

Example 8.2.3 This example is taken from [DQ09] and slightly modified to fulfill the Beavers–Joseph–Saffman interface condition (6.3c). The kinematic viscosity ν,

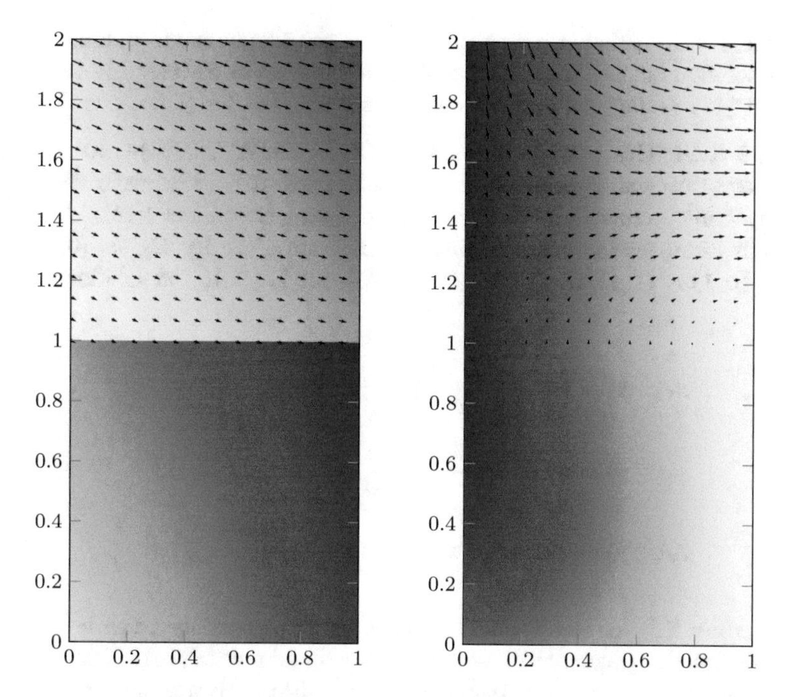

Fig. 8.2 The solution of Examples 8.2.1 (left) and 8.2.3 (right) with parameters $\nu = \mathsf{K} = \alpha = a_i = 1$, $i = 1, \ldots, 9$. The arrows indicate the velocity in the Stokes subdomain Ω_f, while the colors show the pressure in both subdomains ranging from -1.5 to 2.5 (left) and -0.5 to 1.5 (right), respectively. Larger values are represented by darker colors

hydraulic conductivity K, as well as α in the Beavers–Joseph–Saffman condition (6.3c) are set to unity.

As in Example 8.2.1, let $d = 2$, $\Omega_\mathrm{p} = (0, 1)^2$ and $\Omega_\mathrm{f} = (0, 1) \times (1, 2)$, i.e., $\Gamma_\mathrm{I} = (0, 1) \times \{1\}$, see Fig. 8.1. The solution is given by

$$\begin{aligned} \boldsymbol{u}_\mathrm{f}(x, y) &= \begin{pmatrix} -\sin\left(\frac{\pi}{2}x\right)\cos\left(\frac{\pi}{2}y\right) \\ \cos\left(\frac{\pi}{2}x\right)\sin\left(\frac{\pi}{2}y\right) \end{pmatrix}, \\ p_\mathrm{f}(x, y) &= 0.5 - x, \\ \varphi_\mathrm{p}(x, y) &= 0.5 - x + (1 - y)\cos\left(\frac{\pi}{2}x\right), \end{aligned}$$

see also Fig. 8.2 on the right. Note that, restricted to the interface Γ_I, the two pressures p_f and φ_p coincide. Furthermore, the right-hand side f_p in the Darcy subdomain is nonzero. As in the previous example, Example 8.2.1, boundary conditions can be chosen almost arbitrarily.

Note that this solution is a superposition of the simple one from Remark 8.2.2 and a solution obtained from a separation of variables as discussed in Sect. 8.1.1, setting $f_1(x) = \sin(\frac{\pi}{2}x)$, $g_2(y) = \sin(\frac{\pi}{2}y)$, and $h_2(y) = 1 - y$, i.e., $\zeta = \frac{\pi}{2} = \sqrt{\sigma}$.

Example 8.2.4 This example is used in [DQ09, DQV07, CJW14] to study the behavior of several algorithms for different values of the kinematic viscosity ν and hydraulic conductivity K as well as different refinement levels. As in the previous examples the computational domains are given by $\Omega_\mathrm{p} = (0,1)^2$ and $\Omega_\mathrm{f} = (0,1) \times (1,2)$, i.e., $\Gamma_\mathrm{I} = (0,1) \times \{1\}$, see Fig. 8.1. The solution is set to be

$$\boldsymbol{u}_\mathrm{f}(x,y) = \begin{pmatrix} y^2 - 2y + 1 \\ x^2 - x \end{pmatrix},$$

$$p_\mathrm{f}(x,y) = 2\nu\,(x + y - 1) + \frac{1}{3\mathsf{K}},$$

$$\varphi_\mathrm{p}(x,y) = 2\nu x + \frac{1}{\mathsf{K}}\left(x(1-x)(y-1) + \frac{y^3}{3} - y^2 + y\right),$$

see also Fig. 8.3. As this example includes only polynomials it can be obtained from the function `parametrized_polynomial` implemented in Listing 2 with `polynomial_order=3` and suitably choosing the free parameters. Instead of the Beavers–Joseph–Saffman condition (6.3c), here the simplified equation $\boldsymbol{u}_\mathrm{f} \cdot \boldsymbol{\tau} = 0$ is used.

Example 8.2.5 This example is used in [CGHW11] to study the alternative Robin–Robin (C-RR) method on the subdomains $\Omega_\mathrm{p} = (0,\pi) \times (-1,0)$ and $\Omega_\mathrm{f} = (0,\pi) \times (0,1)$, i.e., $\Gamma_\mathrm{I} = (0,\pi) \times \{0\}$. The solution is given as

$$\boldsymbol{u}_\mathrm{f}(x,y) = \begin{pmatrix} \cos(x)v'(y) \\ \sin(x)v(y) \end{pmatrix},$$

$$p_\mathrm{f}(x,y) = 0,$$

$$\varphi_\mathrm{p}(x,y) = \sin(x)e^y,$$

with $v(y) = -K - \frac{1}{2\nu}y + \left(\frac{\mathsf{K}}{2} - \frac{1}{4\alpha\nu^2}\right)y^2$, which can be derived via separation of variables as in Eq. (8.2) setting $f_1 = \cos$, $g_2 = v$, $h_2(y) = e^y$, hence $\zeta = \vartheta = -1$ and $\sigma = 1$. In [CGHW11] the Beavers–Joseph–Saffman condition is formulated with α^{-1} compared to Eq. (6.3c), leading to a slightly modified function v here.

Example 8.2.6 This example is motivated by [CW07a, CW07b] and models a unidirectional flow over a porous media bed with a non-straight interface Γ_I and is used in [CJW14]. The domain is $\Omega = [0, 2L] \times [0, D_\mathrm{p} + D_\mathrm{f}]$ where L is the length of one dune, D_f the depth of the Stokes subdomain Ω_f, and D_p the depth of the Darcy subdomain Ω_p at $x \in \{0, L, 2L\}$. The interface is composed of two

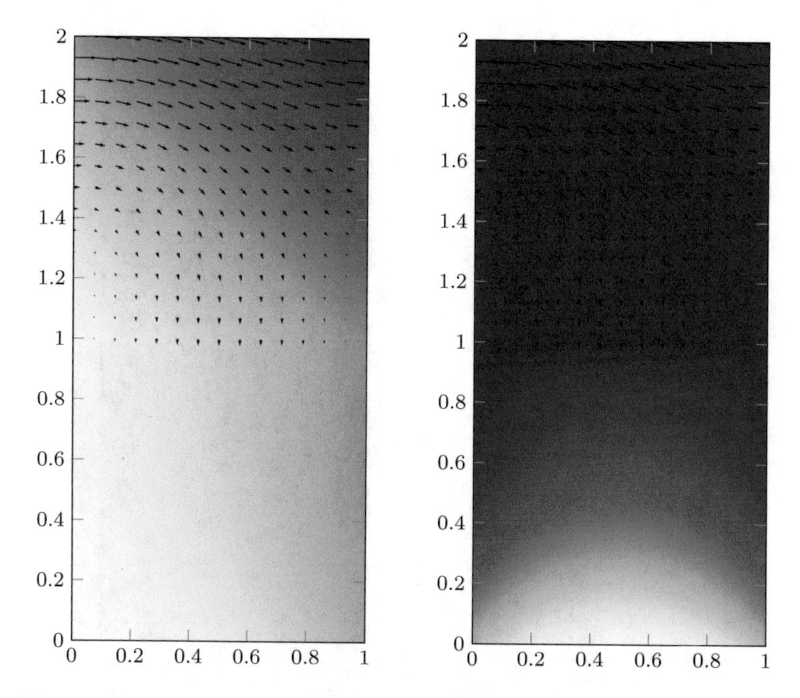

Fig. 8.3 The solution of Example 8.2.4 with parameters $\alpha = 1$ and $\nu = \mathsf{K} = 1$ (left) and $\nu = \mathsf{K} = 10^{-3}$ (right). The arrows indicate the velocity in the Stokes subdomain Ω_f, while the colors show the pressure in both subdomains ranging from 0 to $4\nu + 1/(3\mathsf{K}) = 13/3$ (left) and $-\frac{1}{K}(K\nu - 0.5)^2 \approx -250$ to $4\nu + 1/(3\mathsf{K}) \approx 1000/3$ (right), respectively. Larger values are represented by darker colors

triangles, whose highest points are located at $x = \{l_D, l_D + L\}$ and $y = D_\mathrm{p} + h_D$, see Fig. 8.4.

The solution shall describe a flow from left to right without any inner forces, i.e., $\mathbf{f}_\mathrm{f} = \mathbf{0}$, $f_\mathrm{p} = 0$. Hence, the only driving forces are at the boundary. The domain can be thought of as section of a long river and its ground underneath with a periodically shaped riverbed. Thus the solution is expected to be periodic as well and only driven by a prescribed pressure drop $p_0 = 10^{-3}$:

$$
\begin{aligned}
\boldsymbol{u}_\mathrm{f}(0, y) &= \boldsymbol{u}_\mathrm{f}(2L, y), & y &\in [D_\mathrm{p}, D_\mathrm{p} + D_\mathrm{f}],\\
\mathsf{T}(\boldsymbol{u}_\mathrm{f}, p_\mathrm{f})(0, y) \cdot \boldsymbol{n} &= -\mathsf{T}(\boldsymbol{u}_\mathrm{f}, p_\mathrm{f})(2L, y) \cdot \boldsymbol{n} + p_0 \boldsymbol{n}, & y &\in [D_\mathrm{p}, D_\mathrm{p} + D_\mathrm{f}],\\
\varphi_\mathrm{p}(0, y) &= \varphi_\mathrm{p}(2L, y) + p_0, & y &\in [0, D_\mathrm{p}],\\
-K\nabla\varphi_\mathrm{p}(0, y) \cdot \boldsymbol{n} &= \mathsf{K}\nabla\varphi_\mathrm{p}(2L, y) \cdot \boldsymbol{n}, & y &\in [0, D_\mathrm{p}],
\end{aligned}
$$

where $\boldsymbol{n}$ is the normal pointing out of Ω_f and Ω_p respectively. The bottom boundary is considered impermeable, $\mathsf{K}\nabla\varphi_\mathrm{p}(x, 0) \cdot \boldsymbol{n} = 0$, and a no-slip condition is imposed at the top, $\boldsymbol{u}_\mathrm{f}(x, D_\mathrm{p} + D_\mathrm{f}) = \mathbf{0}$, $x \in [0, 2L]$. With these boundary conditions, the

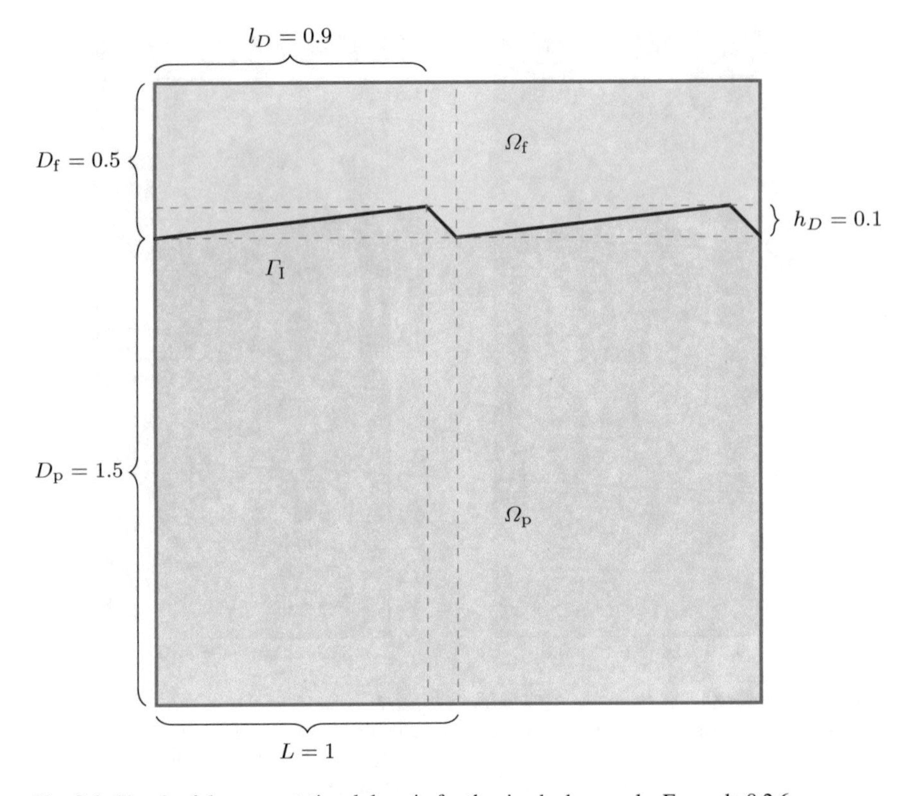

Fig. 8.4 Sketch of the computational domain for the riverbed example, Example 8.2.6

pressure is unique only up to an additive constant, which is fixed in all simulations by forcing one pressure degree of freedom to be zero.

A quantity of interest to hydrologists is the depth and area of the *interfacial exchange zone*, which is described as that part of the Darcy subdomain Ω_p which interacts with the interface, i.e., where the streamlines eventually touch Γ_I. The rest of the porous medium is then called *underflow* area, which is the majority in the computed solutions, see also the middle picture in Fig. 8.5. Additionally, this example can be used to study a Navier–Stokes–Darcy coupled system, where in the free flow subdomain Ω_f the additional nonlinear term $(\boldsymbol{u}_f \cdot \nabla)\boldsymbol{u}_f$ enters Eq. (6.1a). With this model the depth and area of the interfacial exchange zone is considerably bigger, see [Hof13]. Since the focus of this example is to study the algorithms for various ν, K, γ_f, and γ_p, the depth of the interfacial exchange zone is not computed in this work.

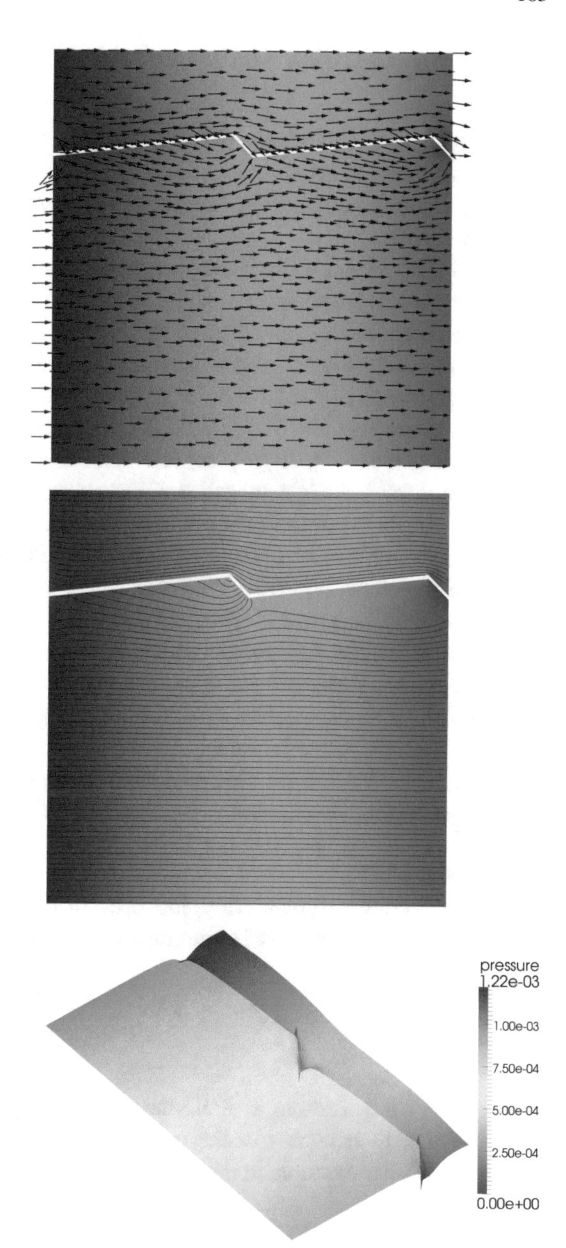

Fig. 8.5 Computed solution for the riverbed example, Example 8.2.6, and $\nu = 10^{-6}, \mathsf{K} = 10^{-7}$. The unscaled arrows (top) and streamlines (middle) in the Darcy subdomain are obtained from the vector field $\boldsymbol{u}_\mathrm{p} = -\mathsf{K}\nabla\varphi_\mathrm{p}$ in a post-processing step, see Eq. (6.1c). The bottom shows the pressure elevation and the color always corresponds to p_f and φ_p. The subdomains in the top two pictures are moved apart slightly to enhance the visibility of the interface Γ_I

8.3 Computations

In this section the discrete and weak solutions are denoted with the same symbol as long as there is no ambiguity, i.e., omitting the index $_h$ where possible. Also norms $\|\cdot\|$ without an index are the Euclidean norm on $\mathbb{R}^N$ with a suitable, typically large, N.

All simulations are performed on matching grids, which means the grids in the respective subdomains Ω_f and Ω_p match on the interface Γ_I. The solutions $(\boldsymbol{u}_\mathrm{f}, p_\mathrm{f}, \varphi_\mathrm{p})$ are always sought in (P_2, P_1, P_2) or (Q_2, Q_1, Q_2) in the respective subdomains, i.e., the inf–sup stable Taylor–Hood pair of finite elements for the Stokes subsystem and second order polynomials for the Darcy pressure. This is the same choice as in [CGHW11, DQ09, DQV07] and on the interface Γ_I it implies Λ_p to be a subspace of P_2 or Q_2, respectively. As noted in Sect. 6.6.2 the same is true for Λ_f only for a straight interface. Otherwise Λ_f allows discontinuities wherever adjacent interface edges/faces are not parallel. This is the case for the riverbed example, Example 8.2.6. The computational grids are unstructured grids obtained from a grid generator.

The subproblems are always solved directly using umfpack, [Dav04], even though in principle any solver which is suitable for the respective subproblem can be applied. Here, all problems are small or medium sized and, since in all the algorithms the respective system matrix does not change during the interface iteration, the LU-factorizations have to be computed only once, leading to fast iterations after the first one. Despite that, the computing times are not used as the primary measure for efficiency. Since all algorithms have a comparable cost per iteration as each subproblem has to be solved,[3] the number of iterations is of most importance.

In the literature typical stopping criteria depend on the absolute or relative differences between successive iterates, i.e., terms of the forms

$$\left\| s^{k+1} - s^k \right\| \quad \text{or} \quad \frac{\left\| s^{k+1} - s^k \right\|}{\left\| s^k \right\|},$$

where s^k is the discrete k-th iterate of $\boldsymbol{u}_\mathrm{f}$, p_f, φ_p, λ, or μ. Additionally, one can measure the accuracy of the mass conservation (6.3a) and balance of normal forces (6.3b), for example in the form of

$$\left\| \boldsymbol{u}_\mathrm{f}^{k+1} \cdot \boldsymbol{n} + \mathsf{K}\nabla\varphi_\mathrm{p}^{k+1} \cdot \boldsymbol{n} \right\|_{0,\Gamma_\mathrm{I}} \quad \text{and} \quad \left\| \boldsymbol{n} \cdot \mathsf{T}(\boldsymbol{u}_\mathrm{f}^{k+1}, p_\mathrm{f}^{k+1}) \cdot \boldsymbol{n} + \varphi_\mathrm{p}^{k+1} \right\|_{0,\Gamma_\mathrm{I}}, \tag{8.5}$$

[3] Iterative solution algorithms for the Steklov–Poincaré equations (6.16) and (6.17) use a preconditioner which adds to the complexity of each iteration, which has to be taken into account, see also at the end of Sect. 7.4.

where the finite element solutions and their derivatives are explicitly evaluated at the interface Γ_{I}, which is not defined for the weak solutions. If not otherwise stated the stopping criteria for some prescribed tolerance eps is met as soon as

$$E_{k+1} < \text{eps}, \tag{8.6}$$

with

$$E_{k+1} = \frac{\left\| \boldsymbol{u}_{\mathrm{f}}^{k+1} - \boldsymbol{u}_{\mathrm{f}}^{k} \right\|}{\left\| \boldsymbol{u}_{\mathrm{f}}^{k} \right\|} + \frac{\left\| p_{\mathrm{f}}^{k+1} - p_{\mathrm{f}}^{k} \right\|}{\left\| p_{\mathrm{f}}^{k} \right\|} + \frac{\left\| \varphi_{\mathrm{p}}^{k+1} - \varphi_{\mathrm{p}}^{k} \right\|}{\left\| \varphi_{\mathrm{p}}^{k} \right\|}, \tag{8.7}$$

where the denominator is omitted whenever it is smaller than one. While such a condition assesses the progress of the iteration, it does not measure the quality of the solution, i.e., it is not able to directly detect whether the computed iterates are close to the solution of the discrete problem or stuck at an early stage. To account for such a situation another measure can be considered, which is based on the residual of the coupled system (6.43):

$$R_k = \left\| \begin{pmatrix} S & C_1 \\ C_2 & D \end{pmatrix} \begin{pmatrix} (\boldsymbol{u}_{\mathrm{f}}^k, p_{\mathrm{f}}^k) \\ \varphi_{\mathrm{p}}^k \end{pmatrix} - b \right\|. \tag{8.8}$$

In all simulations this residual is smaller than the differences, i.e., $R_k < E_k$, so that the stopping criterion (8.6) suffices.

8.3.1 Known Smooth Solution

Consider Example 8.2.3 which prescribes a known, smooth solution and does not consider viscosity and hydraulic conductivity as free parameters, i.e., $\nu = \mathsf{K} = 1$. With this example the behavior of the proposed algorithms with respect to successive grid refinement as well as a good choice of the parameters γ_{f} and γ_{p} are studied.

A family of unstructured triangular grids is considered, see Table 8.1. The simulations are also run on two sequences of grids which originate through successive uniform refinement of a coarse grid, one consisting of two squares, one with 8 triangles. The results are similar or better compared with the unstructured triangular grids possibly due to super convergence phenomena and are not separately presented here.

The considered methods are the Neumann–Neumann and Robin–Robin algorithms as well as the modified Robin–Robin approach, C-RR, (see Chaps. 6 and 7), the discontinuous Robin–Robin, D-RR (see [CJW14] and the end of Sect. 6.7), and two preconditioned Krylov subspace iterations for each of the two Steklov–Poincaré

Table 8.1 Example 8.2.3: Information concerning the grids and finite element spaces

Grid level	Interface edges	Mesh cells Stokes	Mesh cells Darcy	Degrees of freedom Stokes	Degrees of freedom Darcy
2	4	58	40	309	97
3	9	246	224	1205	485
4	19	980	934	4608	1945
5	39	4166	4050	19,145	8257
6	79	16,812	16,422	76,452	33,161

The illustrations on the left indicate the level 2 and 3 unstructured grids. The finer grids are not shown

Table 8.2 Example 8.2.3 on unstructured triangular grids: the number of iterations for different algorithms and grid refinement levels

			Refinement level				
Method	Algorithm	Equation	2	3	4	5	6
C-RR	7.3	(6.41)	19	18	18	18	18
D-RR	7.2	(6.30)	13	14	16	17	18
NN	7.2	(6.20)	12	13	13	13	13
RR	7.2	(6.40)	19	18	18	18	18
cg λ	Conjugate gradients	(6.16)	5	5	5	5	5
cg μ	Conjugate gradients	(6.17)	5	5	5	5	5
gmres λ	Generalized minimal residuals	(6.16)	5	5	5	5	5
gmres μ	Generalized minimal residuals	(6.17)	5	5	5	5	5

equations (6.16) and (6.17), namely conjugate gradients (cg) and generalized minimal residuals (gmres). The parameters γ_{f} and γ_{p} in the Robin–Robin approaches are set to be $\gamma_{\mathrm{f}} = 1/3$ and $\gamma_{\mathrm{p}} = 1$.

The subdomain iterations are stopped whenever the criterion (8.6) is satisfied for eps $= 10^{-10}$ and the Krylov subspace methods are stopped as soon as the residual is below the same threshold. The initial iterate is always chosen to be zero.

Table 8.2 summarizes the numbers of needed iterations until convergence is achieved. All considered methods performed essentially independent of the refinement level in this respect. The Neumann–Neumann scheme needed the fewest iterations compared with the three Robin–Robin approaches. The Krylov subspace methods needed significantly less iterations, but the cost per iteration is higher, because the preconditioner includes an extra solving step in one of the subdomains, see also Sect. 7.4.

The errors of the computed solution components $\boldsymbol{u}_{\mathrm{f}}$, p_{f}, φ_{p} in both L^2 and H^1-semi norm are essentially the same for all methods except D-RR. The convergence rates for the errors in $\boldsymbol{u}_{\mathrm{f}}$ and φ_{p} match the predicted ones, i.e., order 3 in the respective L^2 norm and 2 in the H^1-semi norm. The respective orders for the Stokes pressure p_{f} are 2 and 1. The D-RR approach has slightly larger errors for the L^2-norms of the Stokes velocity and the Darcy pressure. The convergence rates for the errors are also the predicted ones apart from the L^2 error in $\boldsymbol{u}_{\mathrm{f}}$ where it is slightly smaller: $\approx$2.7. Since the D-RR method involves lower order terms, see Sect. 6.7.1, such a decrease in accuracy is not unexpected. For the sake of brevity no plot of the errors is shown.

In order to find values of γ_{f} and γ_{p} for which the Robin–Robin methods converge fast, many simulations are carried out. Since the ratio $\gamma_{\mathrm{f}}/\gamma_{\mathrm{p}}$ determines the convergence speed according to the analysis in Sect. 7.3, various values of γ_{p} around 1 and ratios $\gamma_{\mathrm{f}}/\gamma_{\mathrm{p}}$ are used. Whenever $\gamma_{\mathrm{f}} > \gamma_{\mathrm{p}}$ the iteration does not convergence or is slow, therefore, the considered values for $\gamma_{\mathrm{f}}/\gamma_{\mathrm{p}}$ are smaller than 1.1. In Fig. 8.6 the number of iterations are depicted for the three Robin–Robin methods. With this example most considered pairs of γ_{p} and $\gamma_{\mathrm{f}}/\gamma_{\mathrm{p}}$ lead to more iterations compared with the Neumann–Neumann approach. The three methods have a very similar behavior for smaller ratios, while the D-RR method converges faster for larger ones. Furthermore, all methods perform better for larger γ_{p} and at least if $\gamma_{\mathrm{p}} \geq 1$ for smaller γ_{f}. This observation is expected as in the limit $\gamma_{\mathrm{p}} = \infty$ and $\gamma_{\mathrm{f}} = 0$, all the Robin–Robin methods reduce to the Neumann–Neumann iteration, compare Eqs. (6.20), (6.40), (6.30), and (6.41). Among the Robin–Robin methods D-RR is converging the fastest over all considered parameter values. For larger γ_{p} and smaller γ_{f} the Robin–Robin methods need slightly fewer iterations until convergence compared with the Neumann–Neumann approach. Additionally, all above observations hold on all the used grid levels.

Remark 8.3.1 (Conclusions) Using Example 8.2.3 it is advisable to use Algorithm 7.1 or 7.2 with the Neumann–Neumann approach (6.20). Its number of iterations does not increase on finer grids and it is comparably simple to implement and analyze.

In contrast, the Krylov type methods decrease the iteration count considerably, so that the extra effort to implement the Krylov methods for the Steklov–Poincaré equations (6.16) and (6.17) can be reasonable in practice.

The Robin–Robin methods with $\gamma_{\mathrm{p}} = 1$ and $\gamma_{\mathrm{f}} = 1/3$ need more iterations at a similar cost per iteration and are therefore not advisable for this example. With appropriately chosen parameters however (e.g., $\gamma_{\mathrm{p}} = 10$ and $\gamma_{\mathrm{f}} = 1$), the Robin–Robin methods are competitive and finding such parameters can be done on a coarse grid. While the additional implementation effort for the C-RR and D-RR methods is small, Algorithm 7.2 for the Robin–Robin equation (6.40) is substantially more involved while giving typically worse, at best similar, results compared with D-RR.

8.3.2 Small ν and $\mathbf{K}$

Consider Example 8.2.4 which prescribes a polynomial solution which is not completely in the ansatz space, in particular, the Darcy pressure φ_{p} has polynomial degree 3 while the solution variables in the Stokes subdomain are exactly representable by the chosen Taylor–Hood element. With this example the behavior of the proposed approaches with respect to smaller values of viscosity ν and hydraulic

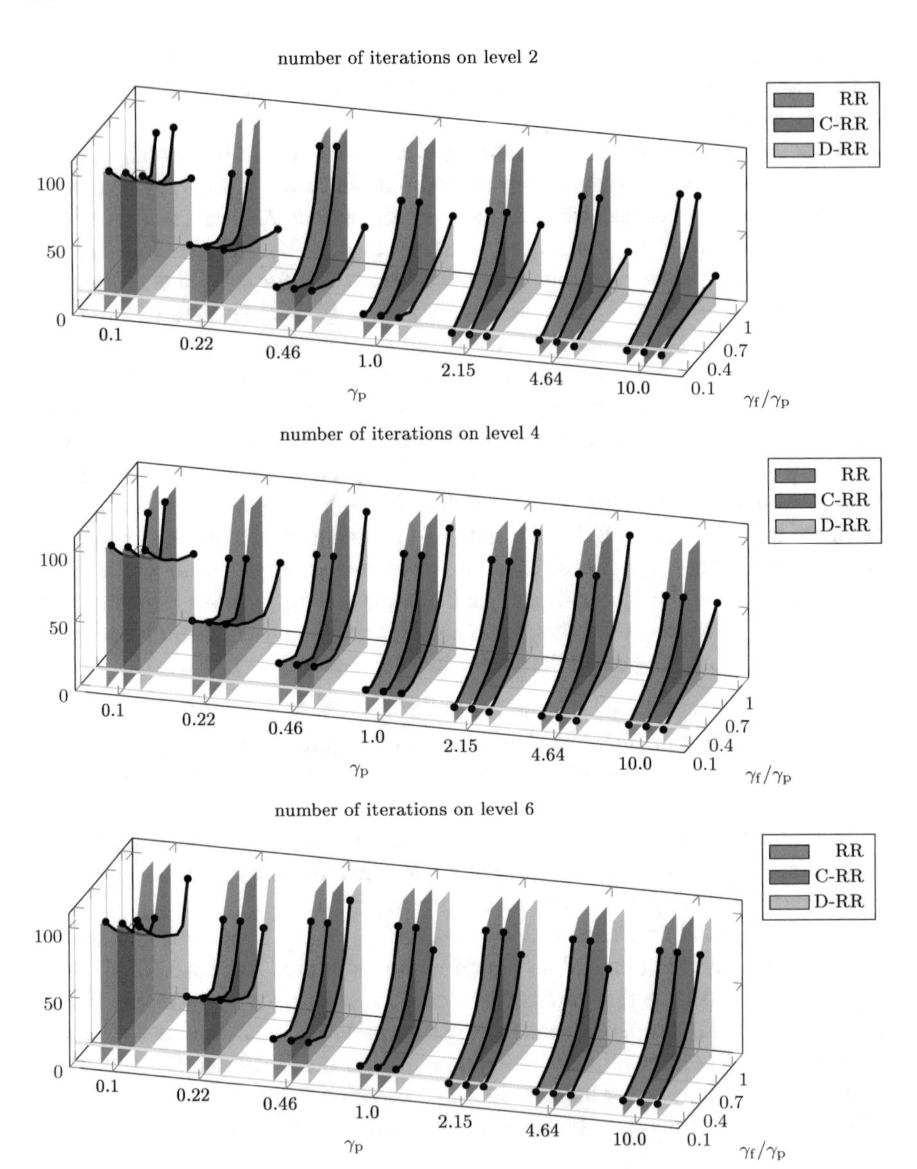

Fig. 8.6 Example 8.2.3: The number of iterations needed for Algorithms 7.2 and 7.3, respectively, for various values of γ_p and ratios γ_f/γ_p on the unstructured triangular grids of Table 8.1. The yellow line indicates the number of Neumann–Neumann iterations (13) as a reference. A value of 100 means that no convergence is achieved within 100 iterations and the black line is absent in this case

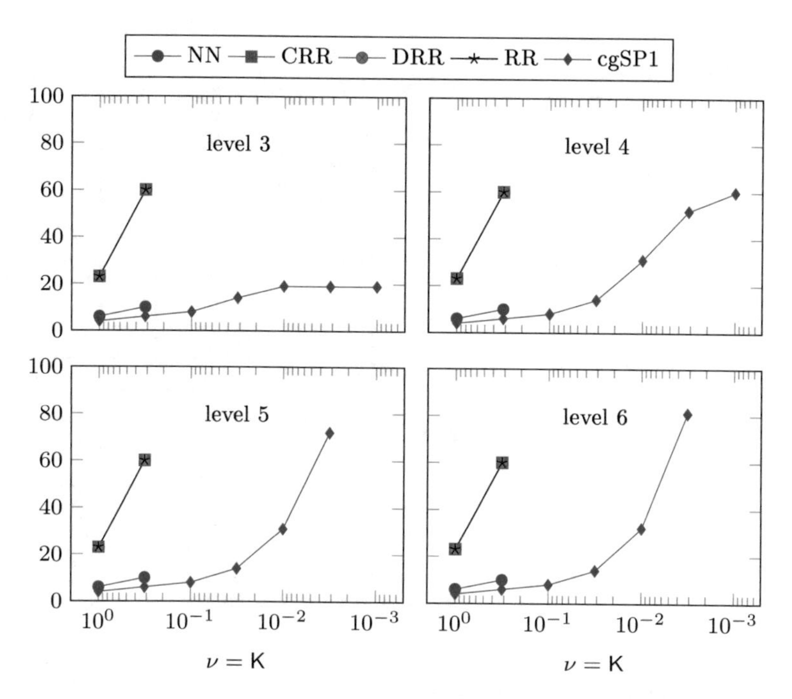

Fig. 8.7 Example 8.2.4: The number of iterations for various values of $\nu = \mathsf{K}$ on different grid levels. All three Robin–Robin methods use $\gamma_\mathrm{p} = 1$, $\gamma_\mathrm{f} = 1/3$ and perform exactly equal for this example, hence their curves are on top of each other. Missing data (for smaller $\nu = \mathsf{K}$) mean that the scheme does not converge within 100 iterations

conductivity K are studied. Since the Krylov iterations for the Steklov–Poincaré equations (6.16) and (6.17) behave comparably, only the method of conjugate gradients for Eq. (6.17) is considered in this example. All simulations use the same unstructured grids as in the previous Sect. 8.3.1, see Table 8.1. Setting $\gamma_\mathrm{p} = 1$ and $\gamma_\mathrm{f} = 1/3$ for the Robin–Robin approaches as in the previous example and slightly smaller values of ν and K lead to results presented in Fig. 8.7. Neither the Robin–Robin methods nor the Neumann–Neumann approach converge within 100 iteration if ν and K are smaller or equal to 10^{-1}. The conjugate gradient method converges even for smaller viscosity and hydraulic conductivity, but the number of iterations depends unfavorably on the grid. On the finer grids also the Steklov–Poincaré iteration does not converge for smaller $\nu = \mathsf{K}$.

In [DQ09, Section 10.4], the C-RR method is studied in more detail for small ν and K. It is proposed to set $\gamma_\mathrm{f} > \gamma_\mathrm{p} > 0$ and to not use large values of these parameters, in particular $\gamma_\mathrm{p} = 0.1$ and $\gamma_\mathrm{f} = 3\gamma_\mathrm{p}$ are considered. For the Robin–Robin and D-RR method the analysis also indicates that $\gamma_\mathrm{f} > \gamma_\mathrm{p}$ is appropriate for smaller ν and K, see Eq. (7.47). Also in [DQ09], the C-RR method is used successfully for realistically small values of ν and K. In order to reproduce these

Table 8.3 The Robin–Robin methods for Example 8.2.4 with $\gamma_\mathrm{p} = 0.1$ and $\gamma_\mathrm{f} = 3\gamma_\mathrm{p}$, various values of ν and K, and with the simplified stopping criterion (8.9)

	Grid	$\nu = \mathsf{K}$				
Method	level	10^{-3}	10^{-4}	10^{-5}	10^{-6}	10^{-7}
RR	2	10	12	13	13	13
	3	10	12	13	13	13
	4	10	12	13	13	13
	5	11	12	13	13	13
	6	11	12	13	13	13
DRR	2	10	12	13	13	13
	3	10	12	13	13	13
	4	10	12	13	13	13
	5	10	12	13	13	13
	6	11	12	13	13	13
CRR	2	–	13	13	13	13
	3	72	12	13	13	13
	4	51	12	13	13	13
	5	36	13	13	13	13
	6	25	13	13	13	13

No convergence within 100 iterations is indicated by '–'

results, the simplified stopping criterion

$$\frac{\left\|\boldsymbol{u}_\mathrm{f}^{k+1} - \boldsymbol{u}_\mathrm{f}^{k}\right\|}{\left\|\boldsymbol{u}_\mathrm{f}^{k}\right\|} < 10^{-6} \tag{8.9}$$

is introduced, compare with Eq. (8.6). A very similar condition is applied in [DQ09]. In Table 8.3 the results for different values of $\nu = \mathsf{K}$ are presented. On coarse grids and comparably large $\nu = \mathsf{K} = 10^{-3}$ the C-RR method does not perform well. Otherwise the number of iterations does not depend on the grid nor on the value of $\nu = \mathsf{K}$ and is acceptable concerning the computational cost for all three Robin–Robin approaches. However, the simplified stopping criterion (8.9) does not imply that a sufficiently accurate solution is obtained in all variables. In particular, the changes in the pressure in both the Stokes and the Darcy subdomain may still be large. The same computations as in Table 8.3 but with the full stopping criterion (8.6) do not converge within 100 iterations for any of the parameters $\nu = \mathsf{K}$. That means even though Eq. (8.9) is fulfilled, the coupled problem is not yet solved with a satisfying tolerance. As an example in Fig. 8.8 it can be seen that the weaker stopping criterion (8.9) holds after only a few iterations (red line) while the sum E_k of the differences of all solution variables, see Eq. (8.7), decreases very slowly, such that even after 10,000 iterations no convergence is achieved yet.

As in the previous study, Sect. 8.3.1, other pairs of $(\gamma_\mathrm{p}, \gamma_\mathrm{f})$ can lead to a faster convergence. Finding such pairs can be done on a coarse grid and is not shown here in detail. For a fixed ratio $\gamma_\mathrm{f}/\gamma_\mathrm{p}$, larger values of γ_p lead to fewer iterations. As an example, the differences E_k for $\gamma_\mathrm{p} = 500$ and $\gamma_\mathrm{f} = 3\gamma_\mathrm{p}$ are shown in Fig. 8.9. With

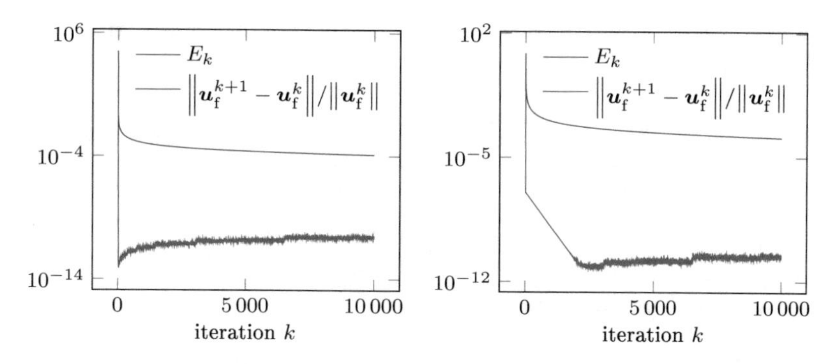

Fig. 8.8 Example 8.2.4 with $\nu = \mathsf{K} = 10^{-4}$: Differences of the iterates for the Robin–Robin and D-RR (left, same behavior) and C-RR (right) method with $\gamma_p = 0.1$ and $\gamma_f = 3\gamma_p$ on the unstructured grid, level 4

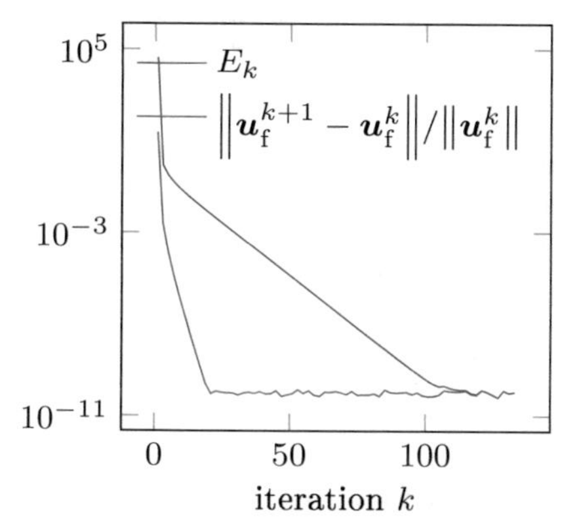

Fig. 8.9 Example 8.2.4 with $\nu = \mathsf{K} = 10^{-4}$: Differences of the iterates for the Robin–Robin and D-RR (same behavior) method with $\gamma_p = 500$ and $\gamma_f = 3\gamma_p$ on the unstructured grid, level 4. The C-RR method diverges for these parameters

this choice the iteration converges even for the full stopping criterion (8.6), yet the number of iterations is rather large.

Remark 8.3.2 (Conclusions) The scaling of the Stokes and Darcy pressure by K^{-1} is a deficiency of this example, nevertheless some conclusion can be drawn. For values of the viscosity ν and hydraulic conductivity K which are of interest in geoscientific applications neither the Neumann–Neumann nor the Krylov type iterations for the Steklov–Poincaré equations lead to feasible algorithms. Also the three Robin–Robin approaches with $\gamma_p > \gamma_f$ are not suitable in this case. Instead considering $\gamma_p < \gamma_f$ leads to converging algorithms. However, the number of iterations is acceptable only for the simplified stopping criterion (8.9) which is based solely on the differences in the Stokes velocity. Using the full stopping criterion (8.6) instead results in slow convergence, at least for $\gamma_p = 0.1$ and $\gamma_f = 3\gamma_p$. Finding better parameters (γ_p, γ_f) can be done on a coarse grid and

Table 8.4 Example 8.2.6: Information concerning the grids and finite element spaces

Grid level	Interface edges	Grid cells		Degrees of freedom	
		Stokes	Darcy	Stokes	Darcy
1	32	277	1009	1392	2098
2	68	1285	4513	6093	9196
3	138	5234	18,322	24,181	36,987

The grid on the left represents level 1

Table 8.5 Example 8.2.6: Information concerning the parameter space in the simulations

Parameter	Set of considered values
ν	$\{10^{-i} \mid i \in \{4, 4.5, 5, 5.5, 6\}\}$
K	$\{10^{-i} \mid i \in \{3, 3.5, 4, 4.5, 5, 5.5, 6, 6.5, 7\}\}$
γ_p	$\{10^{i} \mid i \in \{-1, -0.5, 0, 0.5, 1\}\}$
γ_f/γ_p	$\{1.5, 3, 10, 100, 1000\}$

significantly reduce the number of iterations for the Robin–Robin as well as D-RR method. Larger values of γ_p and γ_f have an effect on the solution which is studied in more detail for the next example.

8.3.3 *Riverbed*

Consider Example 8.2.6 with values of viscosity ν and hydraulic conductivity K which are realistic in geoscientific applications, $\nu \in [10^{-6}, 10^{-4}]$ and $\mathsf{K} \in [10^{-7}, 10^{-3}]$. The Neumann–Neumann method as well as the Krylov iterations are not suitable for such small parameters. Moreover, the C-RR method requires γ_f and γ_p to be very close in this case, such that convergence is slow, see Proposition 7.3.2. Therefore, only the Robin–Robin and the D-RR method are studied with this example. On three different grids, see Table 8.4, various values of $\gamma_p \in [10^{-1}, 10^{1}]$ and ratios $\gamma_f/\gamma_p \in [1.5, 10^3]$ are considered, see also Table 8.5.

The numbers of iterations are summarized in Fig. 8.10[4] for the D-RR method and in Fig. 8.11 for the Robin–Robin approach. Both methods yield comparable results, only larger γ_p and K lead to more iterations of the Robin–Robin iteration.

For the D-RR method and almost all considered values of γ_p the number of iterations is smallest for the largest γ_f, exceptions occur for larger γ_p and K, where the number of iterations grows. Also using smaller γ_p and larger ν leads to an increase in the number of iterations. Otherwise, for each pair (γ_p, γ_f) the surfaces in Fig. 8.10 show only little variation, which means the D-RR method performs

[4]The surfaces sometimes intersect (especially for larger γ_p and larger K) which is not correctly visible in the figure. However the presentation is clearer without these intersections.

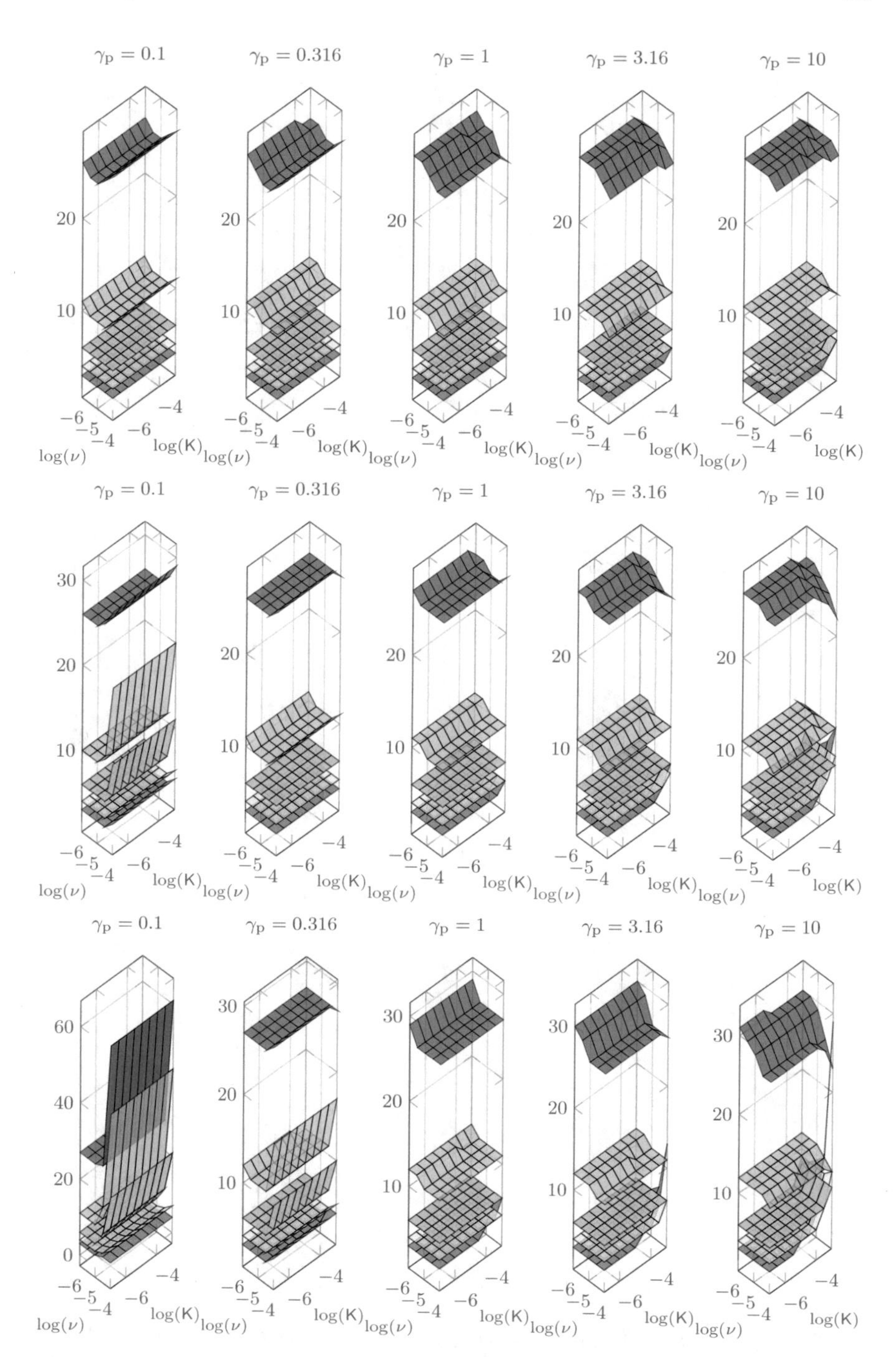

Fig. 8.10 Example 8.2.6: The number of iterations for the D-RR method and different γ_p, ν and K on grid 1 (top), 2 (middle), and 3 (bottom). The surfaces represent the ratios γ_f/γ_p: $\gamma_f = 1.5\gamma_p$ (blue), $\gamma_f = 3\gamma_p$ (cyan), $\gamma_f = 10\gamma_p$ (green), $\gamma_f = 100\gamma_p$ (yellow), $\gamma_f = 1000\gamma_p$ (red). The logarithms are with respect to the base 10

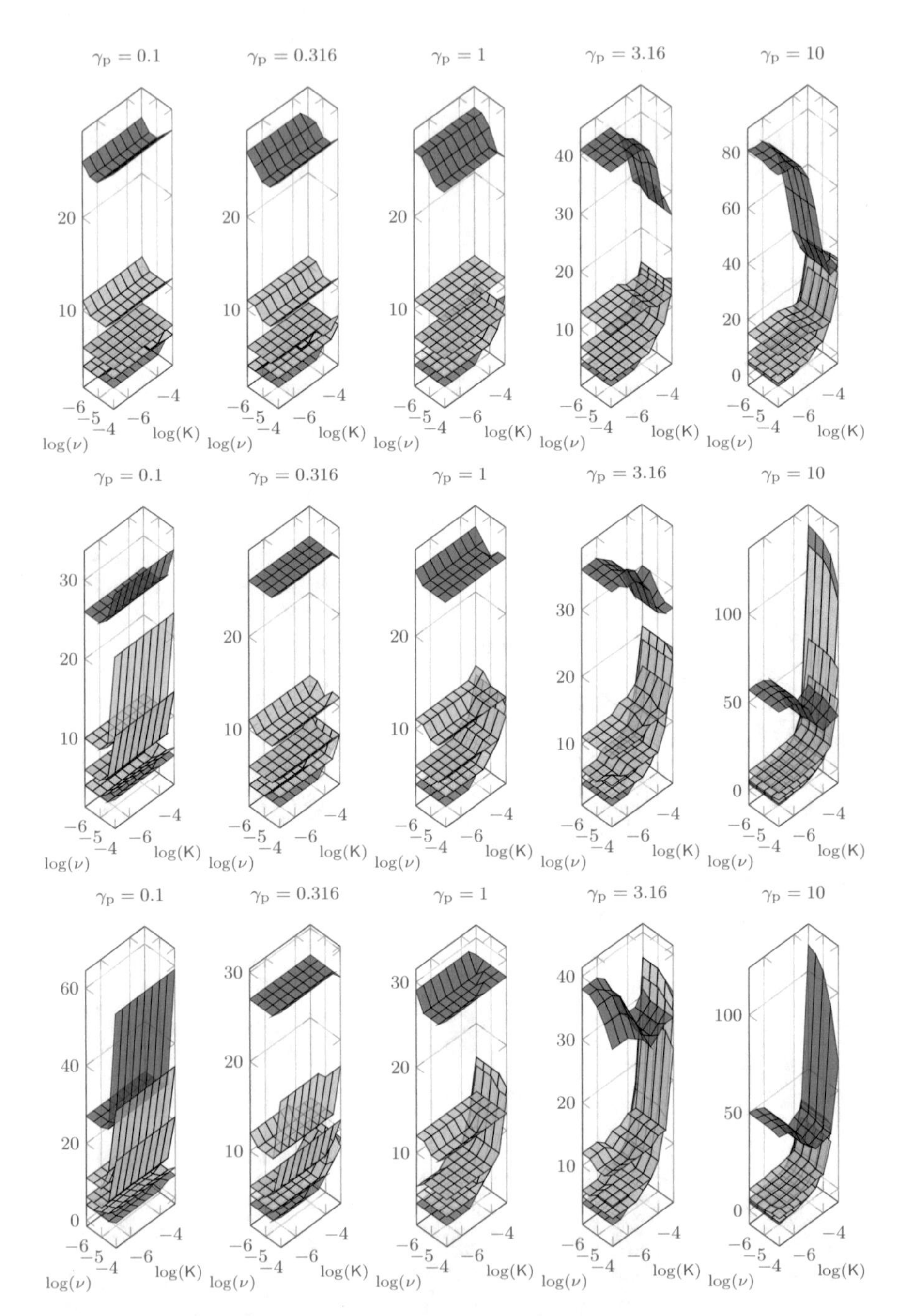

Fig. 8.11 Example 8.2.6: The number of iterations for the Robin–Robin method and different γ_p, ν and K on grid 1 (top), 2 (middle), and 3 (bottom). The surfaces represent the ratios γ_f/γ_p: $\gamma_f = 1.5\gamma_p$ (blue), $\gamma_f = 3\gamma_p$ (cyan), $\gamma_f = 10\gamma_p$ (green), $\gamma_f = 100\gamma_p$ (yellow), $\gamma_f = 1000\gamma_p$ (red). The logarithms are with respect to the base 10

similarly on a wide range of parameters ν and K. The Robin–Robin approach is less robust and performs worse over the same parameter range. Additionally, there is no significant dependence on the grid for both methods at least as long as ν and K are small, $\nu \leq 10^{-5}$, $\mathsf{K} \leq 10^{-4}$.

Despite a fast algorithm, it is of course important to solve the given system with sufficient accuracy. The stopping criterion (8.6) assures this for the equations in the two subdomains but it remains unclear what the effect of different Robin parameters γ_p and γ_f on the mass conservation (6.3a) and the balance of normal stress (6.3b) is. To quantify the errors on the interface the finite element solutions are used to compute

$$I_{\text{mass}} = \left\| \boldsymbol{u}_f \cdot \boldsymbol{n} + \mathsf{K}\nabla\varphi_p \cdot \boldsymbol{n} \right\|_{0,\Gamma_I} \quad \text{and} \quad I_{\text{stress}} = \left\| \boldsymbol{n} \cdot \mathsf{T}(\boldsymbol{u}_f, p_f) \cdot \boldsymbol{n} + \varphi_p \right\|_{0,\Gamma_I},$$

see also Eq. (8.5). Since the modified interface conditions (6.28) used in the Robin–Robin and D-RR approach are linear combinations of the original ones with γ_f and γ_p as factors, it is expected[5] that choosing both gammas large leads to smaller I_{mass} compared with I_{stress}. For small γ_f and γ_p the former is expected to be larger than the latter. In Fig. 8.12 (the logarithm of) the ratio $I_{\text{mass}}/I_{\text{stress}}$ is depicted for all the simulations whose number of iterations is shown in Fig. 8.10. It turns out that, as expected, larger values of γ_f lead to smaller ratios, i.e., $I_{\text{mass}} < I_{\text{stress}}$. The same holds for larger values of γ_p. Furthermore, the ratio $I_{\text{mass}}/I_{\text{stress}}$ does not significantly depend on the grid and except for $\gamma_f/\gamma_p = 1000$ (red) is nearly constant for all considered values of viscosity ν and hydraulic conductivity K.

If for a specific application the accurate approximation of one of the two interface conditions is more important than that of the other, the impact of the choice of the Robin parameters γ_p and γ_f should be taken into account. In this example the errors I_{mass} and I_{stress} are approximately equal if, for example, (γ_p, γ_f) is set to $(0.1, 10)$, $(1, 10)$, or $(3.16, 1.5 \cdot 3.16)$.

Remark 8.3.3 (Conclusions) The D-RR method in general performed equally good or better than the Robin–Robin approach. For small values of viscosity ν and hydraulic conductivity K there are suitable values of the Robin parameters γ_p and γ_f such that the number of iterations remains small, say below 10. Additionally, the interface errors can be tuned somewhat with γ_p and γ_f. The dependence of the grid is observed only at the boundaries of the parameter space (Table 8.5).

The choice $\gamma_p = 1$ and $\gamma_f = 10$ in the D-RR method yields a fast algorithm to solve the coupled problem such that the errors on the interface are almost equal. With this choice the results neither depend on the grid nor on the value of the viscosity ν and hydraulic conductivity K.

[5]As a general similarity let $a, b > 0$. If $\gamma a + b < \varepsilon$ for some ε, then $a < \varepsilon/\gamma$ and $b < \varepsilon$. That means if a and b are equations and γ large, equation a is fulfilled with a higher accuracy compared with b.

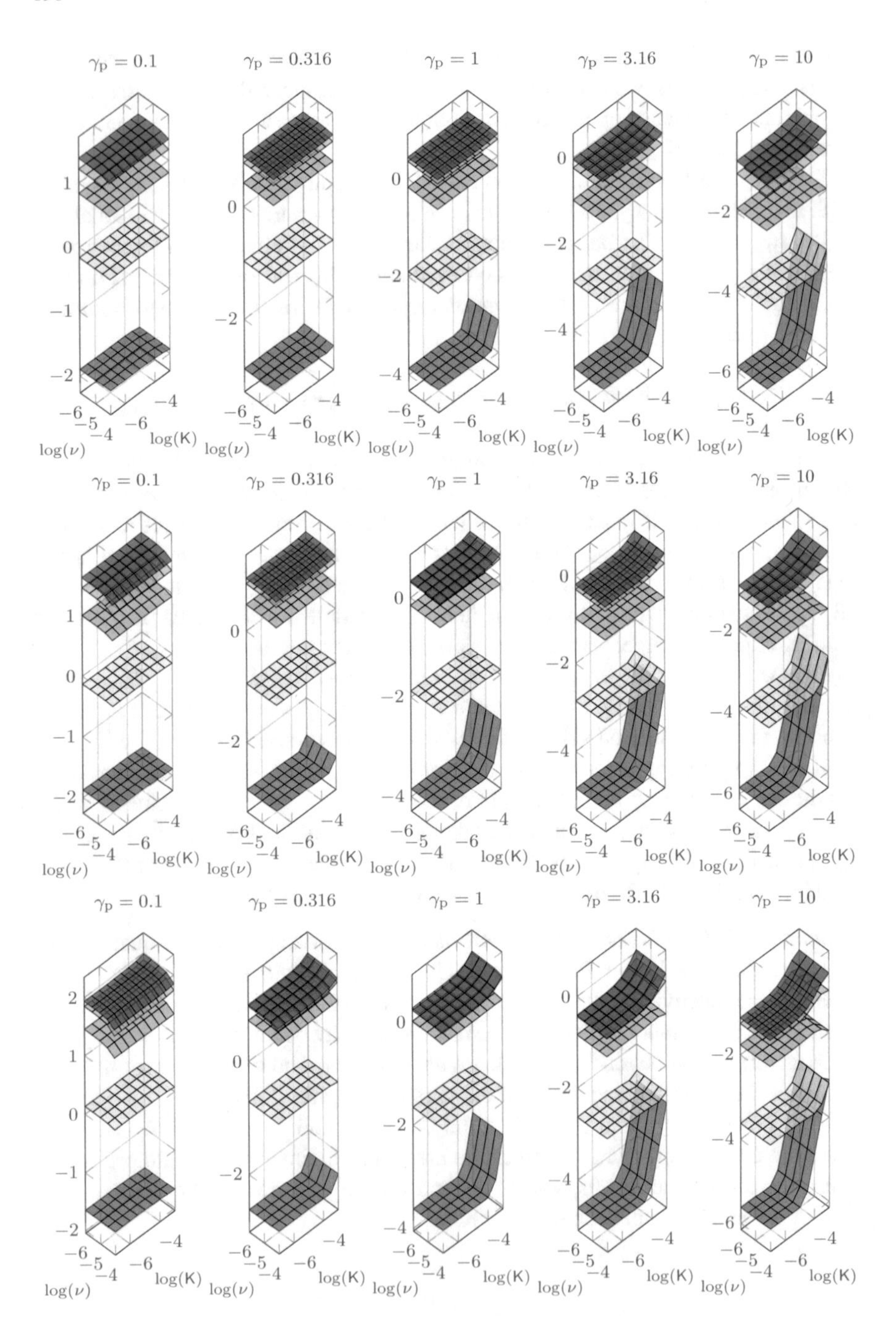

Fig. 8.12 Example 8.2.6: The term $\log(I_{\text{mass}}/I_{\text{stress}})$ for the D-RR method and different γ_p, ν and K on grid 1 (top), 2 (middle), and 3 (bottom). The surfaces represent the ratios γ_f/γ_p: $\gamma_f = 1.5\gamma_p$ (blue), $\gamma_f = 3\gamma_p$ (cyan), $\gamma_f = 10\gamma_p$ (green), $\gamma_f = 100\gamma_p$ (yellow), $\gamma_f = 1000\gamma_p$ (red). The logarithms are with respect to the base 10

8.4 Summary and Conclusions

The numerical tests in the previous section, Sect. 8.3, show that subdomain iterative methods are capable to efficiently solve the Stokes–Darcy coupled problem. The proposed algorithms perform essentially independent of the grid, i.e., they do not degenerate under grid refinement. The presence of tunable parameters γ_{f} and γ_{p} in the Robin–Robin methods is a deficiency which is somewhat compensated by the fact that these parameters can be found cheaply on coarse grids. Additionally, the fulfillment of the mass conservation (6.3a) in relation to the balance of normal stress (6.3b) on the interface can be controlled by these parameters which might as well be an advantage.

In the case of large viscosity $\nu = 1$ and hydraulic conductivity $\mathsf{K} = 1$ all studied algorithms are satisfactory. The ones based on the Neumann–Neumann formulation, see Proposition 7.2.5, do not require any tunable parameters and are therefore recommended in this case. The Steklov–Poincaré iterations do not justify the additional complexity of the implementation, leaving the straightforward Neumann–Neumann iteration to be the approach of choice.

In the case of slightly smaller values of ν and K the Robin–Robin and Steklov–Poincaré methods lead to convergent algorithms. However, it is difficult to find suitable Robin parameters γ_{f} and γ_{p} for the former and the latter suffer from a grid dependency.

The interesting case of very small values of viscosity and hydraulic conductivity for an example of geoscientific interest is solved most efficiently by the D-RR method and to a lesser degree by the Robin–Robin algorithm. All the methods based on the Neumann–Neumann formulation are not suitable in this case. That means in general the best choice is the D-RR method, proposed in Sect. 6.7.1, for physically relevant parameters.

Appendix A
Symbolic Computations to Find Numerical Examples

A.1 Verifying an Example

In case analytical expressions for $\boldsymbol{u}_\mathrm{f}$, p_f, and φ_p are given, it has to be checked if these functions indeed satisfy the Stokes–Darcy coupled problem (6.5). Even though such calculations can be done on paper for each given set of functions, it is much more convenient to have a computer do this. For this purpose so-called computer algebra systems are developed and able to perform calculations symbolically. The `Python` code below uses the library `SymPy` [MSP+17] to define a class `StokesDarcyExample` which is able to check the relevant equations in the case of a horizontal interface in two space dimensions.[1] This class is not very sophisticated but can be of great help when exploring possible analytic solutions to the Stokes–Darcy coupled problem. Furthermore, it might be a starting point to advance these or similar studies, e.g., handle more complicated domains.

```
1   """Copyright 2018, Ulrich Wilbrandt"""
2   import sympy as sy
3
4   # space variables
5   x, y = sy.symbols('x y')
6   # problem parameters
7   nu, K, alpha = sy.symbols('nu K alpha')
8
9
10  # %% class describing a Stokes--Darcy example
11  class StokesDarcyExample:
12      def __init__(self, name, stokes_velocity, stokes_pressure, darcy_pressure,
13                   substitutions=None, bounds=None, BJS=True):
14          """Create a Stokes-Darcy example object and check if it is valid.
15
16          For examples which set nu, K, and alpha to some fixed numbers, the
17          substitutions are needed, eg. {nu: 1, K: 1, alpha: 1}. Additionally,
18          this is needed to draw a plot.
19
```

[1]The used versions are: `Python 3.5.3` and `SymPy 1.0`.

U. Wilbrandt, *Stokes–Darcy Equations*, Advances in Mathematical Fluid Mechanics,
https://doi.org/10.1007/978-3-030-02904-3

```
20          :param name: the name of the example (string)
21          :param stokes_velocity: pair of sympy functions at least depending on
22                                  x and y
23          :param stokes_pressure: sympy function at least depending on x and y
24          :param darcy_pressure: sympy function at least depending on x and y
25          :param substitutions: dictionary which can be passed to 'e.subs' for
26                                sympy expressions 'e'.
27          :param bounds: list of four entries for the domain boundary, the order
28                         is xmin, xmax, ymin, ymax. The interface is then at
29                         (ymax-ymin)/2
30          :param BJS: use the Beavers--Joseph--Saffman condition, otherwise a
31                      zero tangential condition is assumed.
32          """
33          self.name = name
34          self.stokes_velocity = stokes_velocity
35          self.stokes_pressure = stokes_pressure
36          self.darcy_pressure = darcy_pressure
37          self.substitutions = substitutions
38          self.bounds = [0, 1, 0, 2] if bounds is None else bounds
39          self.BJS = BJS
40          self.interface = {y: (self.bounds[2] + self.bounds[3]) / 2}
41          self.check_example()
42
43      def __repr__(self):
44          return self.info(True)
45
46      def info(self, latex=False):
47          """Returns a string with some information on this example."""
48          stokes_u = self.stokes_velocity
49          stokes_p = self.stokes_pressure
50          darcy_p = self.darcy_pressure
51          flux = self.stokes_velocity[1].subs(self.interface).simplify()
52          stress = self.stokes_velocity[1].subs(self.interface).simplify()
53          tangential_u = self.stokes_velocity[0].subs(self.interface).simplify()
54          pressure_mean = self.calculate_stokes_pressure_mean()
55
56          def printer(x):
57              return sy.latex(x) if latex else x
58          ret = "Stokes velocity: " + printer(stokes_u[0]) + "\n"
59          ret += "                  " + printer(stokes_u[1]) + "\n"
60          ret += "Stokes pressure: " + printer(stokes_p) + "\n"
61          ret += "Darcy pressure: " + printer(darcy_p) + "\n"
62          ret += "Darcy  rhs: " + printer(self.calculate_darcy_rhs()) + "\n"
63          ret += "Stokes rhs: " + printer(self.calculate_stokes_rhs()) + "\n"
64          ret += "divergence: " + printer(self.calculate_divergence()) + "\n"
65          ret += "pressure mean value: " + printer(pressure_mean) + "\n"
66          ret += "flux on the interface: " + printer(flux) + "\n"
67          ret += "normal stress on the interface: " + printer(stress) + "\n"
68          ret += "tangential velocity component: " + printer(tangential_u) + "\n"
69          return ret
70
71      def calculate_darcy_rhs(self):
72          """Return the right-hand side for the Darcy equation."""
73          return (- K * (sy.diff(self.darcy_pressure, x, x)
74                  + sy.diff(self.darcy_pressure, y, y))).simplify()
75
76      def calculate_stokes_rhs(self):
77          """Return the right-hand side for the Stokes velocity equations."""
78          x_comp = (-nu * (sy.diff(self.stokes_velocity[0], x, x)
79                           + sy.diff(self.stokes_velocity[0], y, y))
80                    + sy.diff(self.stokes_pressure, x))
81          y_comp = (-nu * (sy.diff(self.stokes_velocity[1], x, x)
82                           + sy.diff(self.stokes_velocity[1], y, y))
83                    + sy.diff(self.stokes_pressure, y))
84          return x_comp.simplify(), y_comp.simplify()
85
86      def calculate_divergence(self):
```

```
87            """Returns the divergence of the Stokes velocity."""
88            return (sy.diff(self.stokes_velocity[0], x)
89                    + sy.diff(self.stokes_velocity[1], y)).simplify()
90
91        def calculate_stokes_pressure_mean(self):
92            """Returns the mean value of the Stokes pressure."""
93            pf = self.stokes_pressure
94            integral = sy.integrate(pf, (x, self.bounds[0], self.bounds[1]),
95                                        (y, self.interface[y], self.bounds[3]))
96            return integral.simplify()
97
98        def check_mass_balance(self):
99            """"Checks if the mass balance on the interface is fulfilled."""
100           pp = self.darcy_pressure
101           ufn = self.stokes_velocity[1]
102           mass_balance = (K * sy.diff(pp, y) + ufn).subs(self.interface)
103           mass_balance = mass_balance.simplify()
104           if not mass_balance.is_zero:
105               if not mass_balance.subs(self.substitutions).is_zero:
106                   print("nonzero mass balance:", mass_balance)
107                   return False
108               else:
109                   print("Warning: mass balance only satisfied after",
110                         "substitutions", self.substitutions)
111           return True
112
113       def check_normal_stress_balance(self):
114           """"Checks if the balance of normal stresses on the interface is
115           fulfilled."""
116           pp = self.darcy_pressure
117           uf2_y = sy.diff(self.stokes_velocity[1], y)
118           pf = self.stokes_pressure
119           normal_stress_balance = (2 * nu * uf2_y - pf + pp).subs(self.interface)
120           normal_stress_balance = normal_stress_balance.simplify()
121           if not normal_stress_balance.is_zero:
122               if not normal_stress_balance.subs(self.substitutions).is_zero:
123                   print("nonzero normal stress:", normal_stress_balance)
124                   return False
125               else:
126                   print("Warning: normal stress only satisfied after",
127                         "substitutions", self.substitutions)
128           return True
129
130       def check_Beavers_Jospeh_Saffman(self):
131           """"Checks if the Beavers-Joseph-Saffman condition on the interface is
132           fulfilled."""
133           uf1 = self.stokes_velocity[0]
134           uf1_y = sy.diff(self.stokes_velocity[0], y)
135           uf2_x = sy.diff(self.stokes_velocity[1], x)
136           bjs = (- alpha * nu * (uf1_y + uf2_x) + uf1).subs(self.interface)
137           bjs = bjs.simplify()
138           if not bjs.is_zero:
139               if not bjs.subs(self.substitutions).is_zero:
140                   print(bjs)
141                   return False
142               else:
143                   print("Warning: Beavers--Joseph--Saffman condition only",
144                         "satisfied after substitutions", self.substitutions)
145           return True
146
147       def check_zero_tangential_velocity(self):
148           """"Checks if the tangential Stokes velocity on the interface is
149           zero."""
150           ut = self.stokes_velocity[0].subs(self.interface)
151           if not ut.is_zero:
152               if not ut.subs(self.substitutions).is_zero:
153                   print("nonzero tangential Stokes velocity:", ut)
```

```
                return False
            else:
                print("Warning: Zero tangential condition only satisfied",
                      "after substitutions", self.substitutions)
        return True

    def check_example(self):
        """Checks all aspects of this example."""
        mb = self.check_mass_balance()
        ns = self.check_normal_stress_balance()
        tc = (self.check_Beavers_Jospeh_Saffman() if self.BJS else
              self.check_zero_tangential_velocity())
        assert (mb and ns and tc), "not a valid Stokes-Darcy example"
```

Listing 1: A Python implementation of a class which can be instantiated to represent a Stokes–Darcy example on a simple domain in two space dimensions

A.2 Finding an Example

The following code illustrates how Example 8.2.1 can be found using the library Sympy [MSP+17]. The function `parametrized_polynomial` returns an object of class `StokesDarcyExample`, see Listing 1. Using the default parameters yields the solution described in Example 8.2.1 while other input may be used to generate more solutions. As with Listing 1, this code is neither optimized nor exceptionally sophisticated, but may serve as a starting point for further experiments and tests.

```
"""Copyright 2018, Ulrich Wilbrandt"""
import sympy as sy
from stokes_darcy import StokesDarcyExample

# space variables
x, y = sy.symbols('x y')
# problem parameters
nu, K, alpha = sy.symbols('nu K alpha')


def parametrized_polynomial(darcy_rhs_zero=False, stokes_rhs_zero=True,
                            polynomial_degree=2, ut0_instead_of_BJS=False):
    """ On the domain $\Omega_p = (0,1)^2$, $\Omega_f = (0,1)\times(1,2)$
        prescribe a Q2-solution for the two Stokes velocity components as
        well as for the Stokes pressure and the Darcy pressure. This yields
        nine free parameters for each solution, 36 for all. Now assume that
        the right-hand sides in the Stokes subdomain, the divergence of the
        Stokes velocity, and the mean value of the Stokes pressure vanish.
        Furthermore the three interface conditions shall be satisfied. This
        will eliminate some of these 36 parameters leaving only nine as free.
        Additionally it is possible to enforce the right-hand side of the
        Darcy subproblem to be zero (see parameter 'darcy_rhs_zero') which
        further reduces the number of free parameters.

        Instead of starting with Q2 solutions any order other than 2 is
        possible, too. However only 1, 2, and 3 have been tested so far.
    """
    # number of coefficients in x for fixed y (and vice versa)
    order = polynomial_degree + 1
    # dimension of Qp (with p = polynomial_degree)
    dim = order*order
    all_coeffs = sy.symbols('b:e:{}:{}'.format(order, order))
    assert len(all_coeffs) == 4 * dim # there are four functions, see below

```

```
35      # %% define functions
36      indices = [(i, j) for i in range(order) for j in range(order)]
37      pp = sy.Poly({indices[i]: all_coeffs[i] for i in range(dim)}, x, y)
38      uf1 = sy.Poly({indices[i]: all_coeffs[i+dim] for i in range(dim)}, x, y)
39      uf2 = sy.Poly({indices[i]: all_coeffs[i+2*dim] for i in range(dim)}, x, y)
40      pf = sy.Poly({indices[i]: all_coeffs[i+3*dim] for i in range(dim)}, x, y)
41      additional_constraints = []
42      interface = {y: 1}
43
44      # %% Stokes constraints (f_f = 0, ∇ · u_f = 0):
45      # divergence
46      g = sy.poly(uf1.diff(x) + uf2.diff(y), x, y)
47      #  right-hand side f_f
48      f1 = uf1.diff(x, x).add(uf1.diff(y, y)).mul(sy.poly(-nu, x, y))
49      f1 = f1.add(pf.diff(x))
50      f2 = uf2.diff(x, x).add(uf2.diff(y, y)).mul(sy.poly(-nu, x, y))
51      f2 = f2.add(pf.diff(y))
52      # mean value of pressure
53      p_mean = sy.integrate(pf, (x, 0, 1), (y, 1, 2))
54      p_mean = sy.poly(p_mean, x, y)
55
56      # %% Darcy constraints
57      # right-hand side f_p
58      fp = pp.diff(x, x).add(pp.diff(y, y)).mul(sy.poly(K, x, y))
59
60      # %% interface constraints
61      mass_balance = sy.diff(pp, y).mul(sy.poly(K, x, y)).add(uf2)
62      mass_balance = mass_balance.subs(interface)
63
64      normal_stress_balance = uf2.diff(y).mul(sy.poly(2*nu, x, y)).add(-pf)
65      normal_stress_balance = normal_stress_balance.add(pp).subs(interface)
66
67      bjs = uf1.diff(y).add(uf2.diff(x)).mul(sy.poly(-alpha*nu, x, y)).add(uf1)
68      bjs = bjs.subs(interface)
69      if ut0_instead_of_BJS:
70          bjs = uf1.subs(interface)
71
72      # %% additional (arbitrary) constraints
73      additional_constraints = []
74      if darcy_rhs_zero:
75          additional_constraints += fp.coeffs()
76      if stokes_rhs_zero:
77          additional_constraints += f1.coeffs() + f2.coeffs()
78
79      # %% solve
80      list_of_coeffs = (g.coeffs() + p_mean.coeffs()
81                        + mass_balance.coeffs() + normal_stress_balance.coeffs()
82                        + bjs.coeffs() + additional_constraints)
83      result_all = sy.solve(list_of_coeffs, *all_coeffs)
84
85      # %% some output:
86      free_symb = [it for it in all_coeffs if it not in result_all.keys()]
87      print("summary: There are", len(all_coeffs), "coefficients.",
88            "The equations elimate", len(result_all), "of them so that",
89            len(free_symb), "remain.")
90
91      # %% update the solutions
92      # rename the ramaining free parameters to 'a' with an index
93      n_free_symbols = len(free_symb)
94      a = sy.symbols('a1:{}'.format(n_free_symbols+1))
95      new_parameters = {free_symb[i]: a[i] for i in range(n_free_symbols)}
96      free_symb = a
97
98      uf1 = sy.simplify(uf1.subs(result_all).subs(new_parameters))
99      uf2 = sy.simplify(uf2.subs(result_all).subs(new_parameters))
100     pf = sy.simplify(pf.subs(result_all).subs(new_parameters))
101     pp = sy.simplify(pp.subs(result_all).subs(new_parameters))
```

```
102
103     # %% create a Stokes--Darcy example from these functions
104     # setting all free symbols to one, this can be changed again for plotting
105     substitutions = {i: 1 for i in free_symb}
106     substitutions.update({K: 1, nu: 1, alpha: 1})
107     sd_ex = StokesDarcyExample("parametrized polynomial", (uf1, uf2), pf, pp,
108                                substitutions, BJS=True)
109     return sd_ex
```

Listing 2: A `Python` implementation to find the space of polynomial solutions of a given order to the Stokes–Darcy coupled problem (6.5) on a rectangular domain

References

[ACM14] Chérif Amrouche, Philippe G. Ciarlet, and Cristinel Mardare. Remarks on a lemma by Jacques-Louis Lions. *C. R. Math. Acad. Sci. Paris*, 352(9):691–695, 2014.

[ACM15] Chérif Amrouche, Philippe G. Ciarlet, and Cristinel Mardare. On a lemma of Jacques-Louis Lions and its relation to other fundamental results. *J. Math. Pures Appl. (9)*, 104(2):207–226, 2015.

[AF03] Robert A. Adams and John J. F. Fournier. *Sobolev spaces*, volume 140 of *Pure and Applied Mathematics (Amsterdam)*. Elsevier/Academic Press, Amsterdam, second edition, 2003.

[Ang11] Philippe Angot. On the well-posed coupling between free fluid and porous viscous flows. *Appl. Math. Lett.*, 24(6):803–810, 2011.

[BBF13] Daniele Boffi, Franco Brezzi, and Michel Fortin. *Mixed finite element methods and applications*, volume 44 of *Springer Series in Computational Mathematics*. Springer, Heidelberg, 2013.

[BC84] Claudio Baiocchi and António Capelo. *Variational and quasivariational inequalities*. A Wiley-Interscience Publication. John Wiley & Sons, Inc., New York, 1984. Applications to free boundary problems, Translated from the Italian by Lakshmi Jayakar.

[BC09] Santiago Badia and Ramon Codina. Unified stabilized finite element formulations for the Stokes and the Darcy problems. *SIAM J. Numer. Anal.*, 47(3):1971–2000, 2009.

[BF91] Franco Brezzi and Michel Fortin. *Mixed and hybrid finite element methods*, volume 15 of *Springer Series in Computational Mathematics*. Springer-Verlag, New York, 1991.

[Bra07a] Dietrich Braess. *Finite Elemente - Theorie, schnelle Löser und Anwendungen in der Elastizitätstheorie*. Springer, 2007.

[Bra07b] Dietrich Braess. *Finite elements*. Cambridge University Press, Cambridge, third edition, 2007. Theory, fast solvers, and applications in elasticity theory, Translated from the German by Larry L. Schumaker.

[Bre02] Kh. Brezis. How to recognize constant functions. A connection with Sobolev spaces. *Uspekhi Mat. Nauk*, 57(4(346)):59–74, 2002.

[Bre11] Haim Brezis. *Functional analysis, Sobolev spaces and partial differential equations*. Universitext. Springer, New York, 2011.

[BS08] Susanne C. Brenner and L. Ridgway Scott. *The mathematical theory of finite element methods*, volume 15 of *Texts in Applied Mathematics*. Springer, New York, third edition, 2008.

[CGH+10] Yanzhao Cao, Max Gunzburger, Xiaolong Hu, Fei Hua, Xiaoming Wang, and Weidong Zhao. Finite element approximations for Stokes-Darcy flow with Beavers-Joseph interface conditions. *SIAM J. Numer. Anal.*, 47(6):4239–4256, 2010.

U. Wilbrandt, *Stokes–Darcy Equations*, Advances in Mathematical Fluid Mechanics,
https://doi.org/10.1007/978-3-030-02904-3

[CGHW10] Yanzhao Cao, Max Gunzburger, Fei Hua, and Xiaoming Wang. Coupled Stokes-Darcy model with Beavers-Joseph interface boundary condition. *Commun. Math. Sci.*, 8(1):1–25, 2010.

[CGHW11] Wenbin Chen, Max Gunzburger, Fei Hua, and Xiaoming Wang. A parallel robin-robin domain decomposition method for the Stokes-Darcy system. *SIAM J. Numerical Analysis*, 49(3):1064–1084, 2011.

[CGHW14] Yanzhao Cao, Max Gunzburger, Xiaoming He, and Xiaoming Wang. Parallel, non-iterative, multi-physics domain decomposition methods for time-dependent Stokes-Darcy systems. *Math. Comp.*, 83(288):1617–1644, 2014.

[Cia02] Philippe G. Ciarlet. *The finite element method for elliptic problems*, volume 40 of *Classics in Applied Mathematics*. Society for Industrial and Applied Mathematics (SIAM), Philadelphia, PA, 2002. Reprint of the 1978 original [North-Holland, Amsterdam; MR0520174 (58 #25001)].

[CJW14] Alfonso Caiazzo, Volker John, and Ulrich Wilbrandt. On classical iterative subdomain methods for the Stokes-Darcy problem. *Comput. Geosci.*, 18(5):711–728, 2014.

[CW07a] M.B. Cardenas and J.L. Wilson. Dunes, turbulent eddies, and interfacial exchange with permeable sediments. *Water Resour. Res.*, 43:08412, 2007.

[CW07b] M.B. Cardenas and J.L. Wilson. Hydrodynamics of coupled flow above and below a sediment–water interface with triangular bedforms. *Adv. Water Resour.*, 30:301–313, 2007.

[Dav04] Timothy A. Davis. Algorithm 832: Umfpack v4.3—an unsymmetric-pattern multifrontal method. *ACM Trans. Math. Softw.*, 30(2):196–199, June 2004.

[DDE12] F. Demengel, G. Demengel, and R. Erné. *Functional Spaces for the Theory of Elliptic Partial Differential Equations*. Universitext. Springer London, 2012.

[DM01] Ricardo G. Durán and Maria Amelia Muschietti. An explicit right inverse of the divergence operator which is continuous in weighted norms. *Studia Math.*, 148(3):207–219, 2001.

[DMQ02] Marco Discacciati, Edie Miglio, and Alfio Quarteroni. Mathematical and numerical models for coupling surface and groundwater flows. *Appl. Numer. Math.*, 43(1–2):57–74, 2002. 19th Dundee Biennial Conference on Numerical Analysis (2001).

[DNPV12] Eleonora Di Nezza, Giampiero Palatucci, and Enrico Valdinoci. Hitchhiker's guide to the fractional Sobolev spaces. *Bull. Sci. Math.*, 136(5):521–573, 2012.

[DQ09] M. Discacciati and A. Quarteroni. Navier-Stokes/Darcy coupling: modeling, analysis and numerical approximation. *Revista Matematica Complutense*, 22(2):315–426, 2009.

[DQV07] Marco Discacciati, Alfio Quarteroni, and Alberto Valli. Robin-Robin domain decomposition methods for the Stokes-Darcy coupling. *SIAM J. Numer. Anal.*, 45(3):1246–1268, 2007.

[DZ11] Carlo D'Angelo and Paolo Zunino. Robust numerical approximation of coupled Stokes and Darcy's flows applied to vascular hemodynamics and biochemical transport. *ESAIM, Math. Model. Numer. Anal.*, 45(3):447–476, 2011.

[EG92] Lawrence C. Evans and Ronald F. Gariepy. *Measure theory and fine properties of functions*. Studies in Advanced Mathematics. CRC Press, Boca Raton, FL, 1992.

[Eva98] Lawrence C. Evans. *Partial differential equations*, volume 19 of *Graduate Studies in Mathematics*. American Mathematical Society, Providence, RI, 1998.

[Eva10] Lawrence C. Evans. *Partial differential equations*, volume 19 of *Graduate Studies in Mathematics*. American Mathematical Society, Providence, RI, second edition, 2010.

[Fol99] Gerald B. Folland. *Real analysis*. Pure and Applied Mathematics (New York). John Wiley & Sons, Inc., New York, second edition, 1999. Modern techniques and their applications, A Wiley-Interscience Publication.

[GOS11a] Gabriel N. Gatica, Ricardo Oyarzúa, and Francisco-Javier Sayas. Analysis of fully-mixed finite element methods for the Stokes-Darcy coupled problem. *Math. Comp.*, 80(276):1911–1948, 2011.

[GOS11b] Gabriel N. Gatica, Ricardo Oyarzúa, and Francisco-Javier Sayas. Convergence of a family of Galerkin discretizations for the Stokes-Darcy coupled problem. *Numer. Methods Partial Differential Equations*, 27(3):721–748, 2011.

[GR86] Vivette Girault and Pierre-Arnaud Raviart. *Finite element methods for Navier-Stokes equations*, volume 5 of *Springer Series in Computational Mathematics*. Springer-Verlag, Berlin, 1986. Theory and algorithms.

[Gri85] P. Grisvard. *Elliptic problems in nonsmooth domains*, volume 24 of *Monographs and Studies in Mathematics*. Pitman (Advanced Publishing Program), Boston, MA, 1985.

[GS07] Juan Galvis and Marcus Sarkis. Non-matching mortar discretization analysis for the coupling Stokes-Darcy equations. *Electron. Trans. Numer. Anal.*, 26:350–384, 2007.

[Hof13] Moritz Hoffmann. The Navier–Stokes–Darcy problem. Bachelor thesis, Freie Universität Berlin, 2013.

[JM00] Willi Jäger and Andro Mikelić. On the interface boundary condition of Beavers, Joseph, and Saffman. *SIAM J. Appl. Math.*, 60(4):1111–1127, 2000.

[Joh16] Volker John. *Finite element methods for incompressible flow problems*. Cham: Springer, 2016.

[Jon73] I.P. Jones. Low Reynolds number flow past a porous spherical shell. *Math. Proc. Cambridge Philos. Soc.*, 73:231–238, 1973.

[Jos05] Jürgen Jost. *Postmodern analysis*. Universitext. Springer-Verlag, Berlin, third edition, 2005.

[KO88] N. Kikuchi and J. T. Oden. *Contact problems in elasticity: a study of variational inequalities and finite element methods*, volume 8 of *SIAM Studies in Applied Mathematics*. Society for Industrial and Applied Mathematics (SIAM), Philadelphia, PA, 1988.

[LI06] Alejandro Limache and Sergio Idelsohn. Laplace form of Navier-Stokes equations: A safe path or a wrong way? *Mecánica Computacional*, 25:151–168, 2006.

[LM72] J.-L. Lions and E. Magenes. *Non-homogeneous boundary value problems and applications. Vol. I*. Springer-Verlag, New York-Heidelberg, 1972. Translated from the French by P. Kenneth, Die Grundlehren der mathematischen Wissenschaften, Band 181.

[LSY02] William J. Layton, Friedhelm Schieweck, and Ivan Yotov. Coupling fluid flow with porous media flow. *SIAM J. Numer. Anal.*, 40(6):2195–2218 (2003), 2002.

[Mac09] Barbara D. MacCluer. *Elementary functional analysis*, volume 253 of *Graduate Texts in Mathematics*. Springer, New York, 2009.

[MM07] Irina Mitrea and Marius Mitrea. The Poisson problem with mixed boundary conditions in Sobolev and Besov spaces in non-smooth domains. *Trans. Amer. Math. Soc.*, 359(9):4143–4182 (electronic), 2007.

[MMS15] Antonio Márquez, Salim Meddahi, and Francisco-Javier Sayas. Strong coupling of finite element methods for the Stokes-Darcy problem. *IMA J. Numer. Anal.*, 35(2):969–988, 2015.

[MSP+17] Aaron Meurer, Christopher P. Smith, Mateusz Paprocki, Ondřej Čertík, Sergey B. Kirpichev, Matthew Rocklin, AMiT Kumar, Sergiu Ivanov, Jason K. Moore, Sartaj Singh, Thilina Rathnayake, Sean Vig, Brian E. Granger, Richard P. Muller, Francesco Bonazzi, Harsh Gupta, Shivam Vats, Fredrik Johansson, Fabian Pedregosa, Matthew J. Curry, Andy R. Terrel, Štěpán Roučka, Ashutosh Saboo, Isuru Fernando, Sumith Kulal, Robert Cimrman, and Anthony Scopatz. Sympy: symbolic computing in python. *PeerJ Computer Science*, 3:e103, January 2017.

[QV99] Alfio Quarteroni and Alberto Valli. *Domain decomposition methods for partial differential equations*. Numerical Mathematics and Scientific Computation. The Clarendon Press Oxford University Press, New York, 1999. Oxford Science Publications.

[RGM15] I.V. Rybak, W.G. Gray, and C.T. Miller. Modeling two-fluid-phase flow and species transport in porous media. *J. Hydrology*, 521:565–581, 2015.

[RM14] Iryna Rybak and Jim Magiera. A multiple-time-step technique for coupled free flow and porous medium systems. *J. Comput. Phys.*, 272:327–342, 2014.

[Rus13] Emmanuel Russ. A survey about the equation $\operatorname{div} u = f$ in bounded domains of $\mathbb{R}^n$. *Vietnam J. Math.*, 41(4):369–381, 2013.

[RY05] Béatrice Rivière and Ivan Yotov. Locally conservative coupling of Stokes and Darcy flows. *SIAM J. Numer. Anal.*, 42(5):1959–1977, 2005.

[Saa03] Yousef Saad. *Iterative methods for sparse linear systems*. Society for Industrial and Applied Mathematics, Philadelphia, PA, second edition, 2003.

[Saf71] P.G. Saffman. On the boundary condition at the interface of a porous medium. *Stud. Appl. Math.*, 50:93–101, 1971.

[Ste70] Elias M. Stein. *Singular integrals and differentiability properties of functions*. Princeton Mathematical Series, No. 30. Princeton University Press, Princeton, N.J., 1970.

[Şuh03] Erdoğan S. Şuhubi. *Functional analysis*. Kluwer Academic Publishers, Dordrecht, 2003.

[Tar07] Luc Tartar. *An introduction to Sobolev spaces and interpolation spaces*, volume 3 of *Lecture Notes of the Unione Matematica Italiana*. Springer, Berlin; UMI, Bologna, 2007.

[Tem77] Roger Temam. *Navier-Stokes equations. Theory and numerical analysis*. North-Holland Publishing Co., Amsterdam-New York-Oxford, 1977. Studies in Mathematics and its Applications, Vol. 2.

[Tri92] Hans Triebel. *Higher analysis*. Hochschulbücher für Mathematik. [University Books for Mathematics]. Johann Ambrosius Barth Verlag GmbH, Leipzig, 1992. Translated from the German by Bernhardt Simon [Bernhard Simon] and revised by the author.

[VY09] Danail Vassilev and Ivan Yotov. Coupling Stokes-Darcy flow with transport. *SIAM J. Sci. Comput.*, 31(5):3661–3684, 2009.

[WBA+17] Ulrich Wilbrandt, Clemens Bartsch, Naveed Ahmed, Najib Alia, Felix Anker, Laura Blank, Alfonso Caiazzo, Sashikumaar Ganesan, Swetlana Giere, Gunar Matthies, and et al. ParMooN—A modernized program package based on mapped finite elements. *Comput. Math. Appl.*, 74(1):74–88, 2017.

[Wer00] Dirk Werner. *Funktionalanalysis*. Springer-Verlag, Berlin, extended edition, 2000.

[Wil13] Michel Willem. *Functional analysis*. Cornerstones. Birkhäuser/Springer, New York, 2013. Fundamentals and applications.

Index

U. Wilbrandt, *Stokes–Darcy Equations*, Advances in Mathematical Fluid Mechanics,
https://doi.org/10.1007/978-3-030-02904-3

Printed in the United States
By Bookmasters